*Civil Engineering Systems
Analysis and Design*

Civil Engineering Systems Analysis and Design

Alan A. Smith

*Department of Civil Engineering,
McMaster University,
Hamilton, Ontario*

Ernest Hinton

and

Roland W. Lewis

*Department of Civil Engineering,
University College of Swansea, Wales*

A Wiley–Interscience Publication

JOHN WILEY & SONS

Chichester · New York · Brisbane · Toronto · Singapore

Copyright © 1983 by John Wiley & Sons Ltd.

All rights reserved.

No part of this book may be reproduced by any means, nor transmitted, nor translated into a machine language without the written permission of the publisher.

Library of Congress Cataloging in Publication Data:

Smith, Alan A.
   Civil engineering systems analysis and design.
   'A Wiley–Interscience publication.'
   Includes index.
   1. Civil engineering. 2. System analysis. 3. System design. I. Hinton, Ernest
II. Lewis, R. W. (Roland Wynne) III. Title.
TA330.S63   1983   624   82-13640
ISBN 0 471 90060 5

British Library Cataloguing in Publication Data:

Smith, Alan A.
   Civil engineering systems analysis and design.
   1. Civil engineering   2. Systems analysis
   I. Title   II. Hinton, Ernest
   III. Lewis, Roland W.
624'.072   TA347.S/
ISBN 0 471 90060 5

Typeset by Mid-County Press, London, S.W.15
Printed by Page Bros. (Norwich) Ltd., Norfolk

Acknowledgements

The authors would like to thank all the students who have helped in the preparation of this book. The text was started about eight years ago and has gone through many proofs. Our students have been most helpful in picking up various errors and commenting on various sections. In particular, we wish to acknowledge Mr Georghiades for his help with the portal frame design in Chapters 3 and 12.

This book would not have been possible without the encouragement of two eminent figures in the field of numerical methods who have been the source of much encouragement over the years and who made it possible for two of us to teach Civil Engineering Systems at the University College of Swansea. We offer our grateful thanks in particular to Dean R. H. Gallagher of the University of Arizona for his continual enthusiasm and support and also to Professor O. C. Zienkiewicz FRS of the University College of Swansea.

Last but by no means least, we thank our families for lost time which can never be regained.

Contents

Preface xi

1 An introduction to systems engineering 1
   1.1 Engineering creativity 1
   1.2 Design methodology 2
   1.3 Systems engineering 7
   1.4 Case study of a systems problem 11
   1.5 Economic aspects of systems evaluation 16
   1.6 Mathematical models 21
   1.7 References 27

2 Optimization by calculus 28
   2.1 Introduction 28
   2.2 Unconstrained functions of a single variable . . . 31
   2.3 Problems involving simple constraints 33
   2.4 Unconstrained functions of several variables . . . 34
   2.5 Treatment of equality constraints 38
   2.6 Extension to multiple equality constraints 43
   2.7 Optimization with inequality constraints 43
   2.8 The generalized Newton–Raphson method 48
   2.9 Exercises 53
   2.10 References 56

3 Linear programming — Part I 57
   3.1 Introduction 57
   3.2 General form of an LP problem 59
   3.3 A two-variable example 60
   3.4 A graphical method of solution 65
   3.5 Introduction to the simplex method 68
   3.6 Basic solutions 69
   3.7 Simplex computation — minimization 72
   3.8 Treatment of maximization problems 81
   3.9 A method for dealing with excess and artificial variables 83
   3.10 Degeneracy 86

3.11	Duality.	88
3.12	Sensitivity analysis	94
3.13	Computer solutions to LP problems	98
3.14	Portal frame design examples — solution	100
3.15	Exercises	108
3.16	References	112

4 Linear programming — Part II — 113
4.1	Classification of special forms of LP problems	113
4.2	Transportation problems	113
4.3	The assignment method	135
4.4	Integer programming	139
4.5	Exercises	145
4.6	References	152

5 Non-linear programming — 154
5.1	Introduction and scope	154
5.2	Effect of non-linearities on the solution	155
5.3	Optimum-seeking strategies	157
5.4	Unconstrained function of a single variable	158
5.5	Unconstrained function of several variables — gradient methods	171
5.6	Unconstrained function of several variables — direct search methods	178
5.7	Pattern search	180
5.8	Constrained function of a single variable	185
5.9	Constrained function of several variables	189
5.10	Exercises	195
5.11	References	198

6 Dynamic programming — 200
6.1	Introduction	200
6.2	A pipeline network problem	201
6.3	Solution of the pipeline network problem	201
6.4	Dynamic programming terminology	204
6.5	The principle of optimality	207
6.6	Allocation processes	215
6.7	Computer solution	223
6.8	Concluding remarks	232
6.9	Exercises	234
6.10	References	236

7 Network analysis — 237
7.1	Introduction	237
7.2	Elementary graph theory	238

	7.3	Network variables and problem types	239
	7.4	Minimum-cost route	246
	7.5	Network capacity problems	256
	7.6	Modification of the directional sense of the network	272
	7.7	Exercises	272
	7.8	References	274

8 Critical path analysis 276
    8.1 Introduction 276
    8.2 Construction of the network 276
    8.3 Formulating the scheduling problem 282
    8.4 The critical path 286
    8.5 Activity times 289
    8.6 The critical time path 290
    8.7 A swimming pool project 291
    8.8 Computer solution 296
    8.9 Resource allocation 297
    8.10 Exercises 306
    8.11 References 308

9 Economic aspects of systems engineering 309
    9.1 Time and money 309
    9.2 Compound interest factors 311
    9.3 Non-uniform series cash flows 316
    9.4 Depreciation and salvage value 320
    9.5 Project appraisal techniques 322
    9.6 Other methods 328
    9.7 Exercises 328
    9.8 References 329

10 Modelling and simulation 330
    10.1 Introduction 330
    10.2 Deterministic models 330
    10.3 Probabilistic models 338
    10.4 Dynamic models 347
    10.5 Exercises 354
    10.6 References 355

11 Decision analysis 356
    11.1 Introduction 356
    11.2 The maximin criterion 356
    11.3 The minimax criterion 358
    11.4 Basic concepts of probability theory 359
    11.5 Bayes' strategies 361
    11.6 Decision trees 363

11.7	Limitations of the expected-value technique.	366
11.8	Concept of utility.	367
11.9	Prior and posterior probabilities.	368
11.10	Pile selection example.	371
11.11	Exercises	376
11.12	References	379

12 Additional worked examples 380

12.1	Introduction.	380
12.2	The Fassbuck Trading Co.	380
12.3	A parking-lot problem.	387
12.4	Design of an insulated pressure vessel.	395
12.5	A new water supply	400
12.6	Minimum weight of a portal frame	413
12.7	References	424

Appendix A   FORTRAN subroutines 425

Index 467

Preface

The book is intended for Civil Engineering undergraduates at the final year or penultimate year level. Although much of the text is necessarily devoted to particular techniques of operations research, one of the principal objectives is to expose the student to the concept and methods of systems analysis and design, to illustrate these notions by means of problem types likely to be encountered in practice, and, most important, to emphasize the full range of socio-economic factors which have a bearing on the solution of multi-disciplinary problems.

The systems approach is shown to be an extension of design methodology in that it usually involves more than one discipline, quite frequently requires the planner or designer to include behavioural policies in the final solution, and almost always defines the objectives in terms of costs or benefits, rather than in technological ideals.

Against the background of the system, characterized by constraints, conflicts, and compromises, the mathematical model is presented as a device to formalize and standardize the treatment of design problems varying widely in both type and scope. The search for an optimal solution provides the motivation for the study of various techniques of operations research. Linear and non-linear programming techniques are discussed in some detail along with dynamic programming, network theory, and decision theory.

The final chapter is devoted to a small number of artificially contrived 'case studies' and worked examples. This serves to illustrate the general methodology of problem-solving and the selection of the appropriate techniques for optimization. Many of the problems are presented in the guise of a series of technical memoranda from imaginary clients. The client's description of the problem is frequently ill-defined and the case studies include the preliminary but vital stages of defining objectives and identifying necessary data.

The order in which the material is presented in a course may differ from the sequence used in the book. For example, there is some advantage in dealing with economic project evaluation (Chapter 9) immediately following Chapter 1, to allow assignments to incorporate present-value calculations. Also, the introduction of modelling and simulation (Chapter 10) early in the course encourages students to become familiar with computing facilities.

In order to allow meaningful examples in mathematical programming to be illustrated and studied by the student, a number of computer subroutines are provided in an Appendix complete with brief documentation. These routines have been selected to be compatible with instructional goals and to encourage discrimination by the user in the use of ready-made software, but are sufficiently versatile to find application in practical problems. Indeed, it may be argued that relatively simple optimization techniques are appropriate for the preliminary analysis and planning of practical problems since in many cases the detailed design is more efficiently completed by simulation techniques or conventional methods.

The FORTRAN routines of Appendix A represent a sub-set of routines in the CIVLIB subroutine library at McMaster University. Readers or instructors who wish to obtain copies of the library or individual routines, in machine readable form, should contact the first-named author for details.

Chapter 1
An Introduction to Systems Engineering

1.1 ENGINEERING CREATIVITY

The function of a Civil Engineer is still well described by the original charter of the Institution of Civil Engineers (1828), viz.

> '... the art of directing the great sources of power in Nature for the use and convenience of man...'

In recent years there has been greater appreciation that 'use' must stop short of abuse and that the 'convenience of man' should not imply advantages for one segment of the community at the expense and inconvenience of another, either now or in the future.

In carrying out this service function for society the engineer is involved in the activities of planning and design, and it is the creative nature of this work that distinguishes the engineer from the scientist. Whereas the latter is concerned with analysing, understanding, and explaining some observed phenomenon, the engineer is required to satisfy a perceived need by creating something which has hitherto not existed. The scientist seeks a unique and comprehensive solution; the engineer, by contrast, must frequently select from a number of alternatives all of which may be imperfect to some degree.

Planning and design may be viewed as distinct but often overlapping parts of the spectrum of creative activity. The planning process involves the setting of goals, feasibility studies, and the balancing of priorities; design, on the other hand, is usually taken to mean the more specific determination of the form of the end product. Both activities are, however, complementary. The planning of a complex project involves preliminary designs of the various components; design often proceeds by a series of stages of progressively increasing refinement and detail up to the point of construction and implementation. In this book, therefore, the terms planning and design will be treated synonymously and used interchangeably.

1.2 DESIGN METHODOLOGY

The process of design may be summarized by the following four sequential activities:

(i) Definition of objectives and identification of design criteria.
(ii) Generation of design alternatives.
(iii) Testing of feasibility of proposals.
(iv) Optimization and refinement of design to maximize effectiveness.

The procedure is illustrated by the flow diagram of Figure 1.1 and the separate steps are discussed in more detail in the following sub-sections.

1.2.1 Definition of objectives

This is often the most difficult or crucial stage in the entire design process. Mistakes or errors of judgement at this point must inevitably have serious

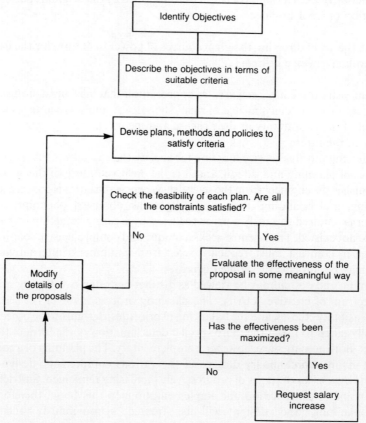

Figure 1.1 Flow diagram of design methodology

consequences on the stages which follow. These statements may appear trite or self-evident, but it is a fact that there exists in the mind of the designer a very strong and natural tendency to focus on the symptoms of the problem rather than on the problem itself. Objectives are frequently not obvious, especially to those who have been closely associated with the problem or with the current 'solution' to the problem, and who have thus become conditioned to thinking along traditional lines. Often considerable pressure may be brought to bear by the client who has a preconception of the solution. Objectives should be stated in the most basic and general terms possible and might typically take the following form:

(a) move traffic,
(b) provide shelter for people,
(c) stop or prevent flooding,
(d) avoid loss of life,
(e) derive profit from an investment,
(f) implement a complex proposal.

In striving to improve on such descriptions of the goals, the designer must exercise lateral thinking by questioning the background to each perceived need in order to build up a picture of the problem environment. Following on the examples of goals listed above one might pose the following questions:

(a) What generates the traffic in the first place?
(b) Where did the unhoused population come from? Is it transient?
(c) Will flood relief at point A cause flood damage at point B to be increased? What is the real cost of flooding?
(d) Why are lives at risk? Is this risk real or imagined?
(e) Who stands to benefit from profit? Is wealth distributed?
(f) Does the proposal require to be so complex? What are the time constraints?

1.2.2 Identification of design criteria

The general objectives having been determined, criteria must be selected whereby these goals may be quantified. Often, these criteria may be readily apparent. With reference to the problem types mentioned previously, the following criteria might be identified to define the objectives more specifically:

(a) traffic volumes, peak ratios, modal split,
(b) population projections, income distribution, group size,
(c) flood levels and discharges, frequency of occurrence,
(d) earthquake probability, code requirements,
(e) limits to investment, competing rates of return, long or short term,
(f) man-hours of effort, quantities of material, maximum and minimum time for activities.

The availability and dependability of data from which actual numbers can be obtained raises queries the answers to which are usually less easy and much more expensive to obtain. Some thought must therefore be given to what constitute *suitable* criteria in terms of the practicality of obtaining a solution within the constraints of time, resources, and technical knowledge.

1.2.3 Generation of design alternatives

The opportunity to develop a wide range of plans and policies is directly dependent on the effort made to define the objective in the most general and fundamental way. For example, the engineer who is instructed to design a new bridge or culvert has considerably less scope for innovation than if the objective had been defined more generally in the following way:

Object: Enable vehicular traffic to cross a watercourse in such a way as to minimize cost and inconvenience to road users, local residents, and taxpayers.

Depending on the scale and nature of this illustrative problem, the designer should be able to question the cost of interrupting the flow of traffic, to examine the use of alternative routes, to consider alternative means of moving vehicles from one side to another, or to estimate the acceptable frequency of flooding on the road. Clearly, as the objective is broadened the nature of the investigation moves through a hierarchy of design situations. With reference to the example mentioned above, Table 1.1 shows the various levels of complexity associated

Table 1.1 Change in scope of the design problem

Level	Objective	Scope of problem
1	Design bridge or culvert of specified dimensions.	Structural and foundation considerations only.
2	Design bridge or culvert for specified discharge.	Hydraulic performance dictates structure dimensions.
3	Design bridge or culvert for 25 year flood.	Hydrology of the catchment included to relate design flood to return period.
4	Design bridge or culvert to minimize inconvenience and construction costs.	Economic considerations required to combine construction costs with cost of traffic interruption, thus defining implicitly the economic return period of the flood.
5	Provide traffic crossing to minimize total costs.	Other alternatives for traffic crossing may now be included in design, e.g. ferry, ford. Economic return period of design flood will differ with each proposal.
6	Provide traffic link across or around watercourse.	Study may now include traffic analysis of network of which the stream crossing is only one link. Re-routing of all or part of the traffic may eliminate or simplify project.

with different statements of the objective. It is obvious that such escalation of the problem may be carried to extremes which are inappropriate for the magnitude of the project. It is equally the responsibility of the engineer to decide the appropriate level of sophistication in the design of a project of known approximate value.

1.2.4 Testing of feasibility

For each alternative proposal which is considered, the first test must be that of feasibility. Technological constraints will usually be carried out routinely in the course of the preliminary design. Other constraints of a social, economic, political, and inter-disciplinary nature may exist and these must be identified and quantified at this stage. As analysis or design of the proposal proceeds, any constraint which is violated will result in the scheme being modified or rejected.

1.2.5 Measures of effectiveness

The most important factors in influencing the nature of the final solution are, in order of importance, the definition of objectives and the selection of an appropriate measure of effectiveness. In its simplest form the effectiveness of a design may be measured in terms of the cost of construction or implementation. In more complex problems account must be taken of the benefits which will accrue following implementation of the solution. When both costs and benefits are variable some function of these must be used, e.g.

$$\text{Effectiveness} = \text{Benefits} - \text{Costs}$$

or

$$\text{Effectiveness} = \text{Benefits}/\text{Costs}.$$

When the benefit–cost ratio is employed care must be taken if certain benefits are labelled as 'negative costs' and vice versa, since this will significantly alter the measure of effectiveness, and may affect the ranking of competing projects.

The real problem, however, is to ensure that all factors have been included in the estimate of costs and benefits.

It is rare that society will judge the outcome of a specific project with a totally unanimous point of view. The greater the complexity of the project the greater the probability that the community will be divided into a number of groups who see the problem and its solution in a different light. As far as possible, the various and possibly conflicting interests of such groups should be expressed by a set of measures of effectiveness which are then combined by means of some weighting process which represents the interest of the various groups as equitably as possible. In attempting to select the necessary weighting factors, the engineer moves rapidly from the area of technology to the arena of politics.

Returning to the problem of the highway bridge or culvert, one may identify at least three 'cost' measures, each of which can be estimated with different degrees of certainty and each of which might be assigned a high priority by different members of the group participating in the decision-making process, e.g.

(i) The actual cost of construction and maintenance can be estimated with reasonable accuracy and would influence strongly the view of the local Council Treasurer.
(ii) The cost arising from interruption to traffic can also be calculated but with less certainty. This factor might figure largely in the thinking of the local Transport officer.
(iii) The inconvenience to local residents might take a number of forms, either during construction, from the generation of heavy vehicular traffic, or the denial of convenient road access. Such 'disutility' costs are rather intangible but could very well be of importance to the local elected Councillor.

The diagram of Figure 1.2 illustrates a less complex problem. The three alternative routes are available between points A and B, the time of travel t_i along

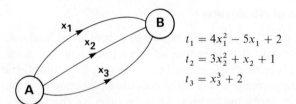

$$t_1 = 4x_1^2 - 5x_1 + 2$$
$$t_2 = 3x_2^2 + x_2 + 1$$
$$t_3 = x_3^3 + 2$$

Figure 1.2 Alternative routes problem

each route being a different function of the number of vehicles per hour x_i assigned to that route.

For the functions given in Figure 1.2, the task is to select the 'best' distribution if the total traffic flow is 100 vehicles per hour. Methods of solving this problem are discussed in Chapter 2 but it is convenient to quote the results here in order to illustrate the point that the solution is sensitive to the measure of effectiveness adopted.

If it is desired to minimize the sum of the three travel times, i.e.

$$t_1 + t_2 + t_3 = \sum_{i=1}^{3} t_i \tag{1.1}$$

the distribution should be $x_1 = 39$, $x_2 = 51$, $x_3 = 10$. On the other hand if the cost to the community in travel time is to be minimized, i.e.

$$\text{cost} = \sum_{i=1}^{3} (x_i t_i) \tag{1.2}$$

the optimum split is found to be $x_1 = 39$, $x_2 = 45$, $x_3 = 16$. Finally, if it is decreed that everyone should spend the same travel time irrespective of route (i.e. $t_1 = t_2 = t_3$) then it can be shown that no opportunity for optimization exists and the unique solution is given by $x_1 = 38.5$, $x_2 = 43.6$, $x_3 = 17.9$ approximately.

1.3 SYSTEMS ENGINEERING

The foregoing discussion on design methodology may be applied almost without qualification to the subject of systems engineering. This is the currently popular name for the engineering processes of planning and design used in the creation of a 'system' or project of considerable complexity. The distinguishing feature between a project design and a system design appears to be in the degree of complexity as measured either by

(i) the number of different components which act together,
(ii) the number or sophistication of the interactions which take place between these components, or
(iii) the fact that the work involves inter-disciplinary teams of specialists to a degree which is uncommon in traditional project design.

In so far as civil engineering comprises several relatively distinct sub-disciplines it may be argued that civil engineers have been practising 'systems engineering' for some considerable time before the term came into vogue. Certain distinguishing characteristics may be noted, however, and it is the purpose of this section to describe these, to relate them to the more basic philosophy of design, and to set the scene for the development of mathematical modelling as one of the principal tools of the systems engineer.

1.3.1 The notion of a system

In the most general sense a system may be defined as a collection of various structural and non-structural (e.g. human) elements which are interconnected and organized in such a way as to achieve some specified objective by the control and distribution of material resources, energy, and information. It is a fundamental characteristic of a properly designed and operated system that the performance achieved by the whole is beyond the total capability of the separate components operating in isolation.

A well-known example is the domestic heating system in which the main objective is the maintenance of a comfortable and healthy environment for human habitation. Auxiliary objectives may be combined with the main system, such as the generation of hot water, provision of cooking facilities, and sometimes the cooling and filtering of the air coupled with control of the humidity. The devices for producing the heat or power are many and varied, employing different sources of energy (coal, oil, gas, electricity, or solar heat) even within the same system. An equally important aspect of the domestic heating system is the minimization of heat loss consistent with acceptable standards of ventilation and natural lighting. Control of the system can be completely automatic or manually adjusted in whole or in part, conditions within the system being monitored either by sophisticated instrumentation or by the human skin. Finally, the behaviour, preferences, and living habits of the residents is of fundamental importance in operating the system at maximum efficiency.

Other examples of systems are easily identified, especially in urban communities. In telecommunications, assembly lines, public transportation, and distribution networks, systems are to be found in every major engineering and technological discipline as well as management and administrative organizations.

1.3.2 The function of the systems engineer

Systems usually comprise a number of distinct elements or components each of which demands a particular type of specialist knowledge. It is the function of the systems engineer to first identify and define the different areas of expertise, then to assemble the necessary skills and knowledge either in the form of personnel or published information, and finally to coordinate the efforts of these various sections of the team to produce the final result. In his capacity of coordinator, the systems engineer must, at each stage in the planning process, maintain an overall view of the whole project and avoid becoming involved in the detailed analysis or design of any particular component. The details of elements and components of the system must be studied *only* in so far as these may affect the performance and cost of the system as a whole.

These activities may be better illustrated with reference to an example. Figure 1.3 illustrates a hypothetical system involving the creation of a reservoir to serve the dual purposes of flood regulation and recreation. The involvement of the systems engineer might typically be summarized as follows:

(a) Identify different components of the system:
   (i) catchment basin,
   (ii) dam and control gates,
   (iii) reservoir,
   (iv) flow data transmission.
(b) Define the areas of expertise likely to be required:
   (i) structural engineering for dam construction,
   (ii) hydrology for flood-frequency prediction,
   (iii) instrumentation engineering for flow data,
   (iv) fresh-water biologist for possible parasite problems,
   (v) economist with special knowledge of recreation parks and auxiliary operations,
   (vi) conservationist — possibly.
(c) Recognize the interactions between the various components and define parameters which affect the interaction between elements and cost effectiveness.
   (i) height of dam affects cost and storage volume,
   (ii) design flood dictates required storage,
   (iii) return frequency of design flood is related to benefits resulting from flood prevention,
   (iv) uncertainty with respect to reservoir level affects cost or use of bathing beaches in summer periods,

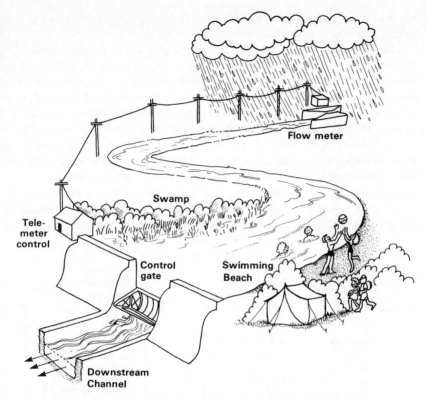

Figure 1.3 A flood control and recreation reservoir

- (v) flooding of low-lying ground may promote parasite breeding which reduces attractiveness of beach,
- (vi) flow measurement telemetered from upstream gauging site affects cost but may give enough advance warning to reduce reservoir size and thus dam costs.

(d) Obtain from literature or from consultation with specialists sufficient information to describe quantitatively the nature of the interactions, cost curves, and benefit functions in terms of the relevant system parameters.
- (i) obtain predictions of probable design floods for various return frequencies,
- (ii) relate storage volume to reduction in flood peak,
- (iii) prepare estimates of damage costs downstream as a function of downstream peak flow,
- (iv) get costs of dam for various storage volumes,
- (v) estimate advance warning obtainable by flow telemetering and corresponding possible drawdown in reservoir by emergency releases,
- (vi) survey potential demand for recreational beach and estimate annual income as function of length of beach,
- (vii) obtain cost estimates for parasite protection.

(e) From knowledge assembled in (d) above develop a model of the system to test interaction of the various components.

The notion of a model is discussed in more detail later in this chapter. For the moment it will suffice to define the process of system modelling as the development of an algorithm or computational procedure whereby the various interactions and trade-off functions can be evaluated for selected trial values of the main input parameters. The algorithm may be implemented by means of a computer program, by analogue devices, by graphical constructions, or by nomograph. Irrespective of the means adopted, the model must have its initial conception as a flow diagram. A typical — but not necessarily complete or ideal — example is illustrated in Figure 1.4.

(f) From the model — as modified and improved in the light of preliminary tests — test the feasibility of the proposal and obtain a preliminary optimization of the main system parameters. At this stage it should be possible to reach a

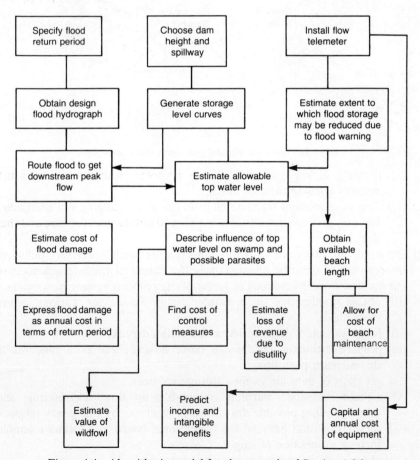

Figure 1.4 Algorithmic model for the example of Section 1.3.2

reasonably firm decision concerning the acceptance or rejection of the proposal.
(g) Assemble and organize the necessary team of talent to carry out more detailed design of the system components.
(h) Maintain an overall view of the development of the design to ensure that the flow of information between the different sections of the team is adequate to ensure interaction without interference, that priorities are maintained, and that the overall objectives of the project are kept in view.

During the design and implementation stage of the project the system engineer must be prepared to assume a more administrative role in coordinating the efforts of the team of specialists, in keeping the various sections together, and also keeping them apart. The team leader must be ready to revise the proposal in the light of any unexpected development in the detailed design in either the economic, technological, or political sector.

1.3.3 Skills required of the systems engineer

The systems engineer concentrates attention on the overall function of the system and on the inter-relationships between the individual components. For this reason his or her primary working medium is mathematics. Physical devices and processes must be represented by mathematical models; performance is analysed and objective measures of costs and benefits are obtained by mathematical approximations; the influence of uncertainty, particularly where human behaviour is involved, must be modelled by probability distributions where necessary. Much of the mathematical simulation is implemented by means of digital computers and, in addition to a basic competence in mathematics, some knowledge of numerical methods and programming is desirable.

Despite the preoccupation with simulation, the systems engineer is more of an engineer than a mathematician. The entire process of system design is essentially practical and, despite the use of modern techniques of operations research, an injection of engineering common sense in the selection of a compromise solution is often necessary. Indeed, it has been said that the process of selecting an optimal policy from a large number of feasible alternatives is as much an art as a science, particularly when the constraints are not merely technological or economic but also social, political, or legal.

The inter-disciplinary nature of most large systems demands that the systems engineer have at least some superficial knowledge of the various areas of expertise on which he or she will rely. A good systems engineer may be thought of as a T-shaped individual with a basic stem of specialization in some discipline to give professional credibility, topped by a lateral cross-piece representing a working knowledge of a relatively large number of related disciplines.

1.4 CASE STUDY OF A SYSTEMS PROBLEM

The problem described in this section can scarcely be claimed to be a systems engineering project since it is relatively simple and falls almost entirely within the single sub-discipline of civil engineering hydraulics. Some of the features of a systems problem are present, however, in that a number of components may be identified between which there are important interactions. The simplicity of the problem, on the other hand, allows the quantitative aspects of the problem to be presented in detail. The reader may visualize elaborations of the situation described here and one of the case studies considered in Chapter 12 is a more complex version of the same problem. The following exchanges of correspondence set the scene and illustrate the process of problem identification and preliminary design.

1.4.1 The Thirstville Project — Preliminary correspondence

To: Max E. Mizer, City Engineer.
From: Rankine Justice, Chairman, Planning Committee.

At the last meeting of the above committee it was decided to proceed with the study of the proposed new development at Thirstville. On the question of water supply, it is estimated that the following fluctuating demand must be provided at an elevation of 300 m:

0900–1800 hours 0.6 m^3/s
1800–0900 hours 0.2 m^3/s.

Please let me have your views on the best method of linking the development with the reservoir at Sparkling Springs.
R.J.

In this initial description of the problem the Planning Committee have not only stated the objective (to provide a supply of water to satisfy a perceived need) but have gone beyond the usual statistics of demographic projections to define the criteria relative to the demand curve which is approximated by a rectangular function. Moreover, the use of an existing reservoir is suggested in the solution. The engineer may have alternative ideas on available sources but, as the following memorandum shows, sets these aside meantime in order to develop preliminary proposals for this source.

To: Chairman, Planning Committee.
From: M. E. Mizer, Engineering.

I have determined that a supply of adequate quantity and quality is available at S.S. Reservoir for the new development at Thirstville. However, the reliable level is only 10 m above the required pressure elevation and it appears likely that pumping may be required. The large variation in demand might best be met by the provision of a balancing reservoir in the town; I assume this would be acceptable. At some stage it might be useful to examine the possibility of lowering the peak demand by load spreading. I shall have some preliminary figures for your next meeting.

M.E.M.

The city engineer's first action is to test the feasibility of the suggested source with respect to quantity, quality, and elevation. He realizes the implication of the low head and also recognizes the hidden cost in the 'peakiness' of the demand. The situation is one which recurs frequently in planning urban service systems and it is wise not to overlook possible economies from spreading transient demands so as to reduce the peaks.

To: Gerrit Startit.
From: Max E. Mizer.

In connection with the Sparkling Springs to Thirstville water main please look up our old contract documents and let me have estimated costs for the following. If possible, provide these as fitted curves.

(i) Cost of pipeline as a function of diameter.
(ii) Cost of balancing reservoir as a function of volume.
(iii) Cost of pump station as a function of horse-power.

Max

Dr Mizer's next action has been to identify the major components which contribute to the performance and cost of the supply system, and also the parameters which characterize these. Being a chief engineer he is able to delegate the task of obtaining cost functions for the three components in terms of the relevant parameters. He will no doubt cast a critical eye over the figures returned by his subordinate to satisfy himself that they are consistent with his considerable experience. Mr Startit responds with the cost functions summarized in the following communication.

Dear Dr Mizer,
Thirstville Project
I note below approximate cost functions which should help in the study of the Thirstville pipeline.

(i) Pipeline $C_1 = 1\,800\,000 D - 180\,000 D^{3/2}$
 D in metres
(ii) Tank $C_2 = 820 + 10.25 V^{1.1}$
 V in cubic metres
(iii) Pump $C_2 = HP(300 + 1.6(10^n))$
 where $n = 3.0 - 0.5 \log(HP)$
 HP in horse-power.

These figures assume that the tank need not be elevated and that the pumps are single stage with normal standby capacity. The figure of 300 in the pump cost represents an estimate of power charges over the life of the system consolidated into a single lump sum.

Yours obediently,
Gerrit

Having assembled most of the necessary data, the chief engineer assigns the job of system optimization to Ms Proffet, a bright and promising young graduate with a flair for computing and numerical analysis.

To: Ms Shona Proffet, Engineering.
From: Max E. Mizer, Chief.

Please find enclosed copies of preliminary correspondence and cost estimates relative to the proposed trunk main serving Thirstville. It seems to me that the scheme may take various forms, e.g.

(i) Gravity supply without balancing tank.
(ii) Gravity supply with balancing tank.
(iii) Pumped supply without balancing tank.
(iv) Pumped supply with balancing tank.

I should like you to analyse these various alternatives and let me have approximate design data for the most economical system.

In addition to the data provided, I have estimated that the friction loss may be approximated from the flow equation,

$$Q = 0.5 D^{8/3} h_f^{1/2}$$

where the pipeline discharge Q is in m³/s and diameter D and headloss h_f are in metres.

Also, power requirements in a pumpstation can be calculated from

$$HP = 17.5 Q (h_f - 10) \quad \text{(horse-power)}$$

since the source reservoir has a minimum reliable elevation 10 m above the required pressure elevation.

Let me have a preliminary report by the 10th of next month.

Max E. Mizer

In his covering letter, Dr Mizer lists what he assesses as the four main options for the scheme, and adds two vital pieces of information of his own. Firstly, he suggests a relationship between discharge, diameter, and headloss, since clearly the choice of resistance law is as much an art as a science and requires the touch of experience. Secondly, an equation relating pump horsepower, discharge, and headloss is defined, not from any uncertainty but because Ms Proffet was trained in a school which persisted in using non-metric units and she is a bit vague about conversion factors.

With that, like a good administrator, he returns to his correspondence tray and puts the project out of his mind until Ms Proffet has come up with some numbers.

1.4.2 Design calculations for the Thirstville Project

It is not known what Ms Proffet's reactions were following receipt of the memorandum and enclosures from the City Engineer. No doubt after the first two readings there was a period of quiet contemplation followed by a careful scrutiny of some of the college notes and texts on hydraulics kept prudently in the desk drawer against just such an eventuality. The calculation sheets which were finally submitted are fortunately carefully preserved and it is instructive to look at the approach to the problem as outlined in the notes illustrated in Figure 1.5.

From the general layout of the system it appears that five design variables sufficiently describe the system, recognizing that both V and HP may be zero. In addition to the interactions suggested by the chief, Ms Proffet recognizes the relationship between the demand curve, the balancing tank storage volume, and the maximum available inflow Q. For the reasoning given, Q and D are selected as decision variables. This means that if values are selected or assumed for Q and D the remaining three variables are uniquely determined, as is the total cost of the scheme. Clearly there are certain values of Q and D — either separately or in combination — which result in an infeasible solution or in unbounded costs. The most obvious constraint of this kind is that Q must be equal at least to the average demand of 0.35 m³/s.

From the flow diagram, the calculation may proceed in a number of ways. The dependence of the system on two variables makes it possible to display the result in either tabular or graphical form, the calculation being carried out by a simple computer program or by hand calculation aided by graphical constructions. Ms Proffet chooses the latter and Figure 1.6 illustrates the layout of the graphical solution.

The solution may be visualized as a cost 'surface' plotted relative to the two independent variables Q and D. Figure 1.7 shows a contour map of this surface and it is apparent that the minimum cost is obtained when the following values are used:

$$Q = 0.35 \text{ m}^3/\text{s}$$
$$D = 0.41 \text{ m}.$$

The cost is then found to be just under \$1 010 000 and the corresponding values of the other design parameters are

$$h_f = 56.93 \text{ m}$$
$$V = 8100 \text{ m}^3$$
$$\text{Power} = 289 \text{ HP}.$$

An important characteristic of the solution is the existence of a local minimum point in the vicinity defined by $Q = 0.54$ m^3/s; $D = 0.5$ m. With this design the storage volume is reduced to $V = 1944$ m^3/s and with $h_f = 47$ m the power requirement is now increased to 350 HP.

Examination of the contour values (or trial calculations using the equations given in Figure 1.5) show that the cost at this local minimum is greater than the global optimum ($Q = 0.35$ m^3/s; $D = 0.41$ m). However, the surface is almost certainly quite sensitive to the cost functions employed and, before making a final recommendation on the case study, it would be desirable to check the effect of small changes in some of the values used to compute the cost. For example, the cost of pumping includes an amount to represent the charges for power (the figure of 300) discounted over the anticipated life of the plant. If this sum were reduced, it is possible that the 'local' minimum would indeed become the true optimum design point. This example therefore illustrates two important facts.

(i) The existence of an apparent minimum (or maximum) is no guarantee that a better solution does not exist. This is true in many practical problems in which non-linear relationships are involved.
(ii) It is often important to test the sensitivity of the solution to changes in one or more of the design or cost parameters in which a measure of uncertainty exists.

Both of these points will be discussed in more detail in certain of the chapters which follow.

The calculation of the surface of Figure 1.7 was obtained using a FORTRAN subroutine indicated by Figure 1.8; alternatively the same calculation can be coded in roughly 100 steps on a programmable pocket calculator. In either case it is important to note that the final cost is the objective to be minimized, but the solutions of interest are the optimal values of the design variables. In the case illustrated, the independent design variables are transferred by means of the subroutine parameters, while other dependent variables are transferred via COMMON block. The advantage of such a procedure will become apparent in Chapter 5 (Non-linear programming).

1.5 ECONOMIC ASPECTS OF SYSTEMS EVALUATION

In the case study of Section 1.4 it was noted that the solution is sensitive to the value used to represent the annual cost of pumping compared with other capital costs. Moreover, a point which was overlooked is the fact that the life of certain components in the system, such as the pipeline or balancing tank, is likely to be much greater than that of rotating machinery such as a pump set. In planning projects it is frequently necessary to make a comparison between proposals involving different proportions of capital investment and operating cost, or to combine in a rational way the contribution in cost or benefit from components which have different life expectancies or which are to be implemented in stages throughout the planned life of the system. In either case it is necessary to be able

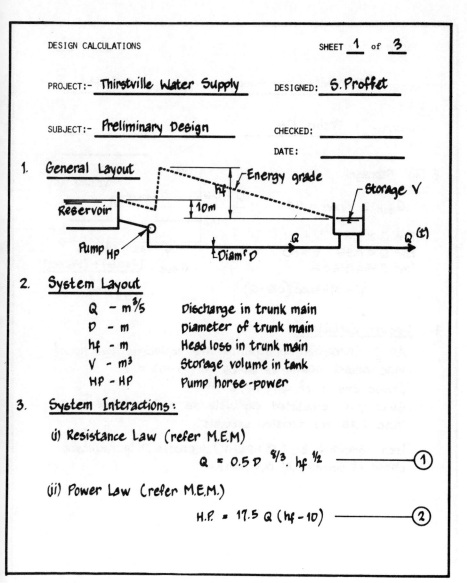

Figure 1.5   Design calculations for Thirstville Water Supply

```
DESIGN CALCULATIONS                      SHEET  2  of  3

PROJECT:- Thirstville Water Supply    DESIGNED: S. Proffet

SUBJECT:- Prelim Des.                 CHECKED: _____
                                      DATE: _____
```

3 (iii) **Storage:**

$Q_{min} \gtrless Q_{ave} = 0.35$

For $Q \geq 0.6$ $V = 0$
For $Q < 0.35$ $V = \infty$
For $0.35 \leq Q \leq 0.6$

$$V = 9 \times 3600 \,(0.6 - Q)$$

$Q_{ave} = \dfrac{(9 \times 0.6) + (15 \times 0.2)}{24} = 0.35 \text{ m}^3/\text{s}$

4. **Design Variables:**

All 3 interaction constraints are equations thus no. of independant design variables = (5-3) = 2
Choose any 2 of Q, D, h_f, V, HP.
Since V is calculated explicitly as a function of Q chose Q as one decision variable.

From remaining 3 (D, h_f, HP) choose D to facilitate choice of commercial pipe sizes.

Figure 1.5 — continued

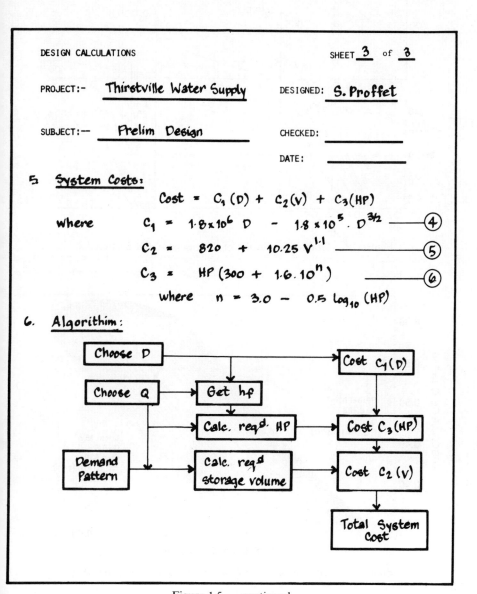

Figure 1.5 — continued

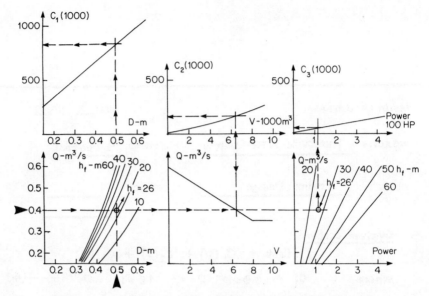

Figure 1.6 Graphical solution of the case study of Section 1.4

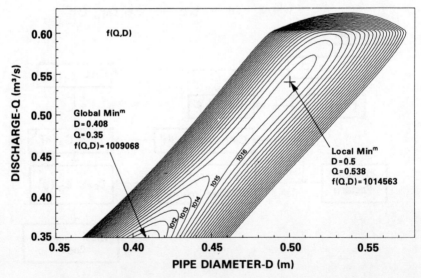

Figure 1.7 Cost surface as a function of diameter and discharge

to relate future and present monetary values in a manner which allows objective comparison. This aspect of project comparison is examined in more detail in Chapter 9.

```
      DIMENSION X(2)
      COMMON /DESIGN/HP,V,HF
   10 PRINT(6,*)"SUPPLY DISCHARGE AND DIAMETER"
      READ(5,*)X(1),X(2)
      IF(X(1).LE.0.0) STOP
      CALL COST(X,2,ANS)
      WRITE(6,40)HP,V,HF,ANS
   40 FORMAT(3F10.1,F15.2)
      GOTO 10
      END
      SUBROUTINE COST(X,N,CST)
C************************************************
C  THE ROUTINE CALCULATES THE TOTAL COST OF A SIMPLE
C  WATER SUPPLY SYSTEM COMPRISING A PIPELINE, PUMP
C  AND BALANCING RESERVOIR.  COST FUNCTIONS AS CODED
C************************************************
      DIMENSION X(N)
      COMMON /DESIGN/HP,V,HF
C  THE COMMON BLOCK RETURNS COMPUTED DESIGN VALUES
      Q=X(1)
      D=X(2)
C  CHECK FOR INFEASIBLE FLOW (.LT. AVERAGE)
      IF(Q.LT.0.35) WRITE(6,20)Q
   20 FORMAT(4H Q=,F10.3,14H  NOT FEASIBLE)
      IF(Q.LT.0.35) STOP
C  CALC. VOLUME OF STORAGE REQUIRED (CHECK FOR V.LT.0.0)
      V=9.0*3600.0*(0.6-Q)
      IF(V.LE.0.0) V=0.0
C  GET HEADLOSS AND HENCE PUMP POWER REQUIRED.
      HF=(2.0*Q/(D**(8.0/3.0)))**2.0
      HPUMP=HF-10.0
      IF(HPUMP.LT.0.0) HPUMP=0.0
      HP=17.5*Q*HPUMP
C  CALC. COSTS OF PIPELINE, RESERVOIR AND PUMP,
      C1=1.8E06*D - 1.8E05*D*SQRT(D)
      C2=0.0
      IF(V.GT.0.0) C2=820.0 + 10.25*V**1.1
      C3=0.0
      IF(HP.LE.0.0) GOTO 30
      EXPT=3.0 - 0.5*ALOG10(HP)
      C3=HP*(300.0 + 1.6*10.0**EXPT)
   30 CONTINUE
      CST=C1 + C2 + C3
      RETURN
      END
```

Figure 1.8   FORTRAN program for case study of Section 1.4

1.6 MATHEMATICAL MODELS

Most engineers are familiar with the idea of using models as an aid in predicting the behaviour of physical devices which may be too costly in time or resources to test in real life. In systems engineering the projects which require to be simulated are not only large and costly, but contain elements such as human behaviour, flow of information, and decision making which cannot be modelled in a physical sense. Moreover, the systems engineer is usually not interested in the detailed working of components but only in the ways in which each component affects the performance and effectiveness of the system as a whole and interacts with other components in the system. For these reasons it is usual to describe the behaviour of a system by means of a mathematical model which is simply a set of equations which describe and represent a real system in terms of its physical, organizational, behavioural, and economic attributes. The model helps to reveal various aspects of the problem, especially those arising as a result of interaction, allows measures of effectiveness to be obtained, and also, on occasions, highlights segments of the system where modification or additional data are required. Of perhaps greatest importance, the mathematical model allows the systems engineer to examine methodically a wide range of system parameters in order to achieve optimum performance and effectiveness.

This process of optimization has already been touched on superficially in the case study of Section 1.4, but more usually the systems engineer will make use of one or more techniques of mathematical programming which have been developed over the last 20 years in the science of operations research. To this end it is helpful to develop certain standard terms and procedures whereby a mathematical model is defined.

1.6.1 Statement of the mathematical model

Each mathematical model usually contains the following elements:

(a) A set of *decision variables*. These are the design variables which may be controlled freely by the designer or decision maker. In many cases these variables will all represent physical quantities such as a load, discharge or dimension. In addition, variables may be numbers representing logical decisions or parameters describing the state of the system. The set of decision variables must be selected in such a way that all other relevant quantities affecting the performance and effectiveness of the system can be evaluated either directly or indirectly. In the mathematical notation used to describe the model the set of decision variables is usually denoted as a vector

$$\mathbf{x} = (x_1, x_2, \ldots, x_n)^\mathrm{T} \tag{1.3}$$

where n is the number of decision variables. The object of the analysis is to determine the best possible set of values with respect to system effectiveness. This is termed the *optimal policy* and is usually denoted by

$$\mathbf{x}^* = (x_1^*, x_2^*, \ldots, x_n^*)^\mathrm{T}. \tag{1.4}$$

(b) An *objective function*. This is the quantity used to measure the effectiveness of a particular policy, and is expressed as a function of the decision variables, i.e.

$$z(\mathbf{x}). \tag{1.5}$$

The aim of the analysis may be either to minimize or maximize the objective function depending on the nature of the problem. In general, however, a problem in minimization may be converted to one of maximization simply by reversing the sign of the objective function. Thus,

$$\text{Minimum } [z(\mathbf{x})] = \text{Maximum } [-z(\mathbf{x})] \qquad (1.6)$$

and problems solved in either form will result in the identical optimal policy.

In some cases the objective function may be defined as being dependent not only on the decision variables $\mathbf{x}$ but also on a set of cost coefficients denoted by a vector

$$\mathbf{c} = (c_1, c_2, \ldots, c_k)^T \qquad (1.7)$$

where k is the number of cost coefficients. For example, in the case study (Section 1.4)

$$\begin{aligned} z &= C_1(D) + C_2(V) + C_3(HP) \\ &= C_1(D) + C_2(\varphi_1(Q)) + C_3(\varphi_2(D, Q)). \end{aligned} \qquad (1.8)$$

These cost coefficients and the decision variables may be combined in any manner whatsoever. In the simplest case the objective function may be defined as the sum of the products of corresponding terms, i.e.

$$z = \sum_{i=1}^{n} (c_i x_i) \qquad (1.9)$$

and where the vectors $\mathbf{c}$ and $\mathbf{x}$ are of the same size. In general, however, the objective function is defined as

$$z = z(\mathbf{c}, \mathbf{x}). \qquad (1.10)$$

(c) A set of *constraints*. Although not mandatory it is usual to find that a mathematical model contains certain conditions which must be satisfied before the set of decision variable values can represent a feasible solution. These *feasibility constraints* are usually expressed as functions of the decision variables and may be either *inequality constraints*, e.g.

$$g_i(\mathbf{x}) \geq 0 \quad i = 1, 2, \ldots, m \qquad (1.11)$$

or *equality constraints*, e.g.

$$h_j(\mathbf{x}) = 0 \quad j = 1, 2, \ldots, p. \qquad (1.12)$$

The functions $g(\mathbf{x})$ and $h(\mathbf{x})$ may involve any number of the decision variables as well as other quantities and numerical constants. If the right-hand sides of the constraints are non-zero, e.g.

$$g(\mathbf{x}) \geq b \qquad (1.13)$$

or
$$h(\mathbf{x}) = b \qquad (1.14)$$

these right-hand-side constants are termed *stipulations*.

In some classes of problem some or all of the design variables are not permitted to take negative values. This is ensured by including in the statement of the model the required number of non-negativity constraints of the form

$$x_i \geq 0. \qquad (1.15)$$

In the case study (Section 1.4) the solution is subject to a number of simple but essential constraints, e.g.

Power: $HP - 17.5Q(h_\mathrm{f} - 10) = 0$
$HP \geq 0$

Flow: $Q - 0.5D^{8/3}h_\mathrm{f}^{1/2} = 0$

Storage: $V - 9 \times 3600(0.5 - Q) = 0$
$V \geq 0$
$Q \geq 0.35.$

Given a set of decision variables, an objective function, and any necessary constraint relationships the mathematical model is expressed as a mathematical programming problem as follows:

$$\underset{\mathbf{x}^*}{\text{Minimize}}\ z(\mathbf{x}, \mathbf{c}) \qquad (1.16)$$

subject to

$$g_i(\mathbf{x}) > 0 \quad i = 1, 2, \ldots, m$$
$$h_j(\mathbf{x}) = 0 \quad j = 1, 2, \ldots, p.$$

In this form the problem of optimization may be tackled in a variety of ways as described in Section 1.6.3.

1.6.2 Classification of mathematical models

Although mathematical models may generally be described in the format defined above, it is useful to classify certain categories of model in order to better understand the applicability of the various techniques available for their solution. The following grouping is not intended to be exhaustive but indicates the types which are frequently encountered.

Linear or non-linear models

If the objective function and all of the constraint functions are linear in terms of the decision variables, the model is said to be *linear* and special conditions may be shown to apply to the solution. If any of the constraints or any part of the objective function contains a non-linearity (e.g. x_1^2, $x_1 x_2$, or sin x_1) the model is said to be *non-linear* and the solution technique which must be employed is generally more complex and expensive. When setting up the model initially it may be advantageous to examine the possibility of eliminating non-linearities in order to obtain the advantage of more efficient algorithms.

Deterministic or probabilistic

A model or an element thereof is said to be *deterministic* if each variable or parameter can be assigned a definite and repeatable value for any given set of conditions. *Probabilistic* models — sometimes termed *stochastic* models — contain variables the values of which are subject to some measure of randomness or uncertainty.

Static or dynamic models

Models which involve time-dependent interactions are said to be *dynamic*; otherwise they are *static*.

Lumped parameter or distributed parameter models

Frequently the characteristics of a component of a system may be described by a single value either because the component is homogeneous with respect to that property, because the value represents an acceptable average, or because insufficient data are available to describe the property in any more detail. In such cases the system is said to be defined by *lumped parameters*. In other cases the property in question — e.g. cross-section, permeability, behaviour, etc. — may vary quite significantly over the domain of the system component, requiring the component to be sub-divided into a number of elements to each of which a specific value of the property is assigned. Such systems and the models used to simulate them are said to possess *distributed parameters*.

1.6.3 Methods of finding an optimal solution

Once a system has been successfully represented by a mathematical model, a number of techniques are available to arrive at the optimal policy which will maximize the effectiveness of the system. The remaining chapters of this book with the exception of the last one, deal with many of these techniques. Once again a complete classification of methods is not given but the following list is broadly representative of the options available. Obviously certain methods are best suited for, or indeed restricted to, particular classes of problem. These points are discussed in some detail in the appropriate chapters and are again touched on in the worked examples of Chapter 12.

Calculus

The classical methods of calculus are discussed in Chapter 2. Although of use in simple problems and of theoretical value in the development of proofs, the methods of calculus become rather impractical when complicated functions are encountered. Even if the objective function and constraints are easily differentiable with respect to the decision variables, the solution frequently degenerates to the solving of systems of non-linear, simultaneous equations which require recourse to be made to numerical methods.

Linear programming

Although at first sight it may appear unlikely that the objective function and constraints can be kept free of non-linearities it is a fact that a very large number of meaningful problems can be expressed as linear models. When this happens it is possible to employ extremely fast and powerful solution techniques. Indeed, for large problems involving many variables linear programming is the only practical way to obtain a solution and linearizing approximations are frequently justified. Chapter 3 introduces the general theory of linear programming and in Chapter 4 a number of applications are described.

Non-linear programming

With the exception of the methods available under the heading of 'methods of calculus', non-linear models must be solved by numerical methods. Probably no other area of numerical methods has received more attention in recent years than the research and development of efficient and reliable algorithms for non-linear programming. Chapter 5 presents an overview of this very large field and concentrates on one or two popular techniques.

Dynamic programming

Many problems of optimization may be viewed as a series of sequential decisions. Although primarily developed for problems which are sequential in time, dynamic programming can be applied profitably to static problems. Dynamic programming is not so much a technique — more a way of looking at the problem. For this reason standard computational routines are less common than in some other areas. Chapter 6 deals with the subject and employs some elementary software for a particular type of problem.

Simulation

When other methods fail due to system complexity or computational difficulty a reasonable attempt at a solution may often be obtained by simulation. Apart from facilitating trial and error design, however, simulation is a valuable technique for studying the sensitivity of system performance to changes in design parameters or operating procedure. Time-sharing computing systems are of particular value in this type of analysis. Some typical examples of simulation are described in Chapter 10.

As well as being concerned with particular techniques of optimization, certain chapters are devoted to studying special classes of problem. Many systems exhibit both the physical and conceptual properties of networks and Chapter 7 deals with the special theory which can be brought to bear on problems of this type.

A particular class of organizational network is the critical path problem and Chapter 8 is devoted to construction management by this technique. Finally, in many systems, although much may be done mathematically to obtain an optimal solution, it frequently happens that decisions have to be made in the face of uncertainty and Chapter 11 discusses decision analysis.

It will be found, not only in studying the remaining chapters but also in putting the ideas contained therein into practice, that computational routines and programs play a large part in systems analysis and design. Many of the techniques described for optimization find expression in standard software designed for application to a wide range of problems of a particular class. To facilitate a start being made in this type of application an Appendix is provided which contains a number of FORTRAN subroutines. These are described, together with complete documentation, so that they may be added to private libraries of routines in teaching institutions or professional offices. It should perhaps be emphasized here that the program coding in many of these routines is designed with clarity of purpose in mind rather than computational efficiency. Obviously whole volumes can be (and are) devoted to software of this type and only a very few representative routines can be included in a text of this scope. Where appropriate the routines are used to demonstrate worked solutions to illustrative problems and no doubt the enthusiastic reader will find many opportunities for modifying or augmenting the routines provided. In so doing it is hoped that the modular nature of subroutine libraries will be adopted, since it is but another expression of the philosophy that is systems engineering.

1.7 REFERENCES

1. Blanchard, B. S., and Fabrycky, W. J. (1981). *Systems Engineering and Analysis*, Prentice-Hall, Englewood Cliffs, New Jersey.
2. de Neufville, R., and Marks, D. (Eds) (1974). *Systems Planning and Design*, Prentice-Hall, Englewood Cliffs, New Jersey.
3. de Neufville, R., and Stafford, J. H. (1971). *Systems Planning for Engineering and Managers*, McGraw-Hill, New York.

Chapter 2
Optimization by Calculus

2.1 INTRODUCTION

The object of the present chapter is to develop some understanding about the properties of the general mathematical model defined in Chapter 1, and to explore the opportunities for solving such problems using classical methods of calculus.

First, we restate the problem:

$$\underset{x^*}{\text{Minimize }} z(\mathbf{x}) \quad \mathbf{x} = (x_1, x_2, \ldots, x_n)^T \tag{2.1}$$

subject to

$$g_i(\mathbf{x}) \geqslant b_i \quad i = 1, 2, \ldots, k \tag{2.2}$$

$$h_j(\mathbf{x}) = 0 \quad j = 1, 2, \ldots, l. \tag{2.3}$$

The functional forms of the $m \, (= k + l)$ constraints may be linear or non-linear. For illustration, consider the following problem involving only 2 variables.

Example 2.1

$$\underset{x_1^*, x_2^*}{\text{Minimize }} z = x_1^2 + 2x_2^2 \tag{2.4}$$

subject to

$$-x_1^2 + x_2 \geqslant 1 \tag{2.5}$$

$$x_1 + x_2 \geqslant 3. \tag{2.6}$$

Since only 2 variables are involved, the problem may be displayed graphically as in Figure 2.1(a).

The inequality constraints (2.5) and (2.6) define respectively the areas above the lines

$$x_2 = x_1^2 + 1$$

and

$$x_2 = -x_1 + 3$$

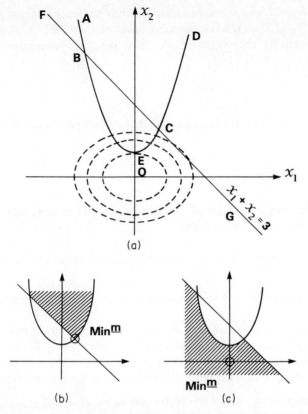

Figure 2.1 (a) Non-linear minimization in 2 variables. (b) Feasible space of Example 2.1. (c) Feasible space of Example 2.2

the intersection of which is shown shaded in Figure 2.1(b). Selection from this feasible space is guided by the objective function which for this case may be graphed as a series of elliptical contours centred on the origin. By inspection the point C ($x_1 = 1$, $x_2 = 2$) is seen to yield the minimum value of $z = 9$.

Example 2.2

Consider an example in which the signs of the constraints (2.5) and (2.6) are reversed, i.e.

$$\underset{x_1^*, x_2^*}{\text{Minimize}}\ z = x_1^2 + 2x_2^2$$

subject to

$$-x_1^2 + x_2 \leq 1 \tag{2.7}$$

$$x_1 + x_2 \leq 3. \tag{2.8}$$

By comparison with the previous example it may be visualized that the feasible space defined by (2.7) and (2.8) is the space below the line FBECG (Figure 2.1(c)). This space includes the origin, which now forms the optimum (minimum) solution.

Example 2.3

If the sign of the x_2 term in (2.5) is changed, the feasible space is now defined by the constraints:

$$x_1^2 + x_2 \leqslant -1 \tag{2.9}$$

$$x_1 + x_2 \geqslant 3.$$

A brief examination shows that this feasible space is non-existent and no solution exists to the minimization problem.

Therefore, it is clear that when a mathematical model includes inequality constraints:

(i) a solution exists only if the constraints intersect to define a finite feasible space,
(ii) the solution may lie on the edge of the feasible space, thus implying that at least one of the constraints is active — i.e. affects the solution,
(iii) the solution may lie in the interior of the feasible space, representing an unconstrained optimization of the non-linear objective function.

Consider now some variation of the previous examples, to examine the significance of equality constraints.

Example 2.4

Assume that (2.5) is changed to an equality constraint. The problem now becomes

$$\underset{x_1^*, x_2^*}{\text{Minimize}} \; z = x_1^2 + 2x_2^2$$

subject to

$$x_1^2 - x_2 + 1 = 0 \tag{2.10}$$

$$x_1 + x_2 \geqslant 3.$$

With reference to Figure 2.1 it may be seen that the feasible 'space' is now reduced to the two arcs AB and CD. Although only line segments, there is still an infinite set of feasible solutions, from which an optimal point may be selected. By inspection, it is apparent that point C is again the minimum.

Example 2.5

Let both (2.5) and (2.6) be converted to equality constraints, to make the problem:

$$\underset{x_1^*, x_2^*}{\text{Minimize}} \; z = x_1^2 + 2x_2^2$$

subject to

$$x_1^2 - x_2 + 1 = 0$$

$$x_1 + x_2 - 3 = 0. \qquad (2.11)$$

The feasible space is now further reduced to the points B and C and the solution is obtained by evaluating the objective functions at these two roots of the simultaneous equations. Optimization in the usual sense is not possible.

Note that if a further equality constraint is added to the problem, e.g. $x_1 = 0$, then, in general, the feasible space is empty and no solution is possible. Only in coincidental cases, e.g. $x_1 = 1$, will a feasible point be defined.

It can be seen that the nature of the solution depends on the number of *variables* n and the number of *equality* constraints l.

For $l > n$ no solution exists in general.
For $l = n$ one or more discrete points (depending on the degree of the constraint equations) are uniquely defined at which the objective function may be evaluated and a solution obtained by inspection.
For $l < n$ the feasible space (if it exists) may contain an infinite number of solutions and an optimal solution, either on the edge or in the interior, must be determined by search or other techniques.

In this connection it may be noted that any inequality constraint may be converted into an equality constraint by the introduction of a slack variable. For example

$$-x_1^2 + x_2 \geqslant 1$$

is equivalent to

$$-x_1^2 + x_2 - x_3 = 1. \qquad (2.12)$$

Thus a problem involving n decision variables, k inequality constraints, and l equality constraints may be converted into a problem with $(n + k)$ variables and $(l + k) = m$ equality constraints. The significance of the relative values of l and n discussed above is therefore unaffected by the value of k.

2.2 UNCONSTRAINED FUNCTIONS OF A SINGLE VARIABLE

A function of a single variable may be examined by means of differential calculus to see if an extreme value (i.e. a maximum or minimum) exists. For this to be possible the function must be:

(i) continuous within the feasible domain of the variable,
(ii) differentiable once to locate critical points, and
(iii) differentiable twice to determine the sense of the critical points.

In general if a function $z = F(x)$ exists then the values of x at which critical points may exist are defined by setting the first derivative to zero, i.e.

$$\frac{dz}{dx} = 0. \tag{2.13}$$

Let one such solution of (2.13) be x^*. Then the nature of the critical point is given by evaluating the second derivative of z at $x = x^*$. Thus for:

$$\frac{d^2z}{dx^2}(x^*) < 0 \quad \text{the critical point is a maximum}$$

$$\frac{d^2z}{dx^2}(x^*) > 0 \quad \text{the critical point is a minimum.}$$

Temporary lapses of memory in recalling this rule may be aided by the picture of Figure 2.2 in which the contents of an inverted or upright wine glass are considered as analogous to the sign of the second derivative.

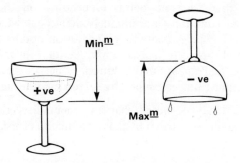

Figure 2.2 Wine-glass analogy

For the case in which the second derivative is zero at the stationary point some further analysis is necessary. Find the lowest derivative which is non-zero at the stationary point, i.e. $d^n z/dx^n(x^*) \neq 0$. If n is odd ($n = 1, 3, 5, \ldots$) the solution found is a point of inflection or a saddle point. If n is even ($n = 2, 4, \ldots$) the rules of the wine-glass analogy may be used as before.

Example 2.6

Traffic passing through a road tunnel is causing a bottleneck and some form of access control or speed control is considered as a means of maximizing the capacity. A survey indicates that the headway (H miles) between vehicles is related to the speed (V mph) by the empirical relation

$$H = \frac{0.4}{50 - V} \tag{2.14}$$

i.e.
$$\text{density } D \text{ (vehicles/mile)} = \frac{1}{H} = \frac{50-V}{0.4} \quad (2.15)$$

or
$$D = 125 - 2.5V.$$

The traffic flow Q (vehicles/hour) is then given by
$$Q = DV = 125V - 2.5V^2. \quad (2.16)$$

The maximum flow is thus given by the critical point defined by
$$\frac{dQ}{dV} = 125 - 5V = 0 \quad (2.17)$$

or
$$V^* = 25 \text{ mph.}$$

Since
$$\frac{d^2Q}{dV^2} = -5 < 0$$

the flow is a maximum when
$$Q = 125 \times 25 - 2.5 \times 25^2 = 1562.5 \text{ vehicles/hour.}$$

Steps might be taken to maintain traffic speed as close to 25 mph as possible (by controlling access). On the other hand, the headway at this speed (84.5 ft) seems somewhat excessive and may be attributable to other factors such as poor illumination or a greasy surface, which could be the real reason for the bottleneck.

2.3 PROBLEMS INVOLVING SIMPLE CONSTRAINTS

Few problems in optimization may be stated as a pure unconstrained mathematical model. The presence of constraints may or may not complicate the method of solution depending on whether:

(i) The constraints are equalities or inequalities.
(ii) The equality constraints allow explicit evaluation of one variable.

Consider first the case of simple equality constraints, such that one variable may be eliminated from the objective function for each equality. This type of manipulation has already been illustrated in the example of Section 1.4. Other examples are given here, in which classical methods of calculus lead to a solution.

Example 2.7

A rather old riveted, steel plate storage tank has an unfortunate habit of 'popping' corroded rivets, issuing a jet of water on the adjacent work area. A small wall is to be located to retain the ponding. If the total depth is H determine the maximum distance x from the tank wall where the jet may strike the ground (Figure 2.3).

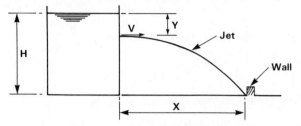

Figure 2.3 Problem of the leaking tank

Let the sprung rivet be at depth Y below the free surface. The jet issues with velocity $V = \sqrt{2gY}$.

The model may be stated as

$$\underset{Y^*}{\text{Maximize}} \ X = f(Y) \tag{2.18}$$

s.t.

$$V = f(Y) \quad \text{or} \quad V = \sqrt{2gY} \tag{2.19}$$

$$X = Vt \tag{2.20}$$

$$(H - Y) = 0.5gt^2. \tag{2.21}$$

The model contains 4 variables: X, Y, V, and t; H and g are known constants (i.e. $l = 3$; $n = 4$; therefore $l < n$).

The trajectory time t is given by (2.21)

$$t = \sqrt{2(H - Y)/g}$$

and substituting in (2.19) and (2.20) gives

$$X = \sqrt{2gY}\sqrt{2(H - Y)/g} = 2\sqrt{Y(H - Y)}. \tag{2.22}$$

The function is now *unconstrained* since the 3 equality constraints have been incorporated in the objective function to eliminate 3 of the 4 variables. A solution may now be found by differentiation.

$$\frac{dX}{dY} = 2\left(\frac{1}{2}\sqrt{\frac{H-Y}{Y}} - \frac{1}{2}\sqrt{\frac{Y}{H-Y}}\right) = 0 \tag{2.23}$$

therefore

$$H - Y = Y$$

or

$$Y = H/2.$$

2.4 UNCONSTRAINED FUNCTIONS OF SEVERAL VARIABLES

Given a function

$$F(\mathbf{x}) \ \text{where} \ \mathbf{x} = (x_1, x_2, \ldots, x_n)^T \tag{2.24}$$

for which all the first partial derivatives exist at all points within the feasible domain of the function, a *necessary* condition for a stationary point is

$$\frac{\partial F}{\partial x_i} = 0 \quad i = 1, 2, \ldots, n. \quad (2.25)$$

Such a stationary point may be a minimum, a maximum, or a point of inflection. Moreover, the possibility must be considered that two or more such stationary points may exist.

In order to test the nature of the stationary point it is necessary to evaluate all the second derivatives at the point, i.e.

$$\frac{\partial^2 F}{\partial x_i \, \partial x_j} \quad i,j = 1, 2, \ldots, n. \quad (2.26)$$

These second derivatives may be assembled in a Hessian matrix, thus:

$$\boldsymbol{H} = \begin{bmatrix} \dfrac{\partial^2 F}{\partial x_1^2} & \dfrac{\partial^2 F}{\partial x_1 \, \partial x_2} & \cdots & \dfrac{\partial^2 F}{\partial x_1 \, \partial x_n} \\ \dfrac{\partial^2 F}{\partial x_2 \, \partial x_1} & \dfrac{\partial^2 F}{\partial x_2^2} & \cdots & \dfrac{\partial^2 F}{\partial x_2 \, \partial x_n} \\ \cdots & \cdots & \cdots & \cdots \\ \dfrac{\partial^2 F}{\partial x_n \, \partial x_1} & \dfrac{\partial^2 F}{\partial x_n \, \partial x_2} & \cdots & \dfrac{\partial^2 F}{\partial x_n^2} \end{bmatrix}. \quad (2.27)$$

It will be noted that $\boldsymbol{H}$ is symmetric since

$$\frac{\partial^2 F}{\partial x_i \, \partial x_j} = \frac{\partial^2 F}{\partial x_j \, \partial x_i}.$$

A *sufficient* condition for the stationary point to be a *minimum* is that all the second derivatives exist at the point, and that all the principal minors are positive. (This implies that $\boldsymbol{H}$ is positive-definite.)

The principal minors of a matrix $\boldsymbol{A}$ are defined as follows:

$$p_1 = a_{11}$$

$$p_2 = \begin{vmatrix} a_{11} & a_{12} \\ a_{21} & a_{22} \end{vmatrix}$$

$$p_3 = \begin{vmatrix} a_{11} & a_{12} & a_{13} \\ a_{21} & a_{22} & a_{23} \\ a_{31} & a_{32} & a_{33} \end{vmatrix} \quad (2.28)$$

$$\cdots\cdots$$

$$p_n = |\boldsymbol{A}|$$

where the symbol $|\ \ |$ means 'the determinant of'.

The corresponding condition that the stationary point is a maximum is that the even-numbered principal minors $(p_2, p_4, \ldots)$ are positive and the odd-numbered principal minors $(p_1, p_3, p_5, \ldots)$ are negative. (This implies that $\bar{H}$ is negative-definite.) If stationarity exists but the sufficient condition is not satisfied, then the nature of the point is indeterminate and must be investigated by other means (e.g. numerical).

Another method of describing the sufficient condition is in terms of the eigenvalues of the Hessian matrix evaluated at the stationary point.

(i) If all eigenvalues of the Hessian are negative at $\mathbf{x}^*$ then the stationary point is a local maximum. Clearly if all the eigenvalues are negative for all possible values of $\mathbf{x}$ the point $\mathbf{x}^*$ is a global maximum.

(ii) If all eigenvalues of the Hessian are positive at $\mathbf{x}^*$ then the stationary point is a local minimum. Similarly, the point is a global minimum if the eigenvalues are positive for all possible values of $\mathbf{x}$.

Example 2.8

A rectangular, open-topped reservoir is to be proportioned. The flowrate, and thus the cost of supplying water to the tank, is inversely proportional to the storage volume provided. Typical figures are as follows:

Volume (m^3)	Flow (m^3/s)	Supply cost ($)
100	0.65	10 000
500	0.50	2000
1000	0.40	1000
2000	0.35	500

The cost of constructing the tank is based on the following rates:

Base $2/m^2
Sides $4/m^2
Ends $6/m^2.

Find the dimensions for least cost.

The supply cost may be approximated by the function

$$c_1 = \frac{10^6}{\text{Volume in m}^3}.$$

Letting the dimensions of the tank in metres be x, y, and z (z being the depth) the construction cost is given by

$$c_2 = 2xy + 8yz + 12xz.$$

Thus the mathematical model may be stated as

$$\text{Minimize } C = 2xy + 8yz + 12xz + \frac{10^6}{xyz}. \tag{2.29}$$

Within the domain of interest defined by $x, y, z > 0$, the necessary conditions for a critical point are given by

$$\frac{\partial C}{\partial x} = 2y + 12z - \frac{10^6}{x^2 yz} = 0 \tag{2.30}$$

$$\frac{\partial C}{\partial y} = 2x + 8z - \frac{10^6}{xy^2 z} = 0 \tag{2.31}$$

$$\frac{\partial C}{\partial z} = 8y + 12x - \frac{10^6}{xyz^2} = 0. \tag{2.32}$$

From these, the following set of simultaneous equations may be obtained:

$$(2.30) \rightarrow 2xy + 12xz = \frac{10^6}{xyz} \tag{2.33}$$

$$(2.31) \rightarrow 2xy + 8yz = \frac{10^6}{xyz} \tag{2.34}$$

$$(2.32) \rightarrow 8yz + 12xz = \frac{10^6}{xyz} \tag{2.35}$$

$$(2.33) \text{ and } (2.34) \rightarrow 12xz = 8yz \tag{2.36}$$

$$(2.33) \text{ and } (2.35) \rightarrow 2xy = 8yz \tag{2.37}$$

$$(2.34) \text{ and } (2.35) \rightarrow 2xy = 12\,xy \tag{2.38}$$

or

$$12x - 8y = 0$$
$$2x - 8z = 0$$
$$ 2y - 12z = 0$$

But

$$\begin{vmatrix} 12 & -8 & 0 \\ 2 & 0 & -8 \\ 0 & 2 & -12 \end{vmatrix} = 0$$

hence no solution is possible. However, from (2.36), (2.37), and (2.38) we have, respectively

$$y = \tfrac{3}{2}x; \quad z = \tfrac{1}{4}x; \quad y = 6z.$$

Using any two of these relations, we may substitute in (2.30), (2.31), or (2.32) to eliminate two of the three variables. For example, eliminating y and z in (2.30) we obtain

$$2 \cdot \tfrac{3}{2}x + 12 \cdot \tfrac{1}{4}x - \frac{10^6}{x^2 \cdot \tfrac{3}{2}x \cdot \tfrac{1}{4}x} = 0 \tag{2.39}$$

or
$$3x + 3x = \frac{(8/3).10^6}{x^4}$$

or
$$x^5 = \frac{8 \times 10^6}{3 \times 6}$$

giving $x^* = 13.48$ m, $y^* = 20.21$ m, and $z^* = 3.37$ m. The total cost is then
$$C = 544.86 + 544.86 + 545.12 + 1089.22$$
$$= 2724.07.$$

To confirm the nature of the critical point we examine the Hessian matrix:

$$H = \begin{bmatrix} \left(\frac{2 \times 10^6}{x^3 yz}\right) & \left(2 + \frac{10^6}{x^2 y^2 z}\right) & \left(12 + \frac{10^6}{x^2 yz^2}\right) \\ \left(2 + \frac{10^6}{x^2 y^2 z}\right) & \left(\frac{2 \times 10^6}{xy^3 z}\right) & \left(8 + \frac{10^6}{xy^2 z^2}\right) \\ \left(12 + \frac{10^6}{x^2 yz^2}\right) & \left(8 + \frac{10^6}{xyz^3}\right) & \left(\frac{2 \times 10^6}{xyz^3}\right) \end{bmatrix} \quad (2.40)$$

therefore

$$H(x^*, y^*, z^*) = \begin{bmatrix} 12 & 6 & 36 \\ 6 & 5.33 & 24 \\ 36 & 24 & 192 \end{bmatrix}. \quad (2.41)$$

The principal minors are:

$p_1 = 12$

$p_2 = (12 \times 5.33) - (6 \times 6) = 27.96$

$p_3 = 12[(5.33 \times 192) - (24 \times 24)] - 6[(6 \times 192) - (36 \times 24)]$
$\quad + 36[(6 \times 24) - (36 \times 5.33)] = 1916.64.$

Since p_1, p_2, and p_3 are all positive, the matrix $H(x^*, y^*, z^*)$ is positive-definite and the critical point is a minimum.

2.5 TREATMENT OF EQUALITY CONSTRAINTS

In Section 2.3 it was demonstrated that problems involving simple constraints may be reduced to an unconstrained form by the method of substitution. This has the double advantage of eliminating one of the constraints and also reducing, by one, the number of variables to be optimized.

In the present section a more general technique is examined which may be applied to constraints of a more complex functional form. In common with the rest of the chapter, however, solutions may be obtained only when the objective function and the constraint function are differentiable with respect to the variables being optimized.

The modified problem may be restated thus:

$$\text{Minimize}_{\mathbf{x}^*} z = F(\mathbf{x}), \quad \mathbf{x} = (x_1, x_2, \ldots, x_n)^T$$

subject to

$$h_j(\mathbf{x}) = 0 \quad j = 1, 2, \ldots, l.$$

For the moment, consider a problem in which $n = 2$ and $l = 1$. Thus

$$\text{Minimize}_{x_1^*, x_2^*} z = F(x_1, x_2) \tag{2.42}$$

subject to

$$h(x_1, x_2) = 0. \tag{2.43}$$

A new function L — called the Lagrangian — is created such that

$$L(x_1, x_2, \lambda) = F(x_1, x_2) + \lambda[h(x_1, x_2)] \tag{2.44}$$

where the term λ is the *Lagrangian multiplier*.

It will be noted, however, that the number of variables is now *increased* by one, compared with the original problem of (2.42) and (2.43).

The inclusion of the additional term in (2.44) will have no effect on the result as long as the constraint is satisfied, since for

$$h(x_1, x_2) = 0$$

then

$$L(x_1, x_2, \lambda) = F(x_1, x_2).$$

Now, as described in Section 2.4, the necessary conditions for a stationary point of the Lagrangian L are given by:

$$\frac{\partial L}{\partial x_1} = \frac{\partial L}{\partial x_2} = \frac{\partial L}{\partial \lambda} = 0 \tag{2.45}$$

or by expanding

$$\frac{\partial L}{\partial x_1} = \frac{\partial F}{\partial x_1} + \lambda \frac{\partial h}{\partial x_1} = 0 \tag{2.46}$$

$$\frac{\partial L}{\partial x_2} = \frac{\partial F}{\partial x_2} + \lambda \frac{\partial h}{\partial x_2} = 0 \tag{2.47}$$

$$\frac{\partial L}{\partial \lambda} = h(x_1, x_2) = 0. \tag{2.48}$$

The solution of these three simultaneous equations will yield the location of one or more stationary points which may be examined to obtain a minimum for L and thus z.

2.5.1 Optimum sizing of a sedimentation tank

Example 2.9

A sedimentation tank is circular in plan with vertical sides above ground and a conical hopper bottom below ground, the slope of the conical part being 3 vertically to 4 horizontally. Determine the proportions to hold a volume of 4070 m³ for minimum area of the bottom and sides (see Figure 2.4).

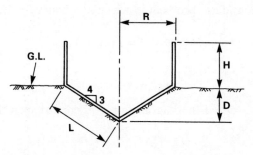

Figure 2.4 Hopper-bottomed sedimentation tank

From Figure 2.4 it is clear that

$$D = 0.75R$$

and

$$L = 1.25R.$$

Thus the volume is given by

$$V = \pi R^2 H + \tfrac{1}{3} D \pi R^2$$
$$= \pi R^2 H + 0.25\pi R^3.$$

The cost is similarly a function of R and H.

Cylindrical: $A_1 = 2\pi RH$
Conical: $A_2 = \pi RL = 1.25\pi R^2.$

The model may be set up as follows:

$$\underset{R^*, H^*}{\text{Minimize}} \; A = 2\pi RH + 1.25\pi R^2 \tag{2.49}$$

subject to

$$\pi R^2 H + 0.25\pi R^3 = \text{Volume } V. \tag{2.50}$$

Forming the Lagrangian, we obtain

$$L = 2\pi RH + 1.25\pi R^2 + \lambda(\pi R^2 H + 0.25\pi R^3 - V). \tag{2.51}$$

The necessary conditions are given by

$$\frac{\partial L}{\partial R} = \frac{\partial L}{\partial H} = \frac{\partial L}{\partial \lambda} = 0$$

i.e.

$$\frac{\partial L}{\partial R} = 2\pi H + 2.5\pi R + 2\pi R\lambda H + 0.75\pi R^2\lambda = 0 \quad (2.52)$$

$$\frac{\partial L}{\partial H} = 2\pi R + \pi R^2\lambda = 0 \quad (2.53)$$

$$\frac{\partial L}{\partial \lambda} = \pi R^2 H + 0.25\pi R^3 - V = 0. \quad (2.54)$$

From (2.53)
$$\lambda = -2/R.$$

Substituting for λ in (2.52) gives

$$2\pi H + 2.5\pi R - 4\pi H - 1.5\pi R = 0$$

or
$$H = R/2.$$

Then from (2.54)
$$\pi R^2 (R/2) + 0.25\pi R^2 - V = 0$$

or
$$0.75\pi R^3 = V$$

therefore
$$R^* = (V/0.75\pi)^{1/3}.$$

For the specified volume of $V = 4070 \text{ m}^3$

$$R^* = 12.0 \text{ m}$$

and thus
$$H^* = 6.0 \text{ m}.$$

The corresponding value of $\lambda^* = -0.167$.

The observant reader will have noticed that although the initial problem appears somewhat complex, the constraint is in fact linear in H so that, by direct substitution, the objective function may be reduced to an unconstrained function in a single variable R which is easily solved by differentiation. The Lagrangian technique is used here for illustrative purposes.

It is also useful to examine the problem graphically. Figure 2.5 shows a number of relations with respect to R and H in the neighbourhood of the solution. The following points may be noted.

(i) The solid line indicates the locus of points which satisfy the volume constraint.
(ii) The dotted contours show the surface of the unmodified objective function A to be monotonic and increasing from the origin as R and H increase. It is seen that the constraint line is approximately tangential to the iso-cost lines at the solution, as it should be.
(iii) The chain lines indicate contours of the Lagrangian $L(R, H, \lambda^*)$ plotted for

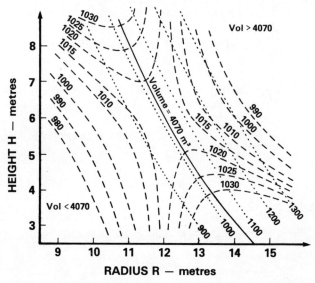

Figure 2.5   Lagrangian and cost surfaces for Example 2.9

$\lambda^* = -1/6$. It is apparent that the surface is a saddle point in the vicinity of the solution.

The last-mentioned point is very significant in that it serves to emphasize that the Lagrangian provides a method of identifying the stationary points and is not itself a method of defining the optimum. The Lagrangian should not be viewed as a modified objective function. The additional terms are not penalty terms (as discussed in Chapter 5), the magnitude of which is proportional to the extent of constraint violation. This is clearly illustrated by the negative sign of the multiplier $\lambda^* = -1/6$.

The magnitude of the Lagrangian multiplier is of significance in interpreting the solution. The fact that λ^* is non-zero is an indication that the constraint is active — i.e. that a lower value of the objective function (area A) could have been located had it not been for the volume constraint. This, of course, is common sense in this problem but may be less obvious in more complex problems. The magnitude of the multiplier may be interpreted as a 'shadow price' — i.e. the extent to which the cost value of the objective function could be 'improved' by relaxing the constraint. For example, if an alternative solution $R = 11.9, H = 5.9$ is examined, the following results are obtained, in comparison with the optimum solution $R^* = 12.0, H^* = 6.0$:

R	H	'cost' A	Volume V
12.0	6.0	1017.876	4071.504
11.9	5.9	997.244	3948.318

The amount by which the constraint is 'relaxed' is 123.186 and the consequent saving, or improvement, in the objective function is -20.632 — a ratio of $-0.167 = \lambda^*$.

2.6 EXTENSION TO MULTIPLE EQUALITY CONSTRAINTS

The treatment of Section 2.5 may be generalized as follows. Consider a problem involving n variables and l equality constraints. The optimization model may be stated (assuming minimization)

$$\underset{x^*}{\text{Minimize}}\ z = F(\mathbf{x}) \quad \mathbf{x} = (x_1, x_2, \ldots, x_n)^T$$

subject to

$$h_j(\mathbf{x}) = 0 \quad j = 1, 2, \ldots, l < n.$$

The Lagrangian must combine the objective function $F(x)$ with every constraint equation, each constraint being modified by an appropriate multiplier λ_j ($j = 1, 2, \ldots, l$). Thus

$$L(\mathbf{x}, \boldsymbol{\lambda}) = F(\mathbf{x}) + \sum_{j=1}^{l} \lambda_j (h_j(\mathbf{x})) \tag{2.55}$$

$$\mathbf{x} = (x_1, x_2, \ldots, x_n)^T$$

$$j = 1, 2, \ldots, l < n.$$

The necessary conditions for a stationary point are then:

$$\frac{\partial L}{\partial x_i} = 0 \quad i = 1, 2, \ldots, n \tag{2.56}$$

$$\frac{\partial L}{\partial \lambda_j} = 0 \quad j = 1, 2, \ldots, l. \tag{2.57}$$

Thus

$$\frac{\partial F}{\partial x_i} + \sum_{j=1}^{l} \lambda_j \, \partial h_j / \partial x_i = 0 \quad i = 1, 2, \ldots, n \tag{2.58}$$

and

$$\sum_{j=1}^{l} h_j \, \partial \lambda_j / \partial \lambda_k = 0 \quad k = 1, 2, \ldots, l \tag{2.59}$$

where

$$\partial \lambda_j / \partial \lambda_k = 1 \quad \text{for } j = k$$
$$= 0 \quad \text{for } j \neq k.$$

2.7 OPTIMIZATION WITH INEQUALITY CONSTRAINTS

Consider a problem involving an inequality constraint and an equality constraint in terms of two variables. The model may be stated in general terms as follows:

$$\text{Minimize } z = F(x_1, x_2) \tag{2.60}$$
$$\underset{x_1^*, x_2^*}{}$$

subject to
$$h(x_1, x_2) = 0 \tag{2.61}$$
and
$$g(x_1, x_2) \geq b. \tag{2.62}$$

It has been demonstrated that the equality constraint (2.61) may be incorporated in a Lagrangian. Equation (2.62) may be converted into a second equality constraint by introducing a new variable θ which is required to be real and which is defined by (2.63)

$$\theta^2 = g(x_1, x_2) - b. \tag{2.63}$$

Obviously if (2.62) is violated then $g(x_1, x_2) < b$ and θ is imaginary. The problem may thus be revised as follows:

$$\text{Minimize } z = F(x_1, x_2)$$
$$\underset{x_1^*, x_2^*}{}$$

subject to
$$h(x_1, x_2) = 0$$
and
$$\theta^2 - g(x_1, x_2) + b = 0.$$

From this model the following Lagrangian may be formed:

$$L(x_1, x_2, \theta, \lambda_1, \lambda_2) = F(x_1, x_2) + \lambda_1 h(x_1, x_2) + \lambda_2[\theta^2 - g(x_1, x_2) + b]. \tag{2.64}$$

As in Section 2.6, the following necessary conditions for stationarity are obtained:

$$\frac{\partial L}{\partial x_1} = \frac{\partial F}{\partial x_1} + \lambda_1 \frac{\partial h}{\partial x_1} - \lambda_2 \frac{\partial g}{\partial x_1} = 0 \tag{2.65}$$

$$\frac{\partial L}{\partial x_2} = \frac{\partial F}{\partial x_2} + \lambda_1 \frac{\partial h}{\partial x_1} - \lambda_2 \frac{\partial g}{\partial x_2} = 0 \tag{2.66}$$

$$\frac{\partial L}{\partial \lambda_1} = \quad h(x_1, x_2) = 0 \tag{2.67}$$

$$\frac{\partial L}{\partial \lambda_2} = \quad \theta^2 - g(x_1, x_2) + b = 0 \tag{2.68}$$

$$\frac{\partial L}{\partial \theta} = \quad 2\lambda_2 \theta = 0. \tag{2.69}$$

The only addition to the case considered in the previous section is equation (2.69). Two possibilities must be examined in solving for the 5 unknowns.

Case 1 $\theta = 0$, in which case the inequality constraint represented by (2.68) reduces to the form

$$g(x_1, x_2) = b.$$

Thus the solution must lie on the constrained boundary (i.e. the constraint is active or binding) and the multiplier λ_2 will be non-zero, representing the 'shadow-price', or the rate at which the objective function may be improved by relaxing the constraint by one unit.

Case 2 $\lambda_2 = 0$, in which case $\theta \neq 0$ and thus

$$g(x_1, x_2) = b + \theta^2 > b$$

i.e. the inequality constraint is inactive and the solution lies in the interior of the feasible space (relative to (2.62)).

Example 2.10

A temporary precast concrete operation is to be set up to produce 70, 180, and 120 units respectively for use in the months of June, July, and August. Due to the need to employ more formwork and overtime labour when production rates are high, the production cost is a quadratic function of the number of units produced in a month, i.e.

$$\text{Production cost/month} = 1.75x^2$$

where x = number of units produced in that month. Some economy can be achieved by spreading production more evenly and storing units. Due to double handling, the cost of storage for one month (or any part) is a linear function of the number of units stored, i.e.

$$\text{Storage cost} = 10x.$$

Determine an optimum production schedule to minimize the total cost.

The model may be set up as follows:

$$\begin{aligned}
\text{Minimize } z = \ & 1.75(x_1^2 + x_2^2 + x_3^2) && \ldots \text{ production} \\
& + 10(x_1 - 70) && \ldots \text{ storage in June} \\
& + 10(x_1 + x_2 - 250) && \ldots \text{ storage in July}
\end{aligned} \quad (2.70)$$

subject to

$$x_1 \geqslant 70 \quad (2.71)$$

$$x_1 + x_2 \geqslant 250 \quad (2.72)$$

$$x_1 + x_2 + x_3 = 370. \quad (2.73)$$

Constraints (2.71) and (2.72) may be rewritten as equality constraints. Thus

$$\theta_1^2 - x_1 + 70 = 0 \quad (2.74)$$

$$\theta_2^2 - x_1 - x_2 + 250 = 0. \quad (2.75)$$

The Lagrangian is then constructed as follows:

$$\begin{aligned}L(x_1, x_2, x_3, \lambda_1, \lambda_2, \lambda_3, \theta_1, \theta_2) = &\ 1.75(x_1^2 + x_2^2 + x_3^2) \\ &+ 10(2x_1 + x_2 - 320) \\ &+ \lambda_1(x_1 + x_2 + x_3 - 370) \\ &+ \lambda_2(\theta_1^2 - x_1 + 70) \\ &+ \lambda_3(\theta_2^2 - x_1 - x_2 + 250).\end{aligned} \quad (2.76)$$

A total of eight simultaneous equations are necessary to define a stationary point:

$$\frac{\partial L}{\partial x_1} = 3.5x_1 + 20 + \lambda_1 - \lambda_2 - \lambda_3 = 0 \quad (2.77)$$

$$\frac{\partial L}{\partial x_2} = 3.5x_2 + 10 + \lambda_1 \quad\quad - \lambda_3 = 0 \quad (2.78)$$

$$\frac{\partial L}{\partial x_3} = 3.5x_3 \quad\quad + \lambda_1 \quad\quad\quad = 0 \quad (2.79)$$

$$\frac{\partial L}{\partial \lambda_1} = x_1 + x_2 + x_3 - 370 \quad\quad = 0 \quad (2.80)$$

$$\frac{\partial L}{\partial \lambda_2} = \theta_1^2 - x_1 + 70 \quad\quad = 0 \quad (2.81)$$

$$\frac{\partial L}{\partial \lambda_3} = \theta_2^2 - x_1 - x_2 + 250 \quad\quad = 0 \quad (2.82)$$

$$\frac{\partial L}{\partial \theta_1} = 2\theta_1 \lambda_2 \quad\quad = 0 \quad (2.83)$$

$$\frac{\partial L}{\partial \theta_2} = 2\theta_2 \lambda_3 \quad\quad = 0. \quad (2.84)$$

Because two inequality constraints have to be considered, there are four cases to be examined. The table below summarizes the possibilities arising from (2.83) and (2.84).

Case	θ_1	θ_2	λ_2	λ_3	Active constraints
1	0	0	finite	finite	Equations (2.71) and (2.72)
2	0	finite	finite	0	Equation (2.71) only
3	finite	0	0	finite	Equation (2.72) only
4	finite	finite	0	0	Neither (2.71) nor (2.72)

Case 1 $\theta_1 = 0;\ \theta_2 = 0$ thus

$$x_1 = 70 \quad\quad \text{therefore } x_1^* = 70$$

and

$$x_1 + x_2 = 250 \quad\quad \text{therefore } x_2^* = 180$$

from (2.80)
$$x_1 + x_2 + x_3 = 370 \quad \text{therefore} \quad x_3^* = 120$$
therefore
$$z = 1.75(70^2 + 180^2 + 120^2)$$
$$= 90\,475.$$

Case 2 $\theta_1 = 0;\ \lambda_3 = 0$ thus $\qquad x_1^* = 70$

$(2.78) \to 3.5x_2 + 10 + \lambda_1 = 0$
$(2.79) \to 3.5x_3 \qquad + \lambda_1 = 0$
$(2.80) \to x_2 + x_3 \qquad = 300$

therefore

$$3.5(300) + 10 + 2\lambda_1 = 0 \qquad \lambda_1^* = -530$$
$(2.77) \to 265 + \lambda_1 - \lambda_2 = 0 \qquad \lambda_2^* = -265$
$(2.78) \to \qquad\qquad\qquad\qquad\qquad x_2^* = 148.57$
$(2.79) \to \qquad\qquad\qquad\qquad\qquad x_3^* = 151.43$
$(2.82) \to \theta_2^2 = x_1 + x_2 - 250 \quad \text{therefore } \theta_2^{2*} = -31.43$

therefore the solution is infeasible.

Case 3 $\theta_2 = 0;\ \lambda_2 = 0$ thus

$(2.82) \to x_1 + x_2 = 250$
$(2.80) \to \qquad\qquad\qquad\qquad x_3^* = 120$
$(2.79) \to \qquad\qquad\qquad\qquad \lambda_1^* = -420$
$(2.77) \to 3.5x_1 - 400 = \lambda_3$
$(2.78) \to 3.5x_2 - 410 = \lambda_3 \quad \text{therefore } \lambda_3^* = 32.5$
$\qquad\qquad\qquad\qquad\qquad\qquad x_1^* = 123.57$
$\qquad\qquad\qquad\qquad\qquad\qquad x_2^* = 126.43$
$(2.81) \to \theta_1^2 = x_1 - 70 = 53.57.$

The solution is feasible and

$$z = 1.75(123.57^2 + 126.43^2 + 120^2) + 10(53.57)$$
$$= 80\,430.36.$$

Case 4 $\lambda_2 = 0;\ \lambda_3 = 0$ thus

$$(2.77) + (2.78) + (2.79) \to 3.5(x_1 + x_2 + x_3) + 30 + 3\lambda_1 = 0$$

therefore
$$\lambda_1^* = -441.67$$
$$x_1^* = 120.48$$
$$x_2^* = 123.33$$
$$x_3^* = 126.19$$

therefore
$$\theta_1^2 = x_1 - 70 = 50.48$$
$$\theta_2^2 = x_1 + x_2 - 250 = -6.19.$$

The solution is infeasible.

The optimum feasible policy is then given by

$$x_1^* = 123.57; \quad x_2^* = 126.43; \quad x_3^* = 120.$$

It will be found that rounding the values of x_1 and x_2 to the nearest integer above or below results in an increase in cost, e.g.

$$x_1 = 123; \quad x_2 = 127; \quad x_3 = 120 \rightarrow z = 80\,431.50$$
$$x_1 = 124; \quad x_2 = 126; \quad x_3 = 120 \rightarrow z = 80\,431.00.$$

2.8 THE GENERALIZED NEWTON–RAPHSON METHOD

From the previous examples it is obvious that, even with relatively simple functional forms in the objective function and constraints, the identification of stationary points by the Lagrangian function frequently results in the solution of sets of simultaneous equations. When these are linear the solution is relatively simple by substitution, but when non-linear equations are encountered some form of numerical solution may be required. One such method is a multi-variable generalization of the Newton–Raphson method, described in almost any book on numerical approximations. Consider the solution of an equation of a single variable of the form

$$F(x) = 0.$$

With reference to Figure 2.6, assume that an initial guess x_0 is used as a starting point. Evaluation of $F(x_0)$ shows the estimate to be in error since $F(x_0) < 0$ (in this sketch).

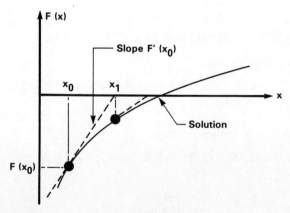

Figure 2.6 The Newton–Raphson method for a function of a single variable

However, the function has a positive slope $F'(x_0)$ at x_0 so that some improvement in the estimate of x may be based on

(i) the error in the function value,
(ii) the slope of the function.

Specifically, a new estimate may be calculated as
$$x_1 = x_0 - F(x_0)/F'(x_0).$$
Successive improvements may be made using the Newton–Raphson iterative formula
$$x_{r+1} = x_r - F(x_r)/F'(x_r). \tag{2.85}$$

Generalization of the approach to functions of several variables is presented here because of the relevance to the solution of systems of simultaneous, non-linear equations.

Let the system of equations be represented by two functions in x and y, i.e.
$$F_1(x, y) = 0$$
$$F_2(x, y) = 0$$
and let an initial approximation to one of the roots be (x_0, y_0). Improvement in the solution may be based on the expansions

$$F_1(x_1, y_1) = F_1(x_0, y_0) + \frac{\partial F_1}{\partial x_0} \delta x_0 + \frac{\partial F_1}{\partial y_0} \delta y_0 \tag{2.86}$$

$$F_2(x_1, y_1) = F_2(x_0, y_0) + \frac{\partial F_2}{\partial x_0} \delta x_0 + \frac{\partial F_2}{\partial y_0} \delta y_0. \tag{2.87}$$

Consider specifically the two equations

$$F_1(x, y) = x^2 - y + 1 = 0 \tag{2.88}$$
$$F_2(x, y) = x - y^2 + 3 = 0. \tag{2.89}$$

Let $(x_0, y_0) = (1, 1)$; then

$$F_1(x_0, y_0) = 1; \quad \frac{\partial F_1}{\partial x} = 2x = 2; \quad \frac{\partial F_1}{\partial y} = -1$$

$$F_2(x_0, y_0) = 3; \quad \frac{\partial F_2}{\partial x} = 1; \quad \frac{\partial F_2}{\partial y} = -2y = -2.$$

Therefore, since the right-hand sides of both F_1 and F_2 are zero,
$$1 + 2\delta x_0 - \delta y_0 = 0$$
$$3 + \delta x_0 - 2\delta y_0 = 0.$$

These linear equations yield a solution
$$\delta x_0 = \tfrac{1}{3}; \quad \delta y_0 = \tfrac{5}{3}.$$
Thus
$$x_1 = x_0 + \delta x_0 = 1.33$$
$$y_1 = y_0 + \delta y_0 = 2.67.$$

The same process repeated with (x_1, y_1) as the starting point yields the next iteration $x_2 = 1.16$, $y_2 = 2.11$, which is converging on the root $(1, 2)$.

For larger systems of equations, the process is more easily described in matrix form. Thus for a set of equations

$$F_i(\mathbf{x}) = 0 \quad i = 1, 2, \ldots, n \quad \text{and} \quad \mathbf{x} = (x_1, x_2, \ldots, x_n)^T$$

the initial estimate of the solution is given by $(\mathbf{x})_0$ where

$$(\mathbf{x}_0) = (x_{10}, x_{20}, \ldots, x_{n0})^T$$

The vector of corrections $\delta \mathbf{x}$ is defined by the equation

$$\mathbf{F}(\mathbf{x}_0) + \mathbf{J}(\mathbf{x}_0).(\delta \mathbf{x})_0 = 0 \tag{2.90}$$

where $\mathbf{J}(\mathbf{x})$ is the Jacobian matrix of first derivatives, i.e.

$$\mathbf{J} = \begin{bmatrix} \partial F_1/\partial x_1 & \partial F_1/\partial x_2 & \ldots & \partial F_1/\partial x_n \\ \partial F_2/\partial x_1 & \partial F_2/\partial x_2 & \ldots & \partial F_2/\partial x_n \\ \ldots\ldots\ldots\ldots\ldots\ldots\ldots\ldots\ldots\ldots\ldots\ldots \\ \partial F_n/\partial x_1 & \partial F_n/\partial x_2 & \ldots & \partial F_n/\partial x_n \end{bmatrix}. \tag{2.91}$$

The vector of corrections is thus given by

$$(\delta \mathbf{x})_0 = -\mathbf{J}^{-1}(\mathbf{x}_0).\mathbf{F}(\mathbf{x}_0) \tag{2.92}$$

in which the system of linear equations in $(\delta \mathbf{x})$ is solved by matrix inversion. Occasionally, the system of linear equations may be solved by substitution. In any event, the iteration must proceed until all of the corrections $(\delta \mathbf{x})_r$ are acceptably small. An illustration of the method is given in the following example.

Example 2.11 A traffic flow problem

Three alternative routes between the origin–destination pair A–B have travel times which as a first approximation may be related to the volume rate of flow (Figure 2.7). If x_i ($i = 1, 2, 3$) represents the number of vehicles per unit time (some arbitrary time interval) the travel times are given by

$$t_1 = 3x_1^2 + 5x_1 - 2$$
$$t_2 = 4x_2^2 - 2x_2 + 1$$
$$t_3 = x_3^3 + 4.$$

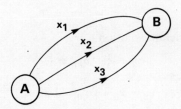

Figure 2.7 The alternative route problem (again!)

If 100 vehicles per unit time leave A, determine the optimal division of traffic between the three routes.

As mentioned in Chapter 1, the solution to a problem such as this is very dependent on the choice of objective function by which the effectiveness of the policy is measured.

One possible objective might be to minimize the sum of the three travel times. This would result in the following model:

subject to
$$\text{Minimize}_{x_1^*, x_2^*, x_3^*} z = t_1 + t_2 + t_3 \tag{2.93}$$

$$t_1 = 3x_1^2 + 5x_1 - 2 \tag{2.94}$$

$$t_2 = 4x_2^2 - 2x_2 + 1 \tag{2.95}$$

$$t_3 = x_3^3 + 4 \tag{2.96}$$

$$x_1 + x_2 + x_3 = 100. \tag{2.97}$$

The first three equality constraints are easily incorporated into the objective function by direct substitution. The fourth equality constraint is linear in each of the variables x_1, x_2, and x_3 and may also be substituted, e.g.

$$x_1 = 100 - x_2 - x_3.$$

This results in an unconstrained function in x_2 and x_3, for which the necessary conditions for a stationary point are obtained as follows:

$$\frac{\partial z}{\partial x_2} = 14x_2 - 607 + 6x_3 = 0 \tag{2.98}$$

$$\frac{\partial z}{\partial x_3} = 3x_3^2 + 6x_3 - 605 + 6x_2 = 0. \tag{2.99}$$

The solution is obtained by substituting for x_2 in the second equation to obtain a quadratic in x_3 which yields the result

$$x_3^* = 10.17$$

from which

$$x_2^* = 39.00$$

and

$$x_1^* = 50.83.$$

A slightly more meaningful objective — and a more complex problem — is obtained by attempting to minimize the total disutility in commuter-hours, i.e.

$$\text{Minimize}_{x_1^*, x_2^*, x_3^*} z = x_1 t_1 + x_2 t_2 + x_3 t_3 \tag{2.100}$$

subject to
$$x_1 + x_2 + x_3 = 100. \tag{2.101}$$

Once again, substitution would be possible, but for purposes of illustration a solution will be obtained by means of a Lagrangian, i.e.

$$L(x_1, x_2, x_3, \lambda) = F_1(x_1) + F_2(x_2) + F_3(x_3)$$
$$+ \lambda(x_1 + x_2 + x_3 - 100) \qquad (2.102)$$

where
$$F_1(x_1) = 3x_1^3 + 5x_1^2 - 2x_1 \qquad (2.103)$$

$$F_2(x_2) = 4x_2^3 - 2x_2^2 + x_2 \qquad (2.104)$$

and
$$F_3(x_3) = x_3^4 + 4x_3. \qquad (2.105)$$

The necessary conditions for a stationary point are given by:

$$\partial L/\partial x_1 = \partial F_1/\partial x_1 + \lambda \quad = f_1 = 0 \qquad (2.106)$$

$$\partial L/\partial x_2 = \partial F_2/\partial x_2 + \lambda \quad = f_2 = 0 \qquad (2.107)$$

$$\partial L/\partial x_3 = \partial F_3/\partial x_3 + \lambda \quad = f_3 = 0 \qquad (2.108)$$

$$\partial L/\partial \lambda = x_1 + x_2 + x_3 - 100 = f_4 = 0 \qquad (2.109)$$

from which values may be obtained for x_1, x_2, x_3, and λ.

Solution of these four simultaneous non-linear equations is somewhat daunting and — as in many such cases — recourse must be made to numerical techniques. Here we may conveniently use the generalized Newton–Raphson method since all of the functions are easily differentiated.

The process of iteration starts from the initial approximation $(x_i)_0$ ($i = 1, 2, 3$) and λ_0.

Improved estimates of the variables are then obtained by computing the changes δx_i ($i = 1, 2, 3$) and $\delta \lambda$ from the set of equations represented by (2.90), i.e.

$$\begin{bmatrix} f_1' & 0 & 0 & 1 \\ 0 & f_2' & 0 & 1 \\ 0 & 0 & f_3' & 1 \\ 1 & 1 & 1 & 0 \end{bmatrix} \begin{bmatrix} \delta x_1 \\ \delta x_2 \\ \delta x_3 \\ \delta \lambda \end{bmatrix} = - \begin{bmatrix} f_1 \\ f_2 \\ f_3 \\ f_4 \end{bmatrix}. \qquad (2.110)$$

The Jacobian of (2.110) is of a particularly simple form since the functions f_1, f_2, and f_3 contain terms in the single variable x_1, x_2, and x_3 respectively, in addition to λ. Solutions for δx_1, δx_2, δx_3, and $\delta \lambda$ may be obtained by row operations relatively simply instead of having to invert the Jacobian. The following steps illustrate the procedure.

(i) The first three rows are divided respectively by f_i' ($i = 1, 2, 3$) following which the fourth row is simplified by reversing the sign and adding all four rows.

This yields

$$\begin{bmatrix} 1 & 0 & 0 & 1/f'_1 \\ 0 & 1 & 0 & 1/f'_2 \\ 0 & 0 & 1 & 1/f'_3 \\ 0 & 0 & 0 & \sum_{i=1}^{3}(1/f'_i) \end{bmatrix} \begin{bmatrix} \delta x_1 \\ \delta x_2 \\ \delta x_3 \\ \delta \lambda \end{bmatrix} = - \begin{bmatrix} f_1/f'_1 \\ f_2/f'_2 \\ f_3/f'_3 \\ \sum_{i=1}^{3}(f_i/f'_i) - f_4 \end{bmatrix}. \quad (2.111)$$

(ii) The fourth column of the Jacobian is simplified by reducing the fourth row element to unity by division and then subtracting proportional parts from the first three rows. Thus we obtain

$$\begin{bmatrix} 1 & 0 & 0 & 0 \\ 0 & 1 & 0 & 0 \\ 0 & 0 & 1 & 0 \\ 0 & 0 & 0 & 1 \end{bmatrix} \begin{bmatrix} \delta x_1 \\ \delta x_2 \\ \delta x_3 \\ \delta \lambda \end{bmatrix} = - \begin{bmatrix} b_1 \\ b_2 \\ b_3 \\ b_4 \end{bmatrix} \quad (2.112)$$

where

$$b_i = f_i/f'_i - b_4/f'_i \quad (i = 1, 2, 3)$$

and

$$b_4 = \left(\sum_{i=1}^{3}(f_i/f'_i) - f_4\right) \bigg/ \sum_{i=1}^{3}(1/f'_i).$$

Equation (2.112) yields the corrections to be applied to x_1, x_2, x_3, and λ. Using the revised values the process is repeated by re-evaluating f_i and f'_i ($i = 1, 2, 3, 4$) for the new values of the variables. The iteration is continued until δx_i ($i = 1, 2, 3$) and $\delta \lambda$ are all acceptably small.

Using arbitrary values for x_1, x_2, x_3, and λ the solution converges in 4 iterations as shown below.

Iteration	1	2	3	4
x_1	33.00	40.88	43.41	44.33	44.41
x_2	33.00	35.38	38.20	38.95	39.02
x_3	34.00	23.74	18.39	16.72	16.57
λ	1.00	−14 886	−17 337	−18 119	−18 193

Programming of the calculation is well within the scope of the more advanced programmable pocket calculators.

2.9 EXERCISES

2.9.1 The workability of an experimental paving material is found to be a function of time. Experimental measures of density may be approximated by the expression

$$z = -3t^4 + 4t^3 + 2.$$

Determine the optimal time.

2.9.2 The percentage removal of impurities in a water treatment process can be improved by the addition of two agents. The percent removed can be expressed as

$$r = 60 + 8x_1 + 2x_2 - x_1^2 - 0.5x_2^2$$

where x_1 and x_2 are the percentage doses of the two agents. Determine the optimal level of dose for each agent.

2.9.3 Determine the position and nature of the stationary points of the following function:

$$z = x_1^3 - x_1 x_2 + x_2^2 - 2x_1 + 3x_2 - 4.$$

2.9.4 Profit from the manufacture and sale of two products is given by the function

$$z = -x_1^2 - x_2^2 + 2x_1 x_2 + 6x_2$$

where x_1 and x_2 are, respectively, the numbers of each product. Resources limit the total number of units (i.e. product 1 plus product 2) to 300. Determine the best distribution.

2.9.5 Solve Exercise 2.9.2 subject to the constraint

$$40x_1 + 20x_2 + 20 \leq 160.$$

2.9.6 Solve Exercise 2.9.5 subject to the additional constraint

$$50x_1 + 35x_2 + 40 \leq 240.$$

2.9.7 A bridge over a ravine comprises three spans and two intermediate columns as illustrated in Figure 2.8. The total span is 100 m and the

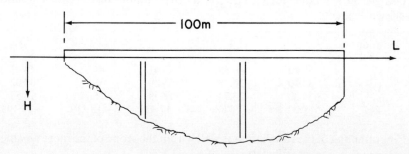

Figure 2.8 Bridge profile for Exercise 2.9.8

topography of the ravine is such that column height H may be expressed as a function of the distance L from the west abutment, thus:

$$H = 2.0 + 0.5L - 0.0045L^2 \text{ (m)}.$$

The cost of beams is given by:

$$C_b = 100x^2$$

where x is the span of the beam. The cost of columns is simply
$$C_c = 5000H.$$
Determine the optimum spans for minimum cost.

2.9.8 A proposed water supply to a factory comprises a pipeline terminating in a balancing tank adjacent to the factory. The purpose of the balancing tank is to allow for a fluctuating demand whereby the factory is to be supplied with 0.28 m³/s over an 8 hour period in each 48 hours. The flow is zero for the remaining 40 hours. Details of the costs and other specifications are given below. Obtain an optimal design by selecting 2 or 3 values of diameter and solving for the other variables. State whether or not the objective function is convex or concave with respect to pipe diameter and state the significance of this on the selection of an optimal solution.

Project Specification
  Pipe length $= 1830$ m
  Flow capacity Q (m³/s) $= 2.75 D^{5/2}$.

The tank is to be square in plan with vertical sides and an open top. Calculate costs in terms of the net internal dimensions with no freeboard allowance.

  Pipe cost/m length $= \$45\sqrt{D}$ (m)
  Tank base cost/m² $= \$10.50$
  Tank wall cost/m² $= \$32.25$.

2.9.4 An open channel cross-section is to be proportioned as a symmetrical triangle such that the section area is 46 m². Also, as the velocity is to be maintained below a value of 2 m/s the wetted perimeter must be not less than 30 m. If excavation costs $C_1 = \$35/\text{m}^3$ and the lining of the inclined banks costs $C_2 = \$21.50/\text{m}^2$, determine the proportions for minimum cost.

2.9.10 Water supply for a community is available from *any or all* of three potential sources, the cost of each supply being described as a quadratic function of the flow rate. Thus:
$$C_1 = 6000 + 100Q_1 + 10Q_1^2$$
$$C_2 = 5000 + 80Q_2 + 20Q_2^2$$
$$C_3 = 7000 + 120Q_3 + 5Q_3^2$$
where C is the cost and Q is the flowrate. Determine the optimal development policy to meet a total demand of 100 units. Discuss briefly how the method of solution and the resulting optimal policy might be modified if the following constraints apply:
$$Q_1 \leqslant 50; \quad Q_2 \leqslant 40; \quad Q_3 \leqslant 65.$$

2.10 REFERENCES

1. Aoki, M. (1971). *Introduction to Optimization Techniques*, MacMillan, New York.
2. Beightler, C. S., Phillips, D. T., and Wilde, D. J. (1978). *Foundations of Optimization*, 2nd Ed., Prentice-Hall, Englewood Cliffs, New Jersey.
3. Daellenbach, H. G., and George, J. A. (1978). *Introduction to Operations Research Techniques*, Allyn and Bacon, Boston.
4. Stark, R. M. and Nicholls, R. L. (1972). *Mathematical Foundations for Design: Civil Engineering Systems*, McGraw-Hill, New York.

Chapter 3
Linear Programming — Part I

3.1 INTRODUCTION

The next two chapters deal with linear programming (LP) problems and associated solution techniques. The present chapter serves as a general introduction to LP and begins with a *graphical method* of solution for problems involving two variables. This approach provides a vital clue to a more general and more powerful method of solution known as the *simplex method* which is then described in detail. The topics of *degeneracy, sensitivity analysis*, and *duality* are then briefly described with the aid of simple examples. Finally, a short but useful computer program based on the simplex method is presented and used to solve several civil engineering problems.

Many civil engineering problems involve the allocation of limited and/or costly resources in order to obtain the best possible results from their use. The term 'best possible use' usually means that the aim is to maximize profits or minimize costs. It also implies that several possible alternatives exist whereby a specific objective may be accomplished. Thus, the problem is to allocate specific amounts of resources to satisfy a given goal in such a way that profits are maximized or costs minimized for the feasible alternatives under consideration.

Linear Programming is probably the best known and most widely used optimization technique for solving certain types of resource-allocation problems. It may be used to solve a wide variety of practical civil engineering problems and can be rapidly programmed for solution on digital computers. Several conditions must be met, however, before LP can be adopted. First, the problem under consideration must be concerned with the specification of non-negative values of a set of variables that optimize a *linear* function expressed in terms of these variables. Secondly, the optimization of this function must also satisfy one or more linear constraints which mathematically describe the availability or requirements of the resources.

The first formal presentation of a linear programming problem and an efficient technique for solving it was made by Dantzig at the US Department of the Air

Force in 1947. The first published work on the topic was by Koopmans[1] in 1951. Dantzig's method, known as the simplex method,[2-5] has subsequently been revised in order to make it computationally more efficient, but the basic approach is the same.

The problems of initial concern were associated with resource allocation and were characterized by very large numbers of decision variables and/or constraints. However, application to civil engineering extends to a broader area of problem solving and design. In problems involving water resources the selection of flows or discharges in a complex network of conduits may often be described by linear relationships. Transportation problems similarly involve multiple applications of the law of continuity which in turn leads to linear constraints. The analysis of forces in a structure can frequently be described in terms of linear algebra (elastic systems involve a basic premise of linearity). Even where strict linearity may not be assumed over the entire range of values of design variables, the power and efficiency of linear programming methods may make justifiable the assumption of approximate linear relationships over a restricted range of variable values.

For the moment, two relatively simple linear programming problems are considered which illustrate the general characteristics of the problem type. The solutions to these and other problems will be discussed in detail later in the chapter.

(a) Concrete Blocks Problem

A concrete products manufacturer makes two types of building blocks: type A and type B. For each set of 100 type A blocks the manufacturer can make a profit of $5 whereas for each set of 100 type B blocks he can make $8. Furthermore, there is a very large market for such blocks nearby and it can be assumed that all blocks which are produced can be sold.

It takes 1 hour to make 100 type A blocks and 3 hours to make 100 type B blocks. Each day there are 12 hours available for block manufacture.

A set of 100 type A blocks requires 2 units of cement, 3 units of fine aggregate, and 4 units of coarse aggregate; whereas a set of 100 type B blocks requires 1 unit of cement and 6 units of coarse aggregate.

Each day, 18 units of cement and 24 units of coarse aggregate are available for block manufacture. There is no restriction on the availability of fine aggregate.

How many type A and type B blocks should be manufactured in order to maximize profits?

(b) Portal Frame Design

The fixed-base rectangular portal frame ABCD shown in Figure 3.1 is subjected to a vertical load V at the centre of the beam BC and a horizontal load H at the top of column CD. The frame is to be designed for minimum weight using plastic hinge methods. The lengths and fully plastic moments of the column (beam) are

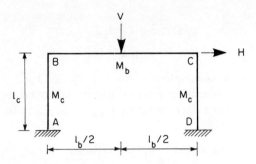

Figure 3.1 Rectangular fixed base portal frame

given as L_c (L_b) and M_c (M_b) respectively. All joints are rigid and it may be assumed that the weight of each frame member is proportional to the fully plastic moment of the member multiplied by its length. Assume that the values of V, H, L_b, and L_c are specified. The problem then reduces to one of finding the plastic moments M_b and M_c which result in a frame of minimum weight.

3.2 GENERAL FORM OF AN LP PROBLEM

In Section 1.6 the concept of a mathematical model and some related notation were introduced. This section describes the special case of a LP problem.

In its general form an LP problem can be written as follows:

Minimize the *objective function*

$$z = c_1 x_1 + c_2 x_2 + \ldots + c_n x_n \tag{3.1}$$

subject to *m constraints* of the form

$$a_{11} x_1 + a_{12} x_2 + \ldots + a_{1n} x_n \leqslant b_1$$
$$a_{21} x_1 + a_{22} x_2 + \ldots + a_{2n} x_n \leqslant b_2$$
$$\ldots\ldots\ldots\ldots\ldots\ldots\ldots\ldots\ldots\ldots\ldots\ldots$$
$$a_{m1} x_1 + a_{22} x_2 + \ldots + a_{mn} x_n \leqslant b_m$$

and subject to the *non-negativity conditions*

$$x_1, x_2, \ldots, x_n \geqslant 0$$

in which c_i are known as the *cost coefficients*, $x_1, \ldots, x_n$ are *structural variables*, b_i are the right-hand sides or *stipulations* of the constraints and are positive by convention, and each a_{ij} is a *structural coefficient* of structural variable x_j in the ith constraint.

Equation (3.1) can be written more compactly as

$$\text{Minimize } z(\mathbf{x}) \tag{3.2}$$

subject to
$$g_i(\mathbf{x}) \leqslant b_i \quad i = 1, 2, \ldots, m$$
with $\mathbf{x} \geqslant 0$

where z and g_i are *linear* functions of $\mathbf{x} = (x_1, x_2, \ldots, x_n)^T$.

The $m \leqslant$ inequality constraints in (3.1) can be converted to equations by the inclusion of non-negative slack variables $x_{n+1}, \ldots, x_{n+m}$ so that it is possible to write

$$\text{Minimize } z = c_1 x_1 + c_2 x_2 + \ldots + c_n x_n \quad (3.3)$$

subject to
$$a_{11} x_1 + a_{12} x_2 + \ldots + a_{1n} x_n + x_{n+1} = b_1$$
$$a_{21} x_1 + a_{22} x_2 + \ldots + a_{2n} x_n + x_{n+2} = b_2$$
$$\cdots\cdots\cdots\cdots\cdots\cdots\cdots\cdots\cdots\cdots\cdots\cdots\cdots$$
$$a_{m1} x_1 + a_{m2} x_2 + \ldots + a_{mn} x_n + x_{n+m} = b_m$$

and
$$x_1, x_2, \ldots, x_{n+m} \geqslant 0.$$

The LP problem is now in the *canonical form* — that is, each constraint is an equation which contains a variable with a unit coefficient. Elsewhere in the other constraints and the objective function this variable has a zero coefficient. (For complete generality the objective function of equation (3.3) should include products for $c_{n+1} x_{n+1}$ to $c_{n+m} x_{n+m}$. In problems such as batch mixes certain resources may be surplus to the finished mix and have some finite cost for their disposal.) Thus, the addition of the slack variables to the $\leqslant$ inequality constraints has automatically produced the canonical form. To convert LP problems with equality and 'greater than or equal to' type constraints into the canonical form, another approach is required. This will be dealt with later in Section 3.8.

It should also be noted that a maximization problem can conveniently be converted to a minimization problem by noticing that maximizing z' is precisely the same as minimizing $z = -z'$.

3.3 A TWO-VARIABLE EXAMPLE

A two-variable example will now be described which will be used later to illustrate both a graphical and the simplex method of solving LP problems.

A local authority is planning to build an industrial waste processing plant at minimum expenditure on land adjacent to two main roads, a canal, and a railway track as shown in Figure 3.2. The unprocessed and processed waste will be transported to and from the plant via the main roads and therefore two access roads will be required. The plant location is given by the coordinates (x_1, x_2) measured from an origin at the junction between the railway and the canal as shown in Figure 3.2. The zone of possible plant locations is limited by several considerations:

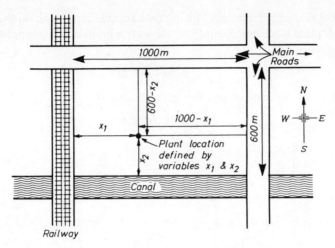

Figure 3.2 Plant location example

(1) The railway track on the western side of the land limits the zone of possible plant locations to

$$x_1 \geqslant 0 \tag{3.4}$$

as shown in Figure 3.3(a).

(2) The canal on the southern side of the land limits the zone to

$$x_2 \geqslant 0 \tag{3.5}$$

as shown in Figure 3.3(b).

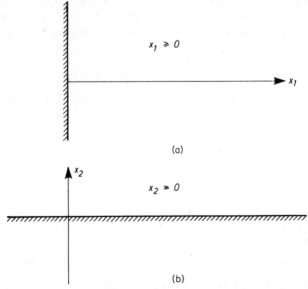

Figure 3.3 (a) Constraint imposed by railway track. (b) Constraint imposed by canal

(3) The local authority boundary line imposes another restriction on the zone of possible plant locations. Figure 3.4† shows this boundary line represented by the equation

$$3x_1 + 5x_2 = 3000 \tag{3.6}$$

thus limiting the zone of possible plant locations to

$$3x_1 + 5x_2 \leq 3000. \tag{3.7}$$

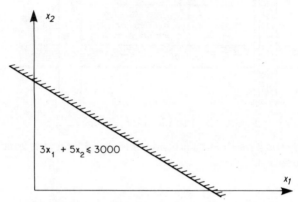

Figure 3.4   Local authority boundary constraint

(4) Precast pile foundations are required for the plant as the soil conditions are very poor on the site, and because of transportation difficulties it has been decided to limit the average pile length to 16 m.

The solid rock lies 6 m below ground level at the S.W. corner of the site and dips 1 in 150 in the East direction and 1 in 50 in the North direction. The depth d to bedrock can thus reasonably be represented by the equation

$$d = \frac{x_1}{150} + \frac{x_2}{50} + 6. \tag{3.8}$$

Since $d \leq 16$ m, a further limit is placed on the zone of possible plant locations, i.e.

$$\frac{x_1}{150} + \frac{x_2}{50} + 6 \leq 16 \tag{3.9}$$

which can be re-written in a more convenient form as

$$x_1 + 3x_2 \leq 1500 \tag{3.10}$$

as shown in Figure 3.5.

† A linear equation $Ax_1 + Bx_2 = C$ can be drawn on the graph as a straight line P_1P_2 as follows:
  (i) Find the point P_1 where the line intersects the x_1-axis; that is put $x_2 = 0$ into the equation and solve for x_1.
  (ii) Find the point P_2 where the line intersects the x_2-axis.
  (iii) Join points P_1 and P_2 with a straight line.

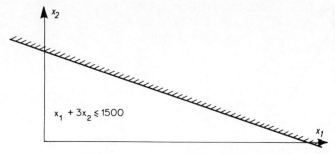

Figure 3.5 Precast pile constraint

(5) The final restriction on the zone of possible plant locations is imposed by the presence of a green belt zone in which the construction of any industrial building is forbidden. The green belt zoning line cuts across the land under consideration and can be represented by the equation

$$3x_1 + 2x_1 = 2100 \qquad (3.11)$$

thus limiting the zone of possible plant locations to

$$3x_1 + 2x_2 \leqslant 2100 \qquad (3.12)$$

as shown in Figure 3.6.

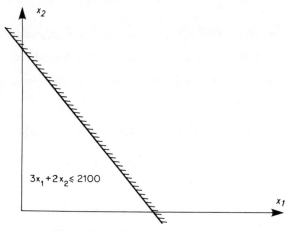

Figure 3.6 Green belt zone constraint

By superimposing all of the constraints the zone of possible plant locations which satisfies all of the constraints can be represented by the shaded zone in Figure 3.7. In LP problems such a zone is known as the *feasible region*.

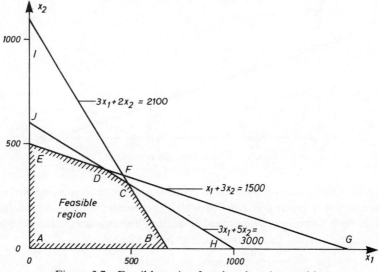

Figure 3.7 Feasible region for plant location problem

The breakdown of the total cost of the industrial waste processing plant is as follows:

$$£405\,000 \quad \text{for the plant structure}$$

$$+ 1500\left(\frac{x_1}{150} + \frac{x_2}{50} + 6\right) \quad \text{for the pile foundations}$$

$$+ 210(1000 - x_1) \quad \text{for the Eastward access road}$$

$$+ 210(600 - x_2) \quad \text{for the Northward access road.}$$

Therefore the objective function for the problem can be expressed as the total cost:

$$z = 405\,000 + 1500\left(\frac{x_1}{150} + \frac{x_2}{50} + 6\right) + 210(1000 - x_1) + 210(600 - x_2)$$

$$= (405\,000 + 9000 + 210\,000 + 126\,000) + (10 - 210)x_1 + (30 - 210)x_2$$

$$= 750\,000 - 200x_1 - 180x_2.$$

The problem can thus be written in the general form described in Section 3.2:

$$\text{Minimize } z = 750\,000 - 200x_1 - 180x_2 \quad \text{(a)} \qquad (3.13)$$

subject to

$$3x_1 + 5x_2 \leqslant 3000 \quad \text{(b)}$$

$$x_1 + 3x_2 \leqslant 1500 \quad \text{(c)}$$

$$3x_1 + 2x_2 \leqslant 2100 \quad \text{(d)}$$

and

$$x_1 \geqslant 0, \quad x_2 \geqslant 0.$$

It should be noted that the objective function and the constraints are all *linear* functions of x_1 and x_2, and also that the non-negativity condition on x_1 and x_2 applies.

(An observant reader may have noticed that two constraints have been omitted:

$$x_1 \leq 1000$$
$$x_2 \leq 600.$$

However, these constraints are automatically satisfied by the three constraints which have been included.)

3.4 A GRAPHICAL METHOD OF SOLUTION

In this method, a two-variable LP problem can be solved graphically; the solution procedure is summarized in Figure 3.8.

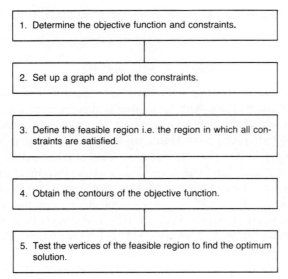

Figure 3.8   Flow chart for the solution of a 2-variable LP problem using a graphical method

Consider the problem outlined in Section 3.3. The feasible region (i.e. the region within which all constraints are satisfied) is shown in Figure 3.9.

Graphically $z = 750\,000 - 200x_1 - 180x_2$ can be drawn as an infinite number of parallel lines corresponding to different values of z, the total cost. For example, if z is arbitrarily set equal to £705 000 then the equation of the corresponding straight line contour is

$$705\,000 = 750\,000 - 200x_1 - 180x_2 \qquad (3.14)$$

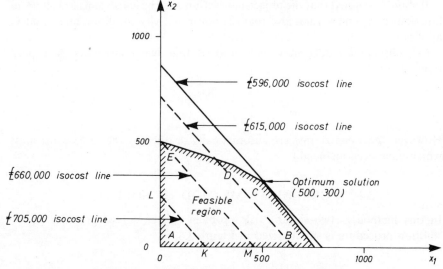

Figure 3.9 A graphical solution to the LP problem

or
$$200x_1 + 180x_2 = 45\,000. \tag{3.15}$$

The intersection of this line with the x_1-axis occurs at point K ($x_1 = 225$, $x_2 = 0$) while its intersection with the x_2-axis is at point L ($x_1 = 0$, $x_2 = 250$). If points K and L are joined together an iso-cost line (or objective function contour) is obtained. Note that the 705 000 iso-cost line is a linear equation that can be expressed as $x_2 = 45\,000/180 - (200/180)x_1$. That is, it has a slope of $-200/180$. All other iso-cost lines have the same slope and are therefore parallel.

Similarly, other iso-cost lines can be drawn yielding different cost levels. For example, line EM in Figure 3.9 represents a 660 000 iso-cost line. Lines EM and KL are parallel but the cost associated with line EM is lower than that of KL. It is possible to continue drawing such iso-cost lines for lower costs as long as the line remains in the feasible region.

Eventually, when either a corner point of the feasible region or one of its boundary lines is reached, no further feasible iso-cost lines can be drawn. In either case the optimal solution(s) has (have) been found.

In the present case if iso-cost lines are drawn on the graph it will be seen that the lowest value of the objective function within the feasible region occurs at vertex C. This point gives the optimal solution which corresponds to $x_1 = 500$ m, $x_2 = 300$ m, and $z = 596\,000$ with constraints 3.13(b) and (d) satisfied as equalities. In fact, it can be proved that, provided the feasible region is convex, the optimal solution will always lie at a vertex (see Figure 3.10 for definition of convexity). To provide extra evidence to support this statement, it is suggested that the reader attempts a graphical solution to the plant location problem with various objective functions. For example, in the unlikely event of the road cost

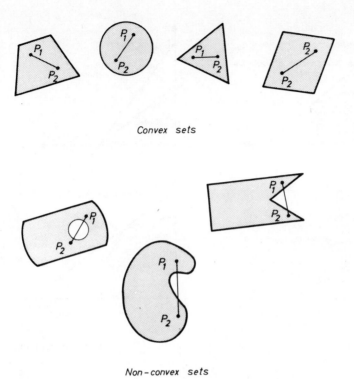

Figure 3.10   Examples of convex and non-convex sets. Convexity in a set of points means that the segment or line joining any two points in the set is also in the set. Here the set of points consists of all those points which satisfy all of the constraints i.e. the feasible region.

being reduced by £90 per metre it may be seen that the solution changes to point B in Figure 3.9. This exercise will also help prepare the reader for the section on sensitivity. Figure 3.11 also gives some further graphical solutions to a number of different LP problems involving two variables. A more rigorous proof that the optimum solution will always lie at a vertex of the feasible region can be obtained from a standard text on linear programming.[2]

The result of the graphical solution to the plant location problem provides a vital clue to the solution of the general LP problem.

The simplex method proceeds by systematically searching adjacent vertices of the feasible region in an attempt to improve and eventually obtain the optimum solution. Thus, the solution would start at point A in Figure 3.7 for the plant location problem and proceed to adjacent point B rather than point E as point B gives a greater improvement in the objective function. The solution would then compare points A and C and subsequently move to point C. At this point no further improvement would be found so that the solution at point C would be taken as optimal.

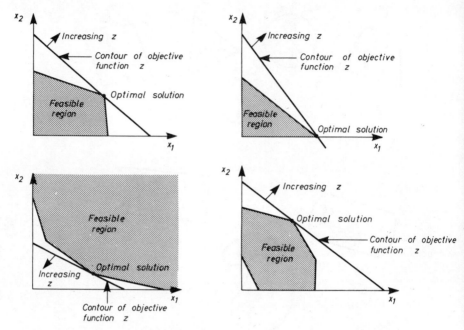

Figure 3.11 Graphical solution to various LP maximization problems

3.5 INTRODUCTION TO THE SIMPLEX METHOD

The simplex method is the most general and powerful of all the various methods of solving LP problems and its main advantage lies in the fact it can provide quick solutions to large problems with the aid of the computer. It not only solves the given LP problem but by simultaneously solving the dual problem (Section 3.11) it provides useful information for sensitivity analysis (Section 3.12).

The simplex method is based on the main theorem of LP which states that

(a) a vertex of the feasible region represents a basic solution, and
(b) one or more vertices will yield an optimal solution.

As mentioned previously, the simplex method can be considered as a systematic means of examining the basic feasible solutions starting with an arbitrary initial basis of m variables where m is the number of constraints. *Slack* and *artificial* variables (Sections 3.7 and 3.9) help in the quick generation of an initial basic feasible solution. If this solution is not optimal the search for an optimal solution continues in an orderly fashion to another vertex with a lower value of the objective function. Such improvement between successive solutions is indicated by the signs of the *simplex criteria* for the non-basic variables (Section 3.7). The solution continues to move from one corner point to another as long as each move results in an improvement (reduction) in the objective function. Since the number of vertices is finite the simplex method guarantees that the optimal

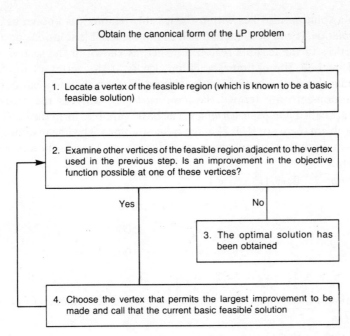

Figure 3.12 General flow chart for the solution of LP problems

solutions will be obtained in a finite number of iterations. The solution procedure is summarized in Figure 3.12.

3.6 BASIC SOLUTIONS

Consider the problem described in Section 3.3. If each of the inequality constraints (excluding the non-negativity condition) is replaced by an equation by introducing slack variables, then the problem can be written as

$$\text{Minimize } z = 750\,000 - 200x_1 - 180x_2 \qquad (3.16)$$

subject to

$$3x_1 + 5x_2 + x_3 \qquad\qquad = 3000$$

$$x_1 + 3x_2 \qquad + x_4 \qquad = 1500$$

$$3x_1 + 2x_2 + \qquad\qquad + x_5 = 2100$$

and

$$x_1 \geqslant 0, \ldots, x_5 \geqslant 0.$$

The significance of the slack variables will be discussed later. It is impossible to solve the constraint equations since there are five unknowns but only three equations. However, if two (i.e. $5 - 3$) variables are set equal to zero it is possible to solve the resulting equations. The solutions obtained in this manner are known as *basic solutions*.

If such a solution also lies within the feasible region, it is known as a *basic feasible solution*. It will be demonstrated that basic feasible solutions occur only at the vertices of the feasible region. Further, as shown in the section on the graphical method, the optimum solution occurs at the vertex of the feasible region and is therefore a basic feasible solution. The variables in a basic solution which are non-zero are termed the *basic variables*, whereas the zero-valued variables are called *non-basic variables*. The basic variables form *the basis*. In the present problem there are 5!/3! 2! = ten basic solutions. The linear simultaneous equations and solutions corresponding to each of the 10 basic solutions will now be examined.

Basic solution 1   Basis (x_1, x_2, x_3) (3.17)

$$3x_1 + 5x_2 + x_3 = 3000$$

$$x_1 + 3x_2 \quad\quad = 1500$$

$$3x_1 + 2x_2 \quad\quad = 2100$$

$$x_1 = 471\tfrac{3}{7}, \quad x_2 = 342\tfrac{6}{7}, \quad x_3 = -128\tfrac{4}{7}, \quad x_4 = 0, \quad x_5 = 0.$$

This solution, which is *infeasible* since x_3 is negative, corresponds to point F in Figure 3.7.

Basic solution 2   Basis (x_1, x_2, x_4) (3.18)

$$3x_1 + 5x_2 \quad\quad = 3000$$

$$x_1 + 3x_2 + x_4 = 1500$$

$$3x_1 + 2x_2 \quad\quad = 2100$$

$$x_1 = 500, \quad x_2 = 300, \quad x_3 = 0, \quad x_4 = 100, \quad x_5 = 0.$$

This solution, which is *feasible*, corresponds to point C in Figure 3.7 and is, in fact, the optimum point.

Basic solution 3   Basis (x_1, x_2, x_5) (3.19)

$$3x_1 + 5x_2 \quad\quad = 3000$$

$$x_1 + 3x_2 \quad\quad = 1500$$

$$3x_1 + 2x_2 + x_5 = 2100$$

$$x_1 = 375, \quad x_2 = 375, \quad x_3 = 0, \quad x_4 = 0, \quad x_5 = 225.$$

This solution, which is *feasible*, corresponds to point D in Figure 3.7.

Basic solution 4   Basis (x_1, x_3, x_4) (3.20)

$$3x_1 + x_3 = 3000$$

$$x_1 + x_4 = 1500$$

$$3x_1 \quad\quad = 2100$$

$$x_1 = 700, \quad x_2 = 0, \quad x_3 = 900, \quad x_4 = 800, \quad x_5 = 0.$$

This solution, which is *feasible*, corresponds to point B in Figure 3.7.

Basic solution 5   Basis (x_1, x_3, x_5) (3.21)

$$3x_1 + x_3 = 3000$$
$$x_1 = 1500$$
$$3x_1 + x_5 = 2100$$

$x_1 = 1500, \quad x_2 = 0, \quad x_3 = -1500, \quad x_4 = 0, \quad x_5 = -2400.$

This solution, which is *infeasible*, corresponds to point G in Figure 3.7.

Basic solution 6   Basis (x_1, x_4, x_5) (3.22)

$$3x_1 = 3000$$
$$x_1 + x_4 = 1500$$
$$3x_1 + x_5 = 2100$$

$x_1 = 1000, \quad x_2 = 0, \quad x_3 = 0, \quad x_4 = 500, \quad x_5 = -900.$

This solution, which is *infeasible*, corresponds to point H in Figure 3.7.

Basic solution 7   Basis (x_2, x_3, x_4) (3.23)

$$5x_2 + x_3 = 3000$$
$$3x_2 + x_4 = 1500$$
$$2x_2 = 2100$$

$x_1 = 0, \quad x_2 = 1050, \quad x_3 = -2250, \quad x_4 = -1650, \quad x_5 = 0.$

This solution, which is *infeasible*, corresponds to point I in Figure 3.7.

Basic solution 8   Basis (x_2, x_4, x_5) (3.24)

$$5x_2 = 3000$$
$$3x_2 + x_4 = 1500$$
$$2x_2 + x_5 = 2100$$

$x_1 = 0, \quad x_2 = 600, \quad x_3 = 0, \quad x_4 = -300, \quad x_5 = 900.$

This solution, which is *infeasible*, corresponds to point J in Figure 3.7.

Basic solution 9   Basis (x_2, x_3, x_5) (3.25)

$$5x_2 + x_3 = 3000$$
$$3x_2 = 1500$$
$$2x_2 + x_5 = 2100$$

$x_1 = 0, \quad x_2 = 500, \quad x_3 = 500, \quad x_4 = 0, \quad x_5 = 1100.$

This solution, which is *feasible*, corresponds to point E in Figure 3.7.

Basic solution 10 Basis (x_3, x_4, x_5) (3.26)

$$x_3 = 3000$$
$$x_4 = 1500$$
$$x_5 = 2100$$

$x_1 = 0, \quad x_2 = 0, \quad x_3 = 3000, \quad x_4 = 1500, \quad x_5 = 2100.$

This solution, which is *feasible*, corresponds to point A in Figure 3.7.

It should be noted that the five basic feasible solutions correspond to the vertices of the feasible region. The remaining five basic solutions violate the nonnegativity condition. All of the basic solutions are summarized in Table 3.1.

Table 3.1 Basic solutions for the plant location problem

Solution number	x_1	x_2	x_3	x_4	x_5	z	Vertex (Refer to Figure 3.7)
1	$471\tfrac{3}{7}$	$342\tfrac{6}{7}$	$-128\tfrac{4}{7}$	0	0		F (infeasible)
2	500	300	0	100	0	596 000	C (feasible)
3	375	375	0	0	225	607 500	D (feasible)
4	700	0	900	800	0	610 000	B (feasible)
5	1500	0	-1500	0	-2400		G (infeasible)
6	1000	0	0	500	-900		H (infeasible)
7	0	1050	-2250	-1650	0		I (infeasible)
8	0	600	0	-300	900		J (infeasible)
9	0	500	500	0	1100	660 000	E (feasible)
10	0	0	3000	1500	2100	750 000	A (feasible)

3.7 SIMPLEX COMPUTATION — MINIMIZATION

As mentioned in an earlier section, the simplex method can be considered as a systematic means of examining the set of basic feasible solutions, starting at an arbitrary initial basis of m variables where m is the number of constraints. If the initial basic feasible solution is not optimal, a neighbouring basis is examined by replacing one of the basic variables, and so forth, until no further improvement can be obtained. This solution procedure for a minimization problem is summarized in Figure 3.12.

The plant location problem introduced in Section 3.3 will now be solved to illustrate the main steps in the simplex method. It should be noted that this problem involves minimization. Problems involving maximization will be discussed in Section 3.8.

Step 1   Formulate the problem

The general LP form of the plant location problem can be written as

$$\text{Minimize } z = 750\,000 - 200x_1 - 180x_2 \qquad (3.27)$$

subject to

$$3x_1 + 5x_2 \leqslant 3000$$
$$1x_1 + 3x_2 \leqslant 1500$$
$$3x_1 + 2x_2 \leqslant 2100$$

and

$$x_1, x_2 \geqslant 0.$$

[Note that we use the notation $1x_1$ instead of x_1 so that we can identify the coefficients in tableaux later.]

Step 2   Convert all constraints to obtain the canonical form

Before proceeding with the simplex method it is necessary to transform the LP problem to its canonical form. This is achieved in this problem by introducing slack variables x_3, x_4, and x_5 so that the problem can be rewritten as

$$\text{Minimize } z = 750\,000 - 200x_1 - 180x_2 \qquad (3.28)$$

subject to

$$3x_1 + 5x_2 + 1x_3 \qquad\qquad\qquad = 3000$$
$$1x_1 + 3x_2 \qquad + 1x_4 \qquad = 1500$$
$$3x_1 + 2x_2 \qquad\qquad\qquad + 1x_5 = 2100$$

$$x_1, x_2, x_3, x_4, x_5 \geqslant 0.$$

A cost coefficient is assigned to each slack variable. Slack variables usually have a physical meaning and their significance in the present case can best be appreciated from the graphical solution. Any solution on a constraint line (i.e. a line on which the constraint is satisfied as an equality) means that the value of the corresponding slack variable is zero and the constraint is fully utilized. Any solution within the limits of the feasible region means that all the slack variables are positive.

Because it is convenient to solve the constraint equations using matrix algebra it is possible to write the constraints in the problem as follows:

$$\begin{bmatrix} 3 & 5 & 1 & 0 & 0 \\ 1 & 3 & 0 & 1 & 0 \\ 3 & 2 & 0 & 0 & 1 \end{bmatrix} \begin{bmatrix} x_1 \\ x_2 \\ x_3 \\ x_4 \\ x_5 \end{bmatrix} = \begin{bmatrix} 3000 \\ 1500 \\ 2100 \end{bmatrix}. \qquad (3.29)$$

Step 3 Construct tableau

After putting the constraints into matrix form, a tableau that summarizes all information relevant to the problem is constructed. The initial tableau shows

(i)   which variable is associated with each column in the matrix,
(ii) the right-hand sides or stipulations,
(iii) the cost coefficients of the basic variables, and
(iv) the basic variables.

The initial tableau, which is still incomplete at this stage, is given as

$c^{(i)}$	Basis	c_j b_i	-200 v_1	-180 v_2	0 v_3	0 v_4	0 v_5	$\dfrac{b_i}{a_{ij}}$
		3000	3	5	1	0	0	
		1500	1	3	0	1	0	
		2100	3	2	0	0	1	

Here the tableau contains the three constraints with slack variables. The five column vectors v_j ($j = 1, 2, \ldots, 5$) denote respectively the column of structural coefficients a_{ij} ($i = 1, 2, 3$) associated with each decision variable x_j. The row vector c_j ($j = 1, 2, \ldots, 5$) defines the cost coefficient associated with each decision variable. Other terms are defined in subsequent steps.

Step 4 Generate an initial basic feasible solution

The iterative process of determining an optimal solution to an LP problem is initiated by selecting a known basic feasible solution. Such a solution can easily be obtained using only slack variables if the values of the other variables (the non-basic variables) are taken as zero. Thus the initial basis consists of x_3, x_4, and x_5 and from (3.28) it can be seen that if $x_1 = 0$ and $x_2 = 0$ then, without actually having to solve the equations,

$$x_3 = 3000, \quad x_4 = 1500, \quad x_5 = 2100, \quad \text{and} \quad z = 750\,000.$$

This initial basic feasible solution is represented by point A in Figure 3.7 and is recorded in the following tableau:

$c^{(i)}$	Basis	c_j b_i	-200 v_1	-180 v_2	0 v_3	0 v_4	0 v_5	$\dfrac{b_i}{a_{ij}}$
	x_3	3000	3	5	1	0	0	
	x_4	1500	1	3	0	1	0	
	x_5	2100	3	2	0	0	1	
		750 000						

The value of the objective function in the initial tableau is given by

$$z = \sum_i c^{(i)} b_i + z_0.$$

The column labelled 'Basis' shows the basic variables for the present solution. The cost coefficients $c^{(i)}$ define the cost coefficient of the ith variable in the basis and are here all equal to zero. To obtain the total cost z for a given solution the cost coefficient $c^{(i)}$ in each row is multiplied by the value of the basic variable associated with that row which is given by the value b_i and added to z_0 which is the constant term in the objective function. In most LP problems z_0 will be equal to zero but in the present case $z_0 = 750\,000$ (see (3.28)). Thus

$$z = z_0 + \sum_i c^{(i)} b_i \qquad (3.30)$$

and the result is given in the last row of the tableau as indicated.

Step 5 Check for optimality and selection of incoming variable

It is now necessary to check whether the present solution is optimal. This is accomplished by calculating a *simplex criterion or coefficient* for each non-basic variable. The criterion relates the net change in the objective function per unit of a non-basic variable hypothetically introduced into the basis. The variable causing the greatest improvement is selected to be the incoming variable. Only one variable will be dropped from the basis at each iteration.

The simplex criterion for non-basic variable x_j is given by

$$\Delta z_j = c_j - \sum_{i=1}^{m} a_{ij} c^{(i)} = c_j - z_j \qquad (3.31)$$

where

$\Delta z_j =$ the value of the simplex criterion for the non-basic variable x_j and represents the net change in the objective function by making $x_j = 1$
$c_j =$ the cost coefficient of the non-basic variable x_j
$c^{(i)} =$ the cost coefficients of the basic variables obtained from the $c^{(i)}$ column of the simplex tableau.

Note that here $c^{(1)}$ is the cost coefficient associated with the basic variable of the first row, $c^{(2)}$ the cost coefficient associated with the basic variable of the second row, etc.

$a_{ij} =$ the coefficient of the ith row and the jth column in the present tableau.

To illustrate the evaluation of the simplex criteria, the effect of introducing a non-basic variable x_1 into the basis is first considered. From (3.28) it can be seen that a movement of the factory by 1 metre in an eastwards direction from its present location at $x_1 = 0$, $x_2 = 0$, to a new location at $x_1 = 1$, $x_2 = 0$, will mean a decrease of £200 in the overall cost. However, this would involve deleting three units of x_3, one unit of x_4, and three units of x_5 in order to balance the equality constraints in (3.28). The contribution of this reduction in x_3, x_4, and x_5 to the total cost is thus

$$a_{11}c^{(1)} + a_{21}c_1^{(2)} + a_{31}c^{(3)} = a_{11}c_3 + a_{21}c_4 + a_{31}c_5$$
$$= 3(0) + 1(0) + 3(0) = 0. \qquad (3.32)$$

Thus the net effect of a unit increase in x_1 (i.e. the simplex criterion for x_1) is given by

$$\Delta z_1 = c_1 - z_1 = \underbrace{1(-200)}_{\text{incoming total cost}} - \underbrace{(3(0) + 1(0) + 3(0))}_{\text{outgoing total cost}} = -200. \qquad (3.33)$$

The current total cost is £750 000 and consequently if x_1 took a value of θ then the new total cost would become

$$z' = \underset{\substack{\uparrow \\ \text{current total cost}}}{z} + \underset{\substack{\uparrow \\ \text{increase in total cost for } x_1 = \theta}}{\Delta z_1 \theta} \qquad (3.34)$$

$$= 750\,000 - 200\theta.$$

Similarly, if the factory is moved one metre in a northwards direction from its initial position the simplex criterion is

$$\Delta z_2 = c_2 - z_2 = c_2 - (a_{12}c^{(1)} + a_{22}c^{(2)} + a_{32}c^{(3)})$$
$$= c_2 - (a_{12}c_3 + a_{22}c_4 + a_{32}c_5) \qquad (3.35)$$
$$= -180 - (5(0) + 3(0) + 2(0))$$
$$= -180.$$

Thus if x_2 took a value of θ then the new total cost would be

$$z' = z + \Delta z_2 \theta$$
$$= 750\,000 - 180\theta. \qquad (3.36)$$

A negative value of Δz_j indicates that a potential improvement can be made by the introduction of one unit of x_j into the basis. Hence the following optimality criteria are adopted:

For *minimization* problems, if one or more simplex criteria Δz_j are negative, the solution is not optimal. If all Δz_j are non-negative the solution is optimal.

It follows that for *maximization* problems, if one or more simplex criteria Δz_j are positive, the solution is not optimal. If Δz_j is negative for all j then the solution is optimal.

Occasionally, a simplex criterion for a non-basic variable will have a value of zero. In this case it is possible to substitute the non-basic variable for one of the basic variables with no change in the objective function. Hence the existence of a zero-valued simplex criteria for a non-basic variable means there are multiple solutions to the problem. In geometric terms, multiple solutions occur when an iso-cost line of the objective function is coincident with (or parallel to) an active constraint so that infinitely many points are optimal solutions to the problem.

Another difficulty can be encountered when two (or more) non-basic variables have identical values of Δz_j. It is suggested that either variable can be chosen arbitrarily to enter the basis when ties are observed.

When one or more of the simplex criteria are negative, an improvement in the objective function can be made by introducing a new variable into the basis. To make the greatest improvement in an objective to be minimized, the entering variable should be the non-basic variable with the smallest (i.e. most negative) Δz_j. In addition this variable must have at least one a_{ij} greater than zero. (The reason for this will be discussed in step 6.)

To bring a new variable into the basis, the column vector v_j associated with the variable must be transformed into a vector with zero coefficients except for a single coefficient of 1. The column vector of the incoming variable will be referred to as the 'pivot column'.

The two negative simplex criteria -200 and -180 indicate respectively the magnitude of the potential improvement in the objective function by the introduction of 1 unit of variable x_1 and x_2 in the new basis. Only one variable can be brought into the basis at one time and since the largest rate of improvement is associated with variable x_1, this variable should be brought into the basis first.

Step 6 Selection of outgoing variable

Having established that x_1 is the incoming variable it is now necessary to select the outgoing variable. The row containing the outgoing variable is called the *pivot row* and its selection is made so that values of all basic variables in the solution are non-negative.

As x_1 increases (with x_2 equal to zero) in (3.28) x_3, x_4, and x_5 become zero respectively when $x_1 = 3000/3 = 1000, x_1 = 1500/1 = 1500$, and $x_1 = 2100/3 = 700$. The variable x_5 becomes equal to zero when $x_1 = 700$ and if x_1 is increased further x_5 will become equal to a value less than zero thus contravening the non-negativity condition. Thus it appears that x_5 should be the variable to leave the basis to make way for the incoming variable x_1. Row 3 of the simplex tableau is associated with variable x_5 and thus becomes the pivot row.

The pivot row can thus be determined by examining the ratios of the stipulations b_i to the corresponding a_{ij} term in the column vector of the incoming variable. A decision is made to remove the variable which has a b_i/a_{ij} with the smallest positive value, where j indexes the terms in the column of the entering variable.

The updated version of the initial tableau can then be written as:

$c^{(i)}$	Basis	c_j b_i	-200 v_1	-180 v_2	0 v_3	0 v_4	0 v_5	$\dfrac{b_i}{a_{ij}}$
0	x_3	3000	3	5	1	0	0	1000
0	x_4	1500	1	3	0	1	0	1500
0	x_5	2100	③	2	0	0	1	700
								x_5 leaves
	z_j Δz_j	750 000	0 -200	0 -180				

pivot x_1 enters

The tableau shows that x_1 (associated with v_1) enters the basis and x_5 (corresponding to v_5), which had been in the basis, is leaving. The coefficient at the intersection of the pivot column and pivot row is normally called the *pivot*. In the present tableau the coefficient in this position is $a_{31} = 3$. The next step is to transform the coefficient matrix so that the pivot equals 1 and all other elements in the pivot column are zero.

Two points should be noted regarding the interpretation of the value of b_i/a_{ij}. Firstly, since the stipulations are non-negative by definition, the algebraic sign of a_{ij} is important. If $a_{ij} < 0$ then it would be possible to increase x_i indefinitely so that the objective function would have a value of $-\infty$ in a minimization problem or a value of $+\infty$ in a maximization problem. (If $a_{ij} = 0$, then b_i/a_{ij} is undefined.) Therefore the coefficient in the column vector of the incoming variable must contain at least one $a_{ij} > 0$ and in identifying the pivot row only positive a_{ij} need be examined.

Secondly, if the maximum value of b_i/a_{ij} occurs at two (or more) rows then an arbitrary choice of the pivot row can result in a condition known as cycling in which there is an endless looping through a set of basic feasible solutions with no convergence towards an optimal solution. A method of avoiding this difficulty is given in Section 3.10.

Step 7   Construction of the simplex tableau for improved solutions

A Gauss–Jordan elimination procedure is used to obtain a pivot column vector with 1 as the new pivot element and zeros elsewhere. The Gauss–Jordan method is summarized in Figure 3.13. The new pivot row is obtained first by dividing the old pivot row by the pivot. Thus in the present case row 3 is divided by the pivot which has a value of 3. To obtain a zero in the element at the junction of the first row and the pivot column, 3 times the new pivot row is subtracted from the first row of the previous tableau. The value 3 is the old row number in the pivot column. Similarly to obtain a zero in the element at the junction of the second row

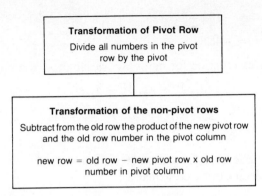

Figure 3.13 Gauss–Jordan method for obtaining simplex tableau for improved solutions

and the pivot column, 1 times the new pivot row is subtracted from the second row of the previous tableau.

In addition, it is necessary to update the entries in the $c^{(i)}$ and basis columns in the tableau. It should be noted that x_1 has replaced x_5 in the basis column. At this stage the new tableau can be written as:

$c^{(i)}$	Basis	c_j b_i	-200 v_1	-180 v_2	0 v_3	0 v_4	0 v_5	$\dfrac{b_i}{a_{ij}}$
0	x_3	3000 $-3(700)$ $=900$	3 $-3(1)$ $=0$	5 $-3(.667)$ $=3$	1 $-3(0)$ $=1$	0 $-3(0)$ $=0$	0 $-3(.333)$ $=-1$	
0	x_4	1500 $-1(700)$ $=800$	1 $-1(1)$ $=0$	3 $-1(.667)$ $=2.333$	0 $-1(0)$ $=0$	1 $-1(0)$ $=1$	0 $-1(.333)$ $=-0.333$	
-200	x_1	2100 $\div 3$ $=700$	3 $\div 3$ $=1$	2 $\div 3$ $=0.667$	0 $\div 3$ $=0$	0 $\div 3$ $=0$	1 $\div 3$ $=0.333$	
		750 000 $-200(700)$ $=610\,000$						

To check whether an optimal solution has been obtained it is necessary to calculate the simplex criteria Δz_j for the non-basic variables. Using (3.31) for non-basic variable x_2, the simplex criterion is given as

$$\Delta z_2 = c_2 - a_{12}c^{(1)} - a_{22}c^{(2)} - a_{32}c^{(3)} = c_2 - z_2$$
$$= -180 - 3(0) - \tfrac{7}{3}(0) - \tfrac{2}{3}(-200)$$

$$= -180 + \tfrac{400}{3}$$
$$= -46.67.$$

Similarly the simplex criterion for non-basic variables x_5 is given as

$$\Delta z_5 = c_5 - a_{15}c^{(1)} - a_{25}c^{(2)} - a_{35}c^{(3)} = c_5 - z_5$$
$$= 0 - (-1)(0) - (-\tfrac{2}{3})(0) - \tfrac{1}{3}(-200)$$
$$= 66.67.$$

Thus the present basic feasible solution is not optimal.
The current tableau can be presented as:

$c^{(i)}$	Basis	c_j b_i	-200 v_1	-180 v_2	0 v_3	0 v_4	0 v_5	$\dfrac{b_i}{a_{ij}}$
0	x_3	900	0	③	1	0	-1	300
0	x_4	800	0	2.333	0	1	-0.333	342.86
-200	x_1	700	1	0.667	0	0	0.333	1050
	z_j	610 000		-133.3			-66.67	
	Δz_j			-46.67			66.67	

x_2 enters

As can be seen from the tableau, to determine the variable leaving the basis the ratios b_i/a_{ij} are calculated. The smallest positive ratio is associated with the first row and hence x_3 leaves the basis to make way for the entering variable x_2. The first row is the pivot row, the second column v_2 is the pivot column, and 3 is the pivot.

The new value of the objective function is given as

$$z = z_0 + c^{(1)}b_1 + c^{(2)}b_2 + c^{(3)}b_3$$
$$= 750\,000 + 0(900) + 0(800) - 200(700)$$
$$= 610\,000.$$

In this solution the factory is sited 700 metres east and 0 metres north of the origin at a cost of £610 000, corresponding to point B in Figure 3.9. Thus the variable coming into the basis is x_2. To determine the variable leaving the basis the ratios b_i/a_{i2} are calculated. The smallest positive ratio is associated with the first row and hence x_3 leaves the basis to make way for the entering variable x_2. The first row is the pivot row, the second column v_2 is the pivot column, and 3 is the pivot.

By applying the Gauss–Jordan procedure once more to obtain a pivot column with 1 as the new pivot element and zeros elsewhere, the following tableau is obtained:

		c_j	-200	-180	0	0	0	$\dfrac{b_i}{a_{ij}}$
$c^{(i)}$	Basis	b_i	v_1	v_2	v_3	v_4	v_5	
-180	x_2	300	0	1	0.333	0	-0.333	
0	x_4	100	0	0	-0.778	1	0.44	
-200	x_1	500	1	0	-0.222	0	0.556	
	z_j	596 000			-15.6		-51.1	
	Δz_j				15.6		51.1	

No negative coefficients

As all the simplex criteria are non-negative, an optimal solution has been obtained,

$$x_1 = 500, \quad x_2 = 300, \quad x_3 = 0, \quad x_4 = 100, \quad x_5 = 0 \quad (3.37)$$

and

$$\begin{aligned} z &= z_0 + c^{(1)}b_1 + c^{(2)}b_2 + c^{(3)}b_3 \\ &= z_0 + c_2 b_1 + c_4 b_2 + c_1 b_1 \\ &= 750\,000 + (-180(300) + 0(100) - 200(500)) \\ &= 596\,000. \end{aligned}$$

In other words the factory should be built 500 m to the east and 300 m to the north of the junction of the canal and the railway which was taken as the origin. In the graphical solution shown in Figure 3.9 this solution corresponds to point C at the junction of the two constraint lines associated with slack variables x_3 and x_5. Hence these variables take zero values as confirmed by the fact that they are non-basic variables in the optimal simplex tableau.

The rules which have been applied in this section are applicable generally for minimization problems and are summarized in Figure 3.14.

Consider Figure 3.15 which shows a three-dimensional view of the feasible region. The progress of the solution from point A to point B and then to C can be compared to the rolling of a ball-bearing down the edge of the feasible region with the greatest slope.

With reference to Figure 3.15 the solution at point A has the alternative of moving from A to B or A to E, the former having the most attractive slope; from point B the alternatives of moving from B back to A or B to C clearly leave only one possible choice for improvement of the solution since BC is the only edge with a negative slope. Finally at point C both the slopes CD and CB are adverse and the optimal solution or lowest point has been found.

3.8 TREATMENT OF MAXIMIZATION PROBLEMS

Many LP problems in civil engineering involve maximization (e.g. of profits) rather than minimization of costs. There are three basic approaches for dealing

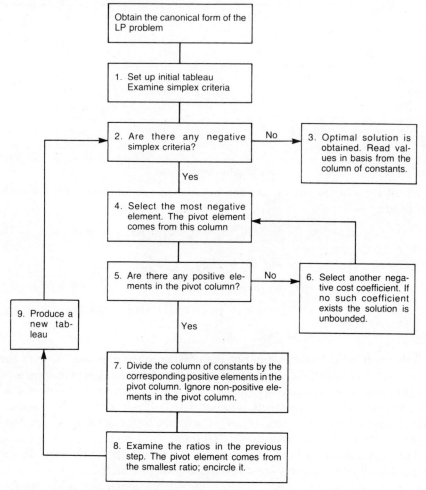

Figure 3.14 Flow chart for the simplex method (minimization)

with maximization problems:

(i) Change the simplex procedure as described in Figure 3.14 to deal with the maximization problem.
(ii) Convert the maximization problem to a minimization problem and solve as indicated by the procedure in Figure 3.14.
(iii) Solve the dual problem (which is a minimization problem) as indicated by the procedure in Figure 3.14.

The significance of the dual problem is much greater than just a means of solving maximization problems. It will be discussed later in Section 3.11 and consequently nothing further will be mentioned regarding method (iii) at present.

As mentioned in Section 3.7 the only difference between solving a

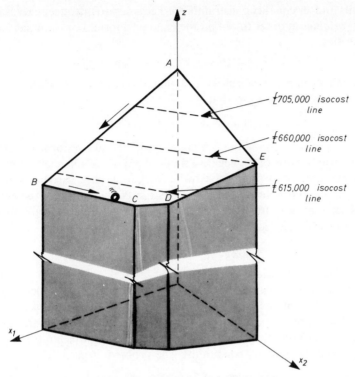

Figure 3.15 Isometric view of the objective function over a two-dimensional feasible space in the plant location problem

maximization and a minimization problem using the simplex method is the sign of the simplex coefficient used in testing for optimality. Thus in method (i), *for a maximization problem if one or more of the simplex criteria Δz_j are positive, the solution is not optimal. If Δz_j is negative for all j the solution is optimal.*

In method (ii) the conversion from maximization to minimization involves three main steps:

(i) Multiply the objective function by -1.
(ii) Solve the problem as a minimization problem.
(iii) Multiply by -1 the optimal value obtained by solving the minimization problem. This gives the optimal value of the original objective function.
   Thus   Minimum (z) = Maximum $(-z)$
   or   Minimum $(-z)$ = Maximum (z). (Refer Section 1.6.1(b).)

3.9 A METHOD FOR DEALING WITH EXCESS AND ARTIFICIAL VARIABLES

Up to this point, only less-than-or-equal-to ($\leqslant$) inequality constraints have been considered. For problems involving greater-than-or-equal-to ($\geqslant$) inequality constraints and/or equality constraints it is necessary to introduce *excess* and/or *artificial variables* in order to obtain the canonical form. For example, consider the constraint

$$a_{i1}x_1 + a_{i2}x_2 + \ldots + a_{in}x_n \geqslant b_i. \qquad (3.38)$$

This inequality can be transformed into an equation by the introduction of a non-negative excess variable x_p so that

$$a_{i1}x_1 + a_{i2}x_2 + \ldots + a_{in}x_n - x_p = b_i. \qquad (3.39)$$

However, a problem containing an equation of the above form is not in the canonical form because of the negative sign in front of the unit coefficient of x_p. A problem containing an equality constraint (without a slack variable) would also not be in the canonical form. For such constraints to conform to the canonical form it is necessary to add an artificial variable to each equation without a slack variable. For example (3.39) would then become

$$a_{i1}x_1 + a_{i2}x_2 + \ldots + a_{in}x_n - x_p + x_q = b_i \qquad (3.40)$$

in which x_q is the artificial variable.

Having obtained the canonical form there still remains the problem of finding an initial basic feasible solution and eliminating the artificial variables as soon as possible. One simple, if inelegant, approach will be illustrated with the following trivial example. (*Note:* this could also be solved by substitution for one of the variables using the equality constraint). The problem is given as

$$\text{Minimize } z = 3x_1 + 2x_2 \qquad \text{(a)} \qquad (3.41)$$

subject to

$$1x_1 + 1x_2 = 10 \quad \text{(b)}$$

$$1x_1 \qquad \geqslant 4 \quad \text{(c)}$$

$$x_1 \geqslant 0, \quad x_2 \geqslant 0.$$

By including an excess variable x_3 in (3.41)(c), the LP problem can be written as

$$\text{Minimize } z = 3x_1 + 2x_2 \qquad \text{(a)} \qquad (3.42)$$

subject to

$$1x_1 + 1x_2 \qquad = 10 \quad \text{(b)}$$

$$1x_1 \qquad - 1x_3 = 4 \quad \text{(c)}$$

$$x_1, x_2 \geqslant 0.$$

To obtain the canonical form it is necessary to introduce artificial variables x_4 and x_5 into both of the constraints so that the LP problem can now be written as

$$\text{Minimize } z = 3x_1 + 2x_2 \qquad \text{(a)} \qquad (3.43)$$

$$1x_1 + 1x_2 \qquad + 1x_4 \qquad = 10 \quad \text{(b)}$$

$$1x_1 \qquad - 1x_3 \qquad + 1x_5 = 4. \quad \text{(c)}$$

Having obtained the canonical form it is now necessary to find an initial basic feasible solution. In this problem, the so-called 'big M method' is employed. First the artificial variables are included in the objective function with large penalty cost coefficients M and the simplex method is used to solve the resulting problem. This has the effect of quickly forcing the artificial variables to zero, at which point they can be discarded. When all of the artificial variables have been eliminated an initial basic feasible solution will result and the solution proceeds in the usual manner.

In the present problem if M is taken as 10 then the problem can be re-written as

$$\text{Minimize } z = 3x_1 + 2x_2 \qquad\qquad + 10x_4 + 10x_5 \qquad \text{(a)} \quad (3.44)$$
$$1x_1 + 1x_2 \qquad + 1x_4 \qquad\qquad = 10 \quad \text{(b)}$$
$$1x_1 \qquad\qquad - x_3 \qquad\qquad + 1x_5 = 4 \quad \text{(c)}$$
$$x_1, x_2 \geqslant 0.$$

The simplex tableaux for the present problem can be then written as:

			c_j	3	2	0	10	10	$\dfrac{b_i}{a_{ij}}$
	$c^{(i)}$	Basis	b_i	v_1	v_2	v_3	v_4	v_5	
I	10	x_4	10	1	1	0	1	0	10
	10	x_5	4	1	0	−1	0	1	4
		z_j		20	10	−10			
		Δz_j	140	−17	−8	10			
II	10	x_4	6	0	1	1	1	−1	6
	3	x_1	4	1	0	−1	0	1	∞
		z_j			10	7		−7	
		Δz_j	72		−8	−7		17	
III	2	x_2	6	0	1	1	1	−1	
	3	x_1	4	1	0	−1	0	1	
		z_j				−1	2	1	
		Δz_j	24			1	8	9	

Thus the solution is $x_1 = 4$, $x_2 = 6$, $x_3 = 0$, $x_4 = 0$, $x_5 = 0$, and $z = 24$. The fact that neither of the artificial variables appears in the final solution confirms that the solution is feasible. The flow chart for the big M method is shown in Figure 3.16.

3.10 DEGENERACY

Degeneracy is a condition sometimes encountered in solving LP problems. To demonstrate degeneracy and a simple procedure for dealing with it, consider the concrete blocks problem described earlier in Section 3.1. If x_1 is the number of sets of 100 type A blocks and x_2 is the number of sets of 100 type B blocks produced each day then the LP problem can be written as

$$\text{Maximize } z' = 5x_1 + 8x_2 \tag{3.45}$$

subject to

$$4x_1 + 6x_2 \leqslant 24$$
$$2x_1 + 1x_2 \leqslant 18$$
$$1x_1 + 3x_2 \leqslant 12$$
$$x_1, x_2 \geqslant 0.$$

The first simplex tableau for this problem is shown below (note that the problem has been transformed to a minimization problem in which $z = -z'$):

			c_j	0	0	0	-5	-8	
	$c^{(i)}$	Basis	b_i	v_3	v_4	v_5	v_1	v_2	$\dfrac{b_i}{a_{ij}}$
I	0	x_3	24	1	0	0	4	6	4
	0	x_4	18	0	1	0	2	1	18
	0	x_5	12	0	0	1	1	3	4
		z_j		0	0	0	0	0	
		Δz_j	0	0	0	0	-5	-8	

The layout of the tableau has been slightly modified to simplify the explanation of the procedure for dealing with degeneracy but does not alter the computational problem to be used in solving the problem. The change is simply that the slack variable columns have been entered in the tableau first. The coefficients of the objective function row indicate that x_2 is the variable which should join the basis in the second tableau. As usual, to obtain the variable leaving the basis, the values b_i/a_{ij} are calculated for column 2. In this case, however, there is a tie between the first and third row. This is the condition known as degeneracy and implies that when the variable x_2 enters the basis it will replace not one but two variables which simultaneously reduce to zero.

The flow chart for dealing with degeneracy is outlined in Figure 3.17. In the present case the first step is performed as follows: moving from left to right, column by column, divide each number in the two tied rows (corresponding to x_3 and x_5) by 6 and 3 respectively (i.e. the values in these rows in the v_2 column). This operation results in the following two sets of ratios corresponding to rows 1 and 3:

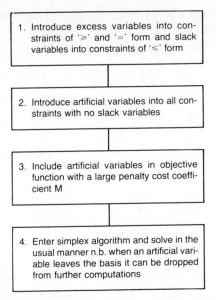

Figure 3.16   Flow chart for the 'Big M Method'

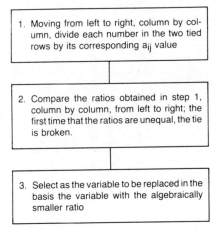

Figure 3.17   Flow chart for procedure for dealing with degeneracy

$$\text{Row 1} \quad \frac{1}{6} \quad \frac{0}{6} \quad \frac{0}{6} \quad \frac{4}{6} \quad \frac{6}{6}$$

$$\text{Row 3} \quad \frac{0}{3} \quad \frac{0}{3} \quad \frac{1}{3} \quad \frac{1}{3} \quad \frac{3}{3}.$$

These ratios are then compared column by column from left to right. The first time that a pair of ratios are found to be unequal the tie is broken and the variable to be replaced in the basis is the variable with the algebraically smaller ratio. Here, the tie is broken with the first pair of ratios. The variable x_2 therefore replaces x_5 in the second tableau as shown below:

	$c^{(i)}$	Basis	b_i	c_j 0 v_3	0 v_4	0 v_5	−5 v_1	−8 v_2	$\dfrac{b_i}{a_{ij}}$
II	0	x_3	0	1	0	−2	2	0	0
	0	x_4	14	0	1	−1/3	5/3	0	42/5
	−8	x_2	4	0	0	1/3	1/3	1	12
		z_j		0	0	−8/3	−8/3	−8	
		Δz_j	−32	0	0	8/3	−7/3	0	

Once the tie has been broken and degeneracy resolved, the technique for completing the second tableau is identical to that given previously in Section 3.7. From the objective function row of the second tableau, it can be seen that x_1 will join the basis in the third tableau. When considering which variable should leave the basis it is usual to choose the variable associated with the smallest non-negative value of b_i/a_{ij}. In the present case, the smallest value is equal to 0 and is associated with basic variable x_3. The next tableau can be completed in the usual manner and is given below:

	$c^{(i)}$	Basis	b_i	c_j 0 v_3	0 v_4	0 v_5	−5 v_1	−8 v_2
III	−5	x_1	0	1/2	0	−1	1	0
	0	x_4	14	−5/6	1	4/3	0	0
	−8	x_2	4	−1/6	0	2/3	0	1
		z_j		−7/6	0	−1/3	−5	−8
		Δz_j	−32	7/6	0	1/3	0	0

Thus the optimal solution implies that no type A blocks should be produced whereas 400 type B blocks should be produced, thus enabling a profit of £32 to be made each day.

Finally, it should be noted that if a tie exists in the simplex criteria row, either of the tied variables may be chosen to join the basis in the next tableau. The choice in some cases may affect the number of iterations required to complete the solution but never the final result.

3.11 DUALITY

The maxim that 'there is more than one way of looking at a problem' holds true in linear programming. Each LP *maximization* problem has its corresponding dual, a *minimization* problem. Conversely, each LP minimization problem has its corresponding dual, a maximization problem. It is possible to solve either the original problem (called the *primal*) or the *dual* to obtain the desired answer.

If the primal problem is given as

$$\text{Maximize } z = (c_1, c_2, \ldots, c_n) \begin{bmatrix} x_1 \\ x_2 \\ \vdots \\ x_n \end{bmatrix} \quad (3.46)$$

subject to

$$\begin{bmatrix} a_{11} & a_{12} & \cdots & a_{1n} \\ a_{21} & a_{22} & \cdots & a_{2n} \\ \vdots & \vdots & & \vdots \\ a_{m1} & a_{m2} & \cdots & a_{mn} \end{bmatrix} \begin{bmatrix} x_1 \\ x_2 \\ \vdots \\ x_n \end{bmatrix} \leqslant \begin{bmatrix} b_1 \\ b_2 \\ \vdots \\ b_m \end{bmatrix}$$

and

$$x_i \geqslant 0 \quad (i = 1, 2, \ldots, n)$$

then the dual problem can be written as

$$\text{Minimize } A = (b_1, b_2, \ldots, b_m) \begin{bmatrix} y_1 \\ y_2 \\ \vdots \\ y_m \end{bmatrix} \quad (3.47)$$

subject to

$$\begin{bmatrix} a_{11} & a_{21} & \cdots & a_{m1} \\ a_{12} & a_{22} & \cdots & a_{m2} \\ \vdots & \vdots & & \vdots \\ a_{1n} & a_{2n} & \cdots & a_{mn} \end{bmatrix} \begin{bmatrix} y_1 \\ y_2 \\ \vdots \\ y_m \end{bmatrix} \geqslant \begin{bmatrix} c_1 \\ c_2 \\ \vdots \\ c_n \end{bmatrix}$$

and

$$y_j \geqslant 0 \quad (j = 1, 2, \ldots, m).$$

The co-existence of the primal and dual problems can be demonstrated by means of the concept of the Lagrangian multiplier, discussed in Section 2.6. Equation (3.46) may be stated more concisely as follows:

$$\text{Maximize } z = \sum_{j=1}^{n} c_j x_j$$

subject to

$$\sum_{j=1}^{n} a_{ij} x_j = b_i \quad \text{for } i = 1, 2, \ldots, m$$

and in which the vector $\mathbf{x} = (x_1, x_2, \ldots, x_n)^T$ now includes the necessary slack variables to convert the '$\leqslant$' constraints to equalities. The Lagrangian may be formed as

$$L(\mathbf{x}, \mathbf{y}) = \sum_{j=1}^{n} c_j x_j + \sum_{i=1}^{m} y_i \left(\sum_{j=1}^{n} a_{ij} x_j - b_i \right) \quad (3.48)$$

in which the vector $\mathbf{y} = (y_1, y_2, \ldots, y_m)$ represents the Lagrangian multipliers. The necessary conditions for an optimum are given by

$$\frac{\partial L}{\partial x_j} = 0 \quad j = 1, 2, \ldots, n \quad (3.49)$$

$$\frac{\partial L}{\partial y_i} = 0 \quad i = 1, 2, \ldots, m. \quad (3.50)$$

Solving (3.49) gives

$$\frac{\partial L}{\partial x_j} = c_j + \sum_{i=1}^{m} y_i a_{ij} = 0$$

or

$$c_j = -\sum_{i=1}^{m} y_i a_{ij} \quad j = 1, 2, \ldots, n. \quad (3.51)$$

Solving (3.50) gives:

$$\frac{\partial L}{\partial y_i} = \sum_{j=1}^{n} a_{ij} x_j - b_i = 0 \quad i = 1, 2, \ldots, m. \quad (3.52)$$

Now substituting (3.52) in (3.48) gives

$$L = \sum_{j=1}^{n} c_j x_j + \sum_{i=1}^{m} y_i(0)$$

$$= \sum_{j=1}^{n} c_j x_j$$

$$= z \quad \text{(the original objective function)}.$$

Also, substitution of (3.51) in (3.48) gives

$$L = -\sum_{j=1}^{n} \sum_{i=1}^{m} y_i a_{ij} x_j + \sum_{i=1}^{m} \sum_{j=1}^{n} y_i a_{ij} x_j - \sum_{i=1}^{m} y_i b_i$$

$$= -\sum_{i=1}^{m} y_i b_i = -A, \quad \text{say}.$$

Therefore by equating the two expressions for L.

$$z = -\sum_{i=1}^{m} b_i y_i = \sum_{j=1}^{n} c_j x_j. \quad (3.53)$$

Equation (3.53) represents the objective function of the dual problem in which y_i

($i = 1, 2, \ldots, m$) are the variables and b_i ($i = 1, 2, \ldots, m$) are cost coefficients. The difference in sign in (3.53) implies that since the primal is a maximization problem, the dual is a minimization problem.

There are several interesting relationships between the optimal solutions for both the primal and dual:

(i) Minimum value of A = Maximum value of z.
(ii) If a slack or excess variable occurs in the kth constraint of either system of equations then the kth variable of its dual vanishes.
(iii) If the kth variable is positive in either system, the kth constraint of its dual is an equation.
(iv) The coefficients of the kth slack or excess variables in the simplex criteria row of the optimal tableau in either system correspond to the optimal values of the kth variable of its dual.

The primal–dual relationships can often lead to a problem formulation which is computationally more convenient than the original problem. For example, the dual formulation might involve fewer constraints and avoid the need for excess and artificial variables. However, the primal–dual relationship can also lead to considerable insight into various aspects of LP problems and their optimal solution and, in this context, the concept of *marginal value* is particularly useful. From the foregoing proof and recalling the discussion following the example of Section 2.5.1, it is clear that the variables in the dual problem represent shadow-prices with respect to the resources of the primal.

To demonstrate the primal–dual relationships consider the primal LP problem

$$\text{Maximize } z = 4.5x_1 + 5x_2 \tag{3.54}$$

subject to

$$3x_1 + 2x_2 \leqslant 24$$

$$4x_1 + 5x_2 \leqslant 40$$

$$x_1, x_2 \geqslant 0.$$

With reference to (3.46) and (3.47), the corresponding dual problem can be written as

$$\text{Minimize } A = 24y_1 + 40y_2 \tag{3.55}$$

subject to

$$3y_1 + 4y_2 \geqslant 4.5$$

$$2y_1 + 5y_2 \geqslant 5$$

$$y_1, y_2 \geqslant 0.$$

The optimum value of a dual variable indicates the rate of change in the value of the objective function if resources associated with the corresponding primal constraint could be increased. For example, if $y_1 = 5/14$ in the optimal dual solution of the above problem it would be possible to increase the primal objective function by 5/14 units for each unit increase in the available resource in the first primal constraint. *Thus the optimal value of the ith variable in the dual reflects the marginal value of the ith resource in the primal.*

In order to solve the primal problem the LP problem expressed by (3.54) must be written in the canonical form, i.e.

$$\text{Minimize } z' = -z = -4.5x_1 - 5x_2 \tag{3.56}$$

subject to

$$3x_1 + 2x_2 + x_3 = 24$$

$$4x_1 + 5x_2 + x_4 = 40$$

$$x_1, x_2, x_3, x_4 \geqslant 0.$$

The simplex tableaux for this problem are given as:

	$c^{(i)}$	Basis	b_i	c_j -4.5 v_1	-5 v_2	0 v_3	0 v_4	$\dfrac{b_i}{a_{ij}}$
I	0 0	x_3 x_4	24 40	3 4	2 5	1 0	0 1	12 8
		z_j Δz_j	0	0 -4.5	0 -5	0 0	0 0	
II	0 -5	x_3 x_2	8 8	1.4 0.8	0 1	1 0	-0.4 0.2	5.71 10
		z_j Δz_j	-40	-4 -0.5			-1 1	
III	-4.5 -5.0	x_1 x_2	5.71 3.43	1 0	0 1	0.71 -0.57	-0.29 0.43	
		z_j Δz_j	-42.8			-0.357 0.357	-0.357 0.857	

Thus, the optimal solution is $z = -z' = 42.8$, $x_1 = 5.71$, and $x_2 = 3.43$. The dual form given by (3.55) can be written in canonical form:

$$\text{Minimize } A = 24y_1 + 40y_2 + My_5 + My_6 \tag{3.57}$$

subject to

$$3y_1 + 4y_2 - y_3 + y_5 = 4.5$$

$$2y_1 + 5y_2 - y_4 + y_6 = 5$$

$$y_1, \ldots, y_6 \geqslant 0$$

where y_3 and y_4 are excess variables and y_5 and y_6 are artificial variables.

Using the big M method described in Section 3.9, the simplex tableaux for this problem are given as:

			c_j	24	40	0	0	M	M	$\dfrac{b_i}{a_{ij}}$
	$c^{(i)}$	Basis	b_i	v_1	v_2	v_3	v_4	v_5	v_6	
I	M	y_5	4.5	3	4	−1	0	1	0	1.25
	M	y_6	5.0	2	5	0	−1	0	1	1.0
		A_j	9.5M	5M	9M	−M	−M			
		ΔA_j		24−5M	40−9M	M	M			
II	M	y_5	0.5	1.4	0	−1	0.8	1	0.8	0.36
	40	y_2	1	0.4	1	0	−1	0	0.2	2.5
		A_j	40+0.5M	16+1.4M		−M	0.8M−8		8−0.8M	
		ΔA_j		8−1.4M		M	8−0.8M		1.8M−8	
III	24	y_1	0.357	1	0	−0.71	0.57	0.71	−0.57	
	40	y_2	0.857	0	1	0.29	−0.43	−0.29	0.43	
		A_j	42.8			−5.71	−3.43	5.71	3.43	
		ΔA_j				5.71	3.43	M−5.71	M−3.43	

It is interesting to compare the optimal solutions to the primal and the dual:

(i) Minimum value of A ($= 42.8$) $=$ Maximum value of z ($= 42.8$).
(ii) In both optimal solutions the slack and excess variables in the optimal solutions are zero, which implies that none of the variables $x_1, x_2, y_1,$ or y_2 vanish.
(iii) In the dual problem y_1 and y_2 are positive and equal to 0.357 (5/14) and 0.857 (6/7) respectively which implies that the first two constraints in the primal problem are satisfied as equalities. In the primal problem x_1 and x_2 are positive and equal to 5.714 (40/7) and 3.429 (24/7) respectively which implies that the first two constraints in the dual problem are satisfied as equalities.
(iv) The simplex criteria of the first and second slack variables of the optimal solution in the primal problem correspond to the optimal values of the first and second variables y_1 and y_2 of the dual and are equal to 0.357 and 0.857 respectively. The simplex criteria of the first and second excess variables of the optimal solution in the dual problem correspond to the optimal values of the first and second variables x_1 and x_2 of the primal and are equal to 5.714 and 3.429 respectively.
(v) Since the optimal dual solution is $y_1 = 0.357$ and $y_2 = 0.857$ the marginal value of resource in the first and second primal constraint is 0.357 and 0.857 respectively. The primal objective function could be improved (i.e. increased)

by these amounts for each additional unit of resource made available in the first and second constraints, respectively.

From (iv) it can be seen that it is not necessary to solve the dual, just to obtain the optimal values of the dual variables; they can be determined directly from the simplex coefficients of the primal solution.

3.12 SENSITIVITY ANALYSIS

In linear programming problems, there are three groups of parameters:

(i) the cost coefficients in the objective function c_j,
(ii) the constants or stipulations in the constraints b_i, and
(iii) the coefficients of the variables in the constraints a_{ij}.

Some of these parameters may be subject to known variations in time or it may be possible to determine them only within certain limits. Consequently, from a practical point of view it is important to explore the sensitivity of the LP solution to changes in these parameters.

One method of estimating the effect of parameter change is to solve a new problem from the beginning whenever one of the coefficients is changed. In this section a different approach known as *sensitivity analysis* will be briefly illustrated.

The important idea associated with sensitivity analysis is that in such an analysis it is not necessary to re-analyse the entire problem from the very beginning each time a check is made on parameter changes.

In this section the sensitivity of the optimal solution to changes in the stipulations b_i and the cost coefficients in the objective function c_j is examined. As no profits are included the interested reader wishing to study sensitivity analysis in greater detail is directed to other texts[2-6] where the topic is given a more comprehensive treatment.

3.12.1 Changes in the cost coefficients

It may frequently be important to consider the effect on the optimal policy (solution) of changes in the objective function. This is particularly significant in situations where cost estimates are subject to uncertainty or fluctuation. Although the value of the objective function is likely to change, the planner or designer is more concerned with the possibility of a change in the optimal policy. The aim then is to determine the limits within which each of the basic variable cost coefficients may vary without altering the optimal policy.

This aspect of sensitivity may be examined by means of the optimal simplex tableau, and reference will be made to the plant location example of Section 3.7 to illustrate the discussion.

For convenience of reference the tableau is reproduced here:

$c^{(i)}$	Basis	b_i	c_j	-200	-180	0	0	0	$\dfrac{b_i}{a_{ij}}$
			v_1	v_2	v_3	v_4	v_5	
-180	x_2	300	0	1	0.333	0	-0.333	
0	x_4	100	0	0	-0.778	1	0.44	
-200	x_1	500	1	0	-0.222	0	0.556	
	z_j	596 000			-15.6		-51.1	
	Δz_j				15.6		51.1	

No negative coefficients

For a problem of minimization, the optimal tableau must contain only positive simplex coefficients Δz_j for the non-basic variables. Before the optimal policy can be changed, one of these simplex coefficients must be driven to a zero or negative value by adjustment of one of the basic variable cost coefficients $c^{(i)}$.

For example, to find the *highest* allowable value of c_1 ($= c^{(3)}$ in the tableau) which will not drive any of the Δz_j less than zero, a search is made for positive values of the structural coefficients a_{3j} (j = non-basic). In this example only a_{35} ($= 0.556$) > 0. Now for $\Delta z_5 = 0$ the requirement which defines Maximum c_1 (or Maximum $c^{(3)}$) is given by

$$c_5 - c^{(1)}a_{15} - c^{(2)}a_{25} - \max c^{(3)}a_{35} = 0$$

i.e.

$$\Delta z_5 + c^{(3)}a_{35} - \max c^{(3)}a_{35} = 0$$

or

$$\max c^{(3)} = c^{(3)} + \frac{\Delta z_5}{a_{35}}$$

e.g.

$$\max c^{(3)} = -200 + \frac{51.1}{0.556} = -108.1.$$

Had there been more than one $a_{3j} > 0$ the value of max $c^{(3)}$ would have been dictated by the non-basic variable with the smallest change. Thus in general:

$$\max c^{(i)} = c^{(i)} + \min \left[\frac{\Delta z_j}{a_{ij}} \right] \quad \text{for } a_{ij} > 0 \quad j = \text{non-basic}. \tag{3.58}$$

The converse argument may be developed to find the *lowest* permissible value of a cost coefficient which will leave the optimal policy unaffected. Thus

$$\min c^{(i)} = c^{(i)} + \max \left[\frac{\Delta z_j}{a_{ij}} \right] \quad \text{for } a_{ij} < 0 \quad j = \text{non-basic}. \tag{3.59}†$$

† In (3.59) 'max' means algebraic maximum and *not* absolute.

For example,

$$\min c_1 = \min c^{(3)} = -200 + \max \left[\frac{15.6}{-0.222} \right] \quad j = 3$$

$$= -200 - 70.3$$

$$= -270.3.$$

The reader should confirm that this value results in a (near) zero value of Δz_3. Thus if c_2 and c_4 remain unchanged, the value of c_1 may lie in the range $-270.3 \leq c_1 \leq -108.1$ without altering the optimal policy $x_1^* = 500; x_2^* = 300; x_4^* = 100$. In the same way the allowable range of c_2 may be calculated as follows:

$$\max c_2 = \max c^{(1)} = -180 + \min \left[\frac{15.6}{0.333} \right] \quad j = 3$$

$$= -180 + 46.8$$

$$= -133.2$$

$$\min c_2 = \min c^{(1)} = -180 + \max \left[\frac{51.1}{-0.333} \right] \quad j = 5$$

$$= -180 - 153.3$$

$$= -333.3.$$

The allowable range is given by $-333.3 \leq c_2 \leq -133.2$. The result may be demonstrated graphically in Figure 3.18 on which are plotted iso-cost lines corresponding to the four objective functions:

$$z = 750\,000 - 270.3x_1 - 180x_2$$

$$z = 750\,000 - 200x_1 - 133.2x_2$$

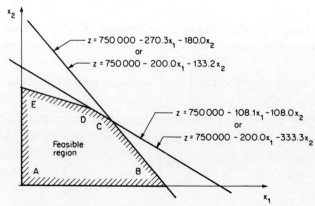

Figure 3.18 Sensitivity analysis with respect to cost coefficients (plant location problem)

$$z = 750\,000 - 108.1x_1 - 180x_2$$
$$z = 750\,000 - 200x_1 - 333.3x_2$$

It will be noted that the first pair of objective functions have the same slope and are thus represented by a single iso-cost line. Similarly the second pair of functions are represented by a second iso-cost line. Each iso-cost line is coincident with one of the edges of the feasible space adjacent to the optimal vertex, thus representing the marginal condition at which the optimal solution is on the point of migrating to an adjacent vertex.

3.12.2 Changes in the stipulations

Any change in the stipulations b_i of the primal is equivalent to a change in the cost coefficient of the dual (see Section 3.11) and may be treated as such. However, such changes can be treated without considering the dual.

Changes in the b_i affect the solution values of the basic variables and the objective function. At the outset it may be helpful to visualize the effect of changes on the graphical solution. In the previous section it was seen that the effect of altering a cost coefficient c_i was to change the slope of the iso-cost line but leave the feasible region unchanged. By contrast, changes to the stipulations do not change the slope of the iso-cost line but alter the shape of the feasible region. Since the values of the simplex criteria are unaffected, the existing basis remains optimal until the solution becomes infeasible. Thus upper and lower limits may be established for each b_i within which the current basis remains both optimal and feasible.

Before the calculation of these limits is described, some definitions are required.

Let the matrix of coefficients in the last (optimal) tableau under the variables of the initial basis (not including the simplex criteria row) be defined as $\boldsymbol{R}$. Thus in the factory location problem where the initial basis is $[x_3, x_4, x_5]$ then

$$\boldsymbol{R} = \begin{bmatrix} r_{11} & r_{12} & r_{13} \\ r_{21} & r_{22} & r_{23} \\ r_{31} & r_{32} & r_{33} \end{bmatrix} = \begin{bmatrix} 0.333 & 0 & -0.333 \\ -0.778 & 1 & 0.44 \\ -0.222 & 0 & 0.556 \end{bmatrix} \begin{matrix} \leftarrow x_2 \\ \leftarrow x_4 \\ \leftarrow x_1 \end{matrix} \quad (3.60)$$

$$\begin{matrix} \uparrow & \uparrow & \uparrow \\ v_3 & v_4 & v_5 \end{matrix}$$

Also, let the optimal values of the basic variables in the last tableau be represented by the vector $\mathbf{q}$. Thus in the factory location problem

$$\mathbf{q} = [q_1, q_2, q_3]^\mathrm{T} = [300, 100, 150]^\mathrm{T}. \quad (3.61)$$

Finally let the vector of stipulations be given as $\mathbf{b}$. Thus in the present problem,

$$\mathbf{b} = [b_1, b_2, b_3]^\mathrm{T} = [3000, 1500, 2100]^\mathrm{T}. \quad (3.62)$$

For the upper limit on b_i for both minimization and maximization cases

$$\max b_i = \min \left\{ b_i - \frac{q_j}{r_{ji}} \right\} \quad \text{for } r_{ji} < 0. \quad (3.63)$$

For the lower limit on b_i

$$\min b_i = \max \left\{ b_i - \frac{q_j}{r_{ji}} \right\} \quad \text{for } r_{ji} > 0. \quad (3.64)$$

Consider the evaluation of the upper limit on the stipulation of the first constraint b_1 in the factory location problem,

$$\max b_1 = \min \begin{bmatrix} b_1 - \frac{q_1}{r_{11}} = 3000 - \frac{300}{0.333} \rightarrow \text{ignore } r_{11} > 0 \\ b_1 - \frac{q_2}{r_{21}} = 3000 - \frac{100}{-0.778} = 3128.6 \\ b_1 - \frac{q_3}{r_{31}} = 3000 - \frac{150}{-0.222} = 3675.0 \end{bmatrix} \quad (3.65)$$

$$= 3128.6.$$

Similarly the lower limit on b_1 is given as

$$b_1 = \max \left\{ \left(3000 - \frac{300}{0.333} \right) = 2250 \right\} \quad (3.66)$$

$$= 2100.0.$$

The effect of changing the stipulation is to move the constraint to a new position parallel to its old position. Thus with reference to Figure 3.19 when $b_1 = 3128.6$, the optimal solution changes to point F. When $b_1 = 2100.0$ the optimal solution changes to point E. While $2100.0 \leqslant b_1 \leqslant 3128.6$ the optimal solution will be at some point D' along line EF. If $2100.0 \leqslant b_1 \leqslant 3000$ then D' will lie on line DE and if $3000 \leqslant b_1 \leqslant 3128.6$ then D' will lie on line DF. This is shown graphically in Figure 3.19. It is left to the reader to examine the effects of changing b_2 and b_3.

A more comprehensive method of dealing with the effects of changes in parameters a_{ij}, b_i, or c_j is known as *parametric analysis*. However, this is beyond the scope of the present text and will not be dealt with here and the interested reader is directed elsewhere.[2-6]

3.13 COMPUTER SOLUTIONS TO LP PROBLEMS

It will be readily apparent that the type of tabular calculation described in Section 3.7 is somewhat impractical for problems of any size. The method described in detail is ideal for implementation in a computer program, thus facilitating the rapid solution of linear problems involving hundreds of variables and

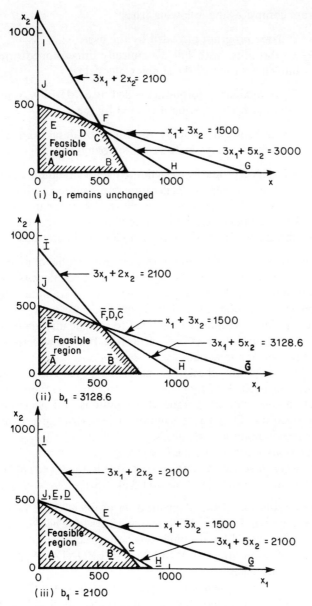

Figure 3.19 Sensitivity analysis of plant location problem — stipulation b_1 (plant location problem)

constraints. The simplex method has been implemented in many such programs and improved algorithms (e.g. revised simplex) have also been developed. One implementation of the standard simplex method is described here and full details of the listing and documentation are given in Appendix A (Subroutine listings).

The program comprises the following parts:

MAIN — a driver program provided by the user.
SIMPLEX — a subroutine with full dynamically dimensioned arrays for the implementation of the simplex algorithm.
TAB
TAB1 — Three auxiliary subroutines called by SIMPLEX to print out the tables at each iteration if desired by the user.
TAB2
LPDATA — a subroutine to simplify the input of data from a time-share terminal. (It may also be used for batch runs.)

The program is designed primarily to facilitate the understanding of the simplex method by providing a means of generating full solutions with tableaux for problems of modest size. The routines are dimensioned for up to 15 variables (including slack and artificial variables) although these statements could be easily modified. The program in addition has the following limitations:

(i) The algorithm is designed for minimization only. Problems of maximization must be re-cast by reversing the sign of the cost coefficients.
(ii) The constraints must be presented in the form of equalities including all slack, surplus, or artificial variables as may be necessary.
(iii) The objective function must be modified to include any artificial variables with appropriate penalty cost coefficients (the 'big M') with the correct sign (e.g. positive for minimization).
(iv) The method does not check for infeasible, degenerate, or unbounded solutions. Infeasible solutions may be indicated by non-zero artificial variables in the solution. Unbounded solutions will generally result in machine overflow. Degenerate solutions will seldom give trouble due to machine representation of numbers.
(i) The stipulations must be specified as positive.
(vi) No provision is made for numerical constants in the objective function (e.g. the 750 000 in the problem of Section 3.3).

The computer solution of the plant location problem could be obtained by means of the program in Figure 3.20. Note that arrays have been dimensioned for the maximum of 15 variables although this is not required here. Refer to the documentation of subroutines SIMPLEX and LPDATA (Appendix A) for definition of the parameters.

The results of a run are shown in Figure 3.21. It will be noted that the objective function of −154 000 is the amount by which the constant of 750 000 is reduced.

3.14 PORTAL FRAME DESIGN EXAMPLES — SOLUTION

The portal frame design problem described in Section 3.1 is now solved using program SIMPLEX. Let the decision variables be $x_1 = M_c$ and $x_2 = M_b$ respectively. Recall that it may be assumed that the weight of each frame member is proportional to the fully plastic moment of the member multiplied by its

```
        DIMENSION A(225),B(15),C(15),X(15),NBASIC(15),SIMCO(15)
        LOGICAL ITAB
        NTAPE=5
        WRITE(6,10)
10      FORMAT(" SUPPLY N,M...2I5")
        READ(5,20)N,M
20      FORMAT(2I5)
        CALL LPDATA(N,M,A,B,C,ITAB,NTAPE)
        CALL SIMPLEX(N,M,A,B,C,X,ANS,ITAB,NBASIC,SIMCO)
        WRITE(6,607)
607     FORMAT(1X,8HVARIABLE,4X,5HVALUE)
        DO 6 I=1,N
        WRITE(6,608)I,X(I)
608     FORMAT(1X,2HX(,I3,4H) = ,F12.3)
6       CONTINUE
        WRITE(6,610)ANS
610     FORMAT(1X,10X,30HOBJECTIVE FUNCTION VALUE IS  ,F15.5)
        STOP
        END
```

Figure 3.20 FORTRAN program for linear programming solution

length.[6,7] Thus the objective is to minimize the frame weight or

$$\text{Minimize } z = \alpha 2l_c x_1 + \alpha l_b x_2. \qquad (3.67)$$

As the constant of proportionality α does not affect the final solution it will be ignored so that the objective can be expressed as

$$\text{Minimize } z = 2l_c x_1 + l_b x_2. \qquad (3.68)$$

The frame will be safe or just on the point of collapse provided that

$$w_i \geqslant w_e \qquad (3.69)$$

where, for each possible collapse mechanism, w_i is the internal plastic work done at the plastic hinges and w_e is the external work done by the applied loads. Figure 3.22 shows the six possible collapse mechanisms for the rectangular portal frame under the loading shown in Figure 3.1.

Thus the objective (3.68) is subject to the following constraints for each mechanism:

$$4x_2 \geqslant Vl_b/2 \qquad \text{Mechanism 1} \qquad (3.70)$$
$$2x_1 + 2x_2 \geqslant Vl_b/2 \qquad \qquad 2$$
$$2x_1 + 2x_2 \geqslant Hl_c \qquad \qquad 3$$
$$4x_1 \geqslant Hl_c \qquad \qquad 4$$
$$2x_1 + 4x_2 \geqslant Hl_c + Vl_b/2 \qquad \qquad 5$$
$$4x_1 + 2x_2 \geqslant Hl_c + Vl_b/2 \qquad \qquad 6$$

and the non-negative condition:

$$x_1, x_2 \geqslant = 0. \qquad (3.71)$$

```
SUPPLY N,M...2I5
?    5    3
ARE TABLES REQUIRED?...YES/NO
? YES
SUPPLY C(I),I = 1, 5...7F10.5
?  -200.0    -180.0      0.0       0.0       0.0
SUPPLY A(I,J),J= 1, 5 AND B(J)...7F10.5
FOR EQUN.( 1)
?    3.0       5.0       1.0       0.0       0.0      3000.0
FOR EQUN.( 2)
?    1.0       3.0       0.0       1.0       0.0      1500.0
FOR EQUN.( 3)
?    3.0       2.0       0.0       0.0       1.0      2100.0

THE INITIAL TABLEAU
                       1          2          3          4          5
OBJ FNCTN          -200.000   -180.000     0.000      0.000      0.000

            3000.000    3.000      5.000      1.000      0.000      0.000
            1500.000    1.000      3.000      0.000      1.000      0.000
            2100.000    3.000      2.000      0.000      0.000      1.000
SIMPLEX C              -200.000   -180.000     0.000      0.000      0.000
            OBJECTIVE FUNCTION VALUE IS       0.00000

TABLEAU NO.   2
            STIP         1          2          3          4          5
X( 3)       900.000      0.000      3.000      1.000      0.000     -1.000
X( 4)       800.000      0.000      2.333      0.000      1.000     -.333
X( 1)       700.000      1.000       .667      0.000      0.000      .333
SIMPLEX C                0.000    -46.667      0.000      0.000     66.667
            OBJECTIVE FUNCTION VALUE IS      -140000.00000

TABLEAU NO.   3
            STIP         1          2          3          4          5
X( 2)       300.000      0.000      1.000       .333      0.000     -.333
X( 4)       100.000      0.000      0.000      -.778      1.000      .444
X( 1)       500.000      1.000      0.000      -.222      0.000      .556
SIMPLEX C                0.000      0.000     15.556      0.000     51.111
            OBJECTIVE FUNCTION VALUE IS      -154000.00000

VARIABLE      VALUE
X( 1) =      500.000
X( 2) =      300.000
X( 3) =        0.000
X( 4) =      100.000
X( 5) =        0.000
            OBJECTIVE FUNCTION VALUE IS  ,   -154000.00000
```

Figure 3.21 Computer solution of the plant location problem

If $l_b = 2$, $l_c = 1$, and $V = H = 1$, then the problem can be expressed as

$$\text{Minimize } z = 2x_1 + 2x_2 \tag{3.72}$$

subject to

$$4x_2 \geqslant 1$$

$$2x_1 + 2x_2 \geqslant 1$$

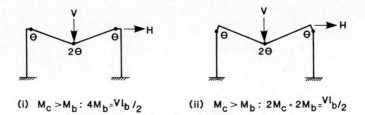

(i) $M_c > M_b$: $4M_b = Vl_b/2$

(ii) $M_c > M_b$: $2M_c + 2M_b = Vl_b/2$

Beam mechanism

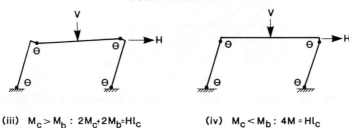

(iii) $M_c > M_b$: $2M_c + 2M_b = Hl_c$

(iv) $M_c < M_b$: $4M = Hl_c$

Sway mechanism

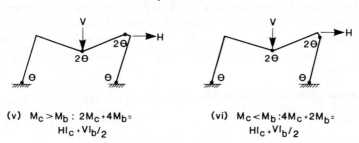

(v) $M_c > M_b$: $2M_c + 4M_b = Hl_c + Vl_b/2$

(vi) $M_c < M_b$: $4M_c + 2M_b = Hl_c + Vl_b/2$

Combined mechanism

Figure 3.22   Collapse mechanisms for rectangular portal frame

$$2x_1 + 2x_2 \geq 1$$
$$4x_1 \geq 1$$
$$2x_1 + 4x_2 \geq 2$$
$$4x_1 + 2x_2 \geq 2$$

and
$$x_1, x_2 \geq 0.$$

This may be solved by a variety of linear programming techniques: graphical solution, simplex, and dual simplex. Figure 3.23 shows the graphical solution which gives the optimal solution A:

$$x_1^* = \tfrac{1}{3}, \quad x_2^* = \tfrac{1}{3}, \quad z^* = \tfrac{4}{3}.$$

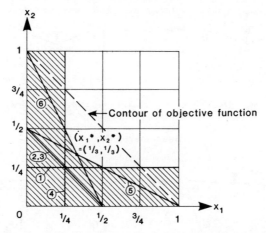

Figure 3.23   Graphical solution for design of rectangular portal frame

Failure in the optimal minimum weight frame takes the form of either mechanism (4) or mechanism (5).

In order to solve this problem using the SIMPLEX program the $\geqslant$ inequalities must be converted by the introduction of excess variables $x_3, x_4, \ldots, x_8$ and artificial variables $x_9, x_{10}, \ldots, x_{14}$. The 'big M' method is then used and $x_9, x_{10}, \ldots, x_{14}$ are introduced into the objective function with a large penalty coefficient M:

$$\text{Minimize } z = 2x_1 + 2x_2 + M(x_9 + x_{10} + x_{11} + x_{12} + x_{13} + x_{14}) \quad (3.73)$$

subject to

$$\begin{aligned}
4x_2 - x_3 & + x_9 &= 1 \\
2x_1 + 2x_2 - x_4 & + x_{10} &= 1 \\
2x_1 + 2x_2 - x_5 & + x_{11} &= 1 \\
4x_1 - x_6 & + x_{12} &= 1 \\
2x_1 + 4x_2 - x_7 & + x_{13} &= 2 \\
4x_1 + 2x_2 - x_8 & + x_{14} &= 2
\end{aligned}$$

with

$$x_1, x_2, \ldots, x_{14} \geqslant 0.$$

The driving program illustrated in Section 3.13 may be used in this example also. Adopting an arbitrary value of $M = 10$, the input and output is as shown in Figure 3.24. From this it can be seen that the optimal solution is identical to that obtained using the graphical solution.

The dual of (3.72) may also be solved using program SIMPLEX and takes the form

```
SUPPLY N, M...2I5
?   14    6
ARE TABLES REQUIRED?...YES/NO
? NO
SUPPLY C(I),I = 1,14...7F10.5
?   2.0       2.0       0.0       0.0       0.0       0.0       0.0
?   0.0      10.0      10.0      10.0      10.0      10.0      10.0
SUPPLY A(I,J),J= 1,14 AND B(J)...7F10.5
FOR EQUN.( 1)
?   0.0       4.0      -1.0       0.0       0.0       0.0       0.0
?   0.0       1.0       0.0       0.0       0.0       0.0       0.0
?   1.0
FOR EQUN.( 2)
?   2.0       2.0       0.0      -1.0       0.0       0.0       0.0
?   0.0       0.0       1.0       0.0       0.0       0.0       0.0
?   1.0
FOR EQUN.( 3)
?   2.0       2.0       0.0       0.0      -1.0       0.0       0.0
?   0.0       0.0       0.0       1.0       0.0       0.0       0 .0
?   1.0
FOR EQUN.( 4)
?   4.0       0.0       0.0       0.0       0.0      -1.0       0.0
?   0.0       0.0       0.0       0.0       1.0       0.0       0.0
?   1.0
FOR EQUN.( 5)
?   2.0       4.0       0.0       0.0       0.0       0.0      -1.0
?   0.0       0.0       0.0       0.0       0.0       1.0       0.0
?   2.0
FOR EQUN.( 6)
?   4.0       2.0       0.0       0.0       0.0       0.0       0.0
?  -1.0       0.0       0.0       0.0       0.0       0.0       1.0
?   2.0
VARIABLE      VALUE
X(  1) =        .333
X(  2) =        .333
X(  3) =        .333
X(  4) =        .333
X(  5) =        .333
X(  6) =        .333
X(  7) =       0.000
X(  8) =       0.000
X(  9) =       0.000
X( 10) =       0.000
X( 11) =       0.000
X( 12) =       0.000
X( 13) =       0.000
X( 14) =       0.000
          OBJECTIVE FUNCTION VALUE IS          1.33333
```

Figure 3.24 Computer solution of the portal frame example by the 'Big M Method'

$$\text{Maximize } z' = y_1 + y_2 + y_3 + y_4 + 2y_5 + 2y_6 \quad (3.74)$$

subject to

$$2y_2 + 2y_3 + 4y_4 + 2y_5 + 4y_6 \leq 2$$

$$4y_1 + 2y_2 + 2y_3 + 4y_5 + 2y_6 \leq 2$$

and

$$y_1, y_2, \ldots, y_6 \geq 0.$$

In order to solve the problem using the SIMPLEX program the $\leq$ inequalities must be converted by the introduction of slack variables y_7 and y_8. The maximization problem is converted to a minimization problem so that the dual may be written as

$$\text{Minimize } A = -y_1 - y_2 - y_3 - y_4 - 2y_5 - 2y_6 \quad (3.75)$$

subject to

$$2y_2 + 4y_3 + 4y_4 + 2y_5 + 4y_6 + y_7 = 2$$

$$4y_1 + 2y_2 + 2y_3 + 4y_5 + 2y_6 + y_8 = 2$$

and

$$y_1, y_2, \ldots, y_8 \geq 0.$$

```
SUPPLY N, M...2I5
?   8    2
ARE TABLES REQUIRED?...YES/NO
? YES
SUPPLY C(I),I = 1, 8...7F10.5
?  -1.0      -1.0      -1.0      -1.0      -2.0      -2.0       0.0
?   0.0
SUPPLY A(I,J),J= 1, 8 AND B(J)...7F10.5
FOR EQUN.( 1)
?   0.0       2.0       2.0       4.0       2.0       4.0       1.0
?   0.0       2.0
FOR EQUN.( 2)
?   4.0       2.0       2.0       0.0       4.0       2.0       0.0
?   1.0       2.0


THE INITIAL TABLEAU
                          1         2         3         4         5
                          6         7         8
OBJ FNCTN              -1.000    -1.000    -1.000    -1.000    -2.000
                       -2.000     0.000     0.000
              2.000     0.000     2.000     2.000     4.000     2.000
                        4.000     1.000     0.000
              2.000     4.000     2.000     2.000     0.000     4.000
                        2.000     0.000     1.000
SIMPLEX C              -1.000    -1.000    -1.000    -1.000    -2.000
                       -2.000     0.000     0.000
            OBJECTIVE FUNCTION VALUE IS         0.00000


TABLEAU NO.    2
            STIP          1         2         3         4         5
                          6         7         8
 X( 7)       1.000    -2.000     1.000     1.000     4.000     0.000
                        3.000     1.000    -.500
 X( 5)        .500     1.000      .500      .500     0.000     1.000
                         .500     0.000      .250
SIMPLEX C              1.000     0.000     0.000    -1.000     0.000
                       -1.000     0.000      .500
            OBJECTIVE FUNCTION VALUE IS        -1.00000
```

```
TABLEAU NO.    3
        STIP        1         2         3         4         5
                    6         7         8
X( 4)       .250   -.500      .250      .250     1.000     0.000
                    .750      .250     -.125
X( 5)       .500   1.000      .500      .500     0.000     1.000
                    .500     0.000      .250
SIMPLEX C           .500      .250      .250     0.000     0.000
                   -.250      .250      .375
            OBJECTIVE FUNCTION VALUE IS          -1.25000


TABLEAU NO.    4
        STIP        1         2         3         4         5
                    6         7         8
X( 6)       .333   -.667      .333      .333     1.333     0.000
                   1.000      .333     -.167
X( 5)       .333   1.333      .333      .333    -.667      1.000
                   0.000     -.167      .333
SIMPLEX C           .333      .333      .333     .333      0.000
                   0.000      .333      .333
            OBJECTIVE FUNCTION VALUE IS          -1.33333
VARIABLE    VALUE
X( 1) =     0.000
X( 2) =     0.000
X( 3) =     0.000
X( 4) =     0.000
X( 5) =      .333
X( 6) =      .333
X( 7) =     0.000
X( 8) =     0.000
            OBJECTIVE FUNCTION VALUE IS          -1.33333
```

Figure 3.25 Computer solution of the portal frame example dual method

Again the driving program illustrated in Section 3.13 may be used in this example also. The input and output is as shown in Figure 3.25. From the output the optimal solution may be interpreted from the final tableau as described earlier. From the positive variables x_5 and x_6 in the dual, only the 5th and 6th constraints are active.

Several assumptions and approximations have been made in this problem:

(i) The weight of each frame member is proportional to the fully plastic moment of the member multiplied by its length.
(ii) There are an infinite number of section sizes available corresponding to any value of plastic moment.
(iii) Although not stated explicitly it is implied that the total cost of the portal frame is heavily dependent on the frame weight.
(iv) The only important constraints are those associated with frame collapse.

In Section 12.6 these approximations and assumptions are discussed further and some of them are examined in detail.

3.15 EXERCISES

3.15.1 Max E. Mizer is planning for a Civil Engineers' conference. Someone has stolen all the beer but he has available the following amounts of alternative liquid refreshments:

Amount (in fluid ounces)	Liquid
84	Vermouth
60	Bourbon
56	Scotch
48	Vodka

In addition, he has an unlimited supply of lemons, limes, orange juice, bitters, and other accompaniments.

Max mixes these masterpieces:

Masterpiece	Ingredients
Manhattan	$\frac{1}{2}$ ounce Vermouth $1\frac{1}{2}$ ounce Bourbon
Rob Roy	$\frac{3}{4}$ ounce Vermouth $1\frac{1}{2}$ ounce Scotch
Scotch-on-the-rocks	2 ounces Scotch
Martini	$\frac{1}{4}$ ounce Vermouth 1 ounce Vodka
Max E. Mizer's appetizer	$1\frac{1}{2}$ ounce Bourbon
Screwdriver	$1\frac{1}{2}$ ounce Vodka

Each drink will sell for $2 and Max knows he can sell as many drinks as he can make up. Which Masterpiece should Max mix in order to maximize his gross income?

3.15.2 The Abacus ('You can count on us') Cement Company has the opportunity to supply various concrete mixes to a nearby construction project. The company, because it is a low-cost operation, has the option of selecting the mixes it finds most profitable to itself, subject only to the restriction that it supply a minimum of 600 cubic metres of 3–4–2 (3 parts cement to 4 parts gravel and 2 parts sand) each day at one site and 200 cubic metres of 3–0–2 each day at another site.

The concrete company will realize a profit contribution of $3.00 per cubic metre from the 3–4–2 mixture and $5.00 per cubic metre from the 3–0–2 mixture.

The plant can obtain for each day's operation a total of 3600 cubic metres of cement, 4000 cubic metres of gravel, and 3000 cubic metres of sand. The plant is situated on a stream so that water is no problem.

The plant manager needs to know what proportion of his production should be the 3–4–2 mix and what proportion of the 3–0–2 mix in order to realize the highest possible profit.

Set up a linear programming model to help the manager determine the optimal production strategy. Explain to the manager all of the information that is made available by this model.

3.15.3 A factory produces two types of concrete block. A ton of the first type requires 2 batches of gravel, 2 bags of cement, and 4 batches of sand. A ton of the second type of block requires 4 batches of gravel and 2 bags of cement but no sand. There are 20 batches of gravel, 12 bags of cement, and 16 batches of sand available. How many tons of each type of block should be manufactured so that maximum profit may be achieved?

It is assumed that the first type of block yields $2 profit per ton and the second type yields $3. Solve this problem using the simplex method and check your solution graphically.

3.15.4 A rectangular fixed based portal frame similar to the one shown in Figure 3.1 has the following dimensions and load intensities: $l_b = 40$, $l_c = 20$, $V = 2$, and $H = 1$. Determine the plastic moment capacities M_b and M_c for a minimum weight frame using

(i) a graphical method,
(ii) the simplex method.

3.15.5 A two-span beam is shown in Figure 3.26. Determine using a graphical

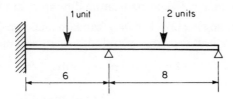

Figure 3.26   Two-span beam (Example 3.15.5)

method the plastic moment capacities in the left-hand and right-hand spans (M_1 and M_2 respectively) for minimum total beam weight. Note that the weight of a beam per unit length is directly proportional to its plastic moment capacity. Give the dual form of the above problem. Why is it more advantageous to use the dual rather than the primal form when using the simplex method to solve problems of the type given above?

3.15.6 The Singleton Concrete Products Company plans production of three of their products A, B, and C. The profits on these products expressed in $100 units are 2, 3, and 1 respectively, and they require two resources — labour and materials. The company's production manager formulates the following linear programming model for determining the optimal product mix:

$$\text{Maximize } z' = 2x_1 + 3x_2 + x_3$$

subject to

$$\tfrac{1}{3}x_1 + \tfrac{1}{3}x_2 + \tfrac{1}{3}x_3 \leqslant 1 \quad \text{labour constraint}$$

$$\tfrac{1}{3}x_1 + \tfrac{4}{3}x_2 + \tfrac{7}{3}x_3 \leqslant 3 \quad \text{material constraint}$$

$$x_1, x_2, x_3 \geqslant 0$$

where x_1, x_2, and x_3 are the number of products A, B, and C produced. This problem is converted to its canonical form and treated as a minimization problem. The optimum solution is given by the following tableau in which x_4 and x_5 are the slack variables:

$c^{(i)}$	Basis	b_i	v_1	v_2	v_3	v_4	v_5
2	x_1	1	1	0	-1	4	-1
3	x_2	2	0	2	2	-1	1
		-8	0	0	3	5	1

(i) Interpret the optimal solution from the above tableau.
(ii) If the profit on product C is increased from 1 to 6 what is the new optimal solution?
(iii) For what range of profits on product A does the solution in the above tableau remain optimal?
(iv) How many additional units of labour must be made available in order to change the current optimal product mix?

3.15.7 Convert the following problem into standard linear programming format and then solve it (i) graphically and (ii) using the simplex method.

$$\text{Maximize } z = 3x_1 + 8x_2$$

subject to

$$3x_1 + 4x_2 \leqslant 20$$

$$x_1 + 3x_2 \geqslant 6$$

$$x_1 \geqslant 0, \quad x_2 \text{ unrestricted in sign.}$$

(*Hint:* For the simplex method solution, let $y_1 = x_1, x_2 = y_2 - y_3$. Thus the problem may be expressed as

$$\text{Minimize } z = -3y_1 - 8y_2 + 8y_3$$

subject to

$$3y_1 + 4y_2 - 4y_3 \leqslant 20$$

$$y_1 + 3y_2 - 3y_3 \geqslant 6$$

$$y_1, y_2, y_3 \geqslant 0.)$$

3.15.8 This example is included to illustrate how techniques such as linear programming may be used to determine the load-carrying capacity of a structure.[8-10]

A symmetric three-bar truss is subjected to a horizontal load x_1 and a vertical load x_2 at point D as shown in Figure 3.27. The cross-sectional

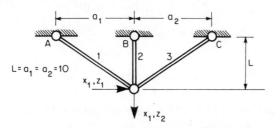

Figure 3.27 Three-bar truss (Example 3.15.8)

areas of bars 1, 2, and 3 are equal to b_1, b_2, and b_3 respectively and $b_1 = b_3$. From the matrix theory of elastic structures it may be shown that the horizontal displacement z_1 and the vertical displacement z_2 at point D may be obtained from the matrix equations

$$\frac{\sqrt{2}E}{40} \begin{bmatrix} (b_1 + b_3) & (b_1 - b_3) \\ (b_1 - b_3) & (b_1 + b_3 + 2\sqrt{2}b_2) \end{bmatrix} \begin{bmatrix} z_1 \\ z_2 \end{bmatrix} = \begin{bmatrix} x_1 \\ x_2 \end{bmatrix}$$

where E is the elastic modulus of the truss member material. Certain constraints are imposed on these displacements and are given as

$$|z_1| - z_1^a \leq 0$$
$$|z_2| - z_2^a \leq 0$$

where z_1^a and z_2^a are specified upper bounds on z_1 and z_2 respectively.

The stress in each member may be calculated by evaluating the strain in terms of the nodal displacements and then by using Hooke's Law to obtain

$$\text{for member 1} \quad \sigma_1 = \frac{E(z_1 + z_2)}{20}$$

$$\text{for member 2} \quad \sigma_2 = \frac{Ez_2}{10}$$

$$\text{for member 3} \quad \sigma_3 = \frac{E(z_2 - z_1)}{20}.$$

Certain constraints are also imposed on these stresses and are given as

$$|\sigma_1| - \sigma_1^a < 0$$
$$|\sigma_2| - \sigma_2^a \leq 0$$
$$|\sigma_3| - \sigma_3^a \leq 0$$

where σ_i^a, $i = 1, 2, 3$ are the allowable stresses specified for each member. Show how linear programming techniques may be used to find the maximum value of P $(= x_1 + x_2)$ which the structure may support. Assume that $x_1, x_2 \geqslant 0$.

Hint: To deal with the absolute value signs in a constraint, say

$$|z_1| - z_1^a \leqslant 0$$

impose two equivalent constraints

$$z_1 \leqslant z_1^a$$

and

$$-z_1 \leqslant z_1^a.$$

3.15.9 Solve Exercises 3.15.1–3.15.8 using program SIMPLEX.

Note that other sources of linear programming applications associated with civil engineering systems may be found in references 5 and 8 to 12.

3.16 REFERENCES

1. Koopmans, T. C. (Ed.) (1951). *Activity Analysis of Production and Allocation*, Wiley, New York.
2. Dantzig, G. B. (1963). *Linear Programming and Extensions*, Princeton University Press, Princeton, N.J.
3. Gass, S. I. (1969). *Linear Programming*, 3rd Ed., McGraw-Hill, New York.
4. Hadley, G. (1962). *Linear Programming*, Addison-Wesley, Reading, Mass.
5. Daellenbach, H. G., and George, J. A. (1978). *Introduction to Operations Research Techniques*, Allyn and Bacon, Boston, Mass. Chapters 2–4.
6. Neal, B. G. (1963). *The Plastic Methods of Structural Analysis*, Chapman and Hall, London.
7. Moy, S. J. (1981). *Plastic Methods for Steel and Concrete Structures*, Macmillan, London.
8. Gallagher, R. H., and Zienkiewicz, O. C. (Eds) (1973). *Optimum Structural Design: Theory and Application*, Wiley, London.
9. Haug, E. J., and Arora, J. S. (1979). *Applied Optimal Design: Mechanical and Structural Systems*, Wiley, New York.
10. Fox, R. L. (1971). *Optimization Methods for Engineering Design*, Addison-Wesley, Reading, Mass.
11. Stark, R. B., and Nicholls, R. L. (1972). *Mathematical Foundations for Design: Civil Engineering Systems*, McGraw-Hill, New York.
12. Au, T., and Stelson, T. E. (1969). *Introduction to Systems Engineering, Deterministic Models*, Addison-Wesley, Reading, Mass.

Chapter 4
Linear Programming – Part II

4.1 CLASSIFICATION OF SPECIAL FORMS OF LP PROBLEMS

Some special types of LP problems can be solved by the use of computational procedures which are more efficient than the simplex method described in the last chapter. *Transportation problems* are a special class of LP problems which frequently occur in civil engineering systems and they can be solved by a purpose-built algorithm known as the *transportation method* which is dealt with in Section 4.2.

An important class of transportation problems, known as assignment problems, can be solved by an even more specialized algorithm known as the *assignment method*. This method is described in Section 4.3.

In some LP problems, only integer-valued variables are allowed and consequently require the specialized techniques of *integer programming* for their solution. One of these techniques, known as the cutting plane method, will be briefly discussed in Section 4.4.

4.2 TRANSPORTATION PROBLEMS

4.2.1 Introduction

In a transportation problem[1-5] it is necessary to assign quantities of a single commodity from various origins to certain destinations. The objective is to find the transportation route which gives minimum costs or maximum profit. The problem is defined by the amount and location of the available supplies and the quantities demanded. Furthermore, a value (usually a cost) associated with the effort required to transport supplies from their origin to their destination must also be determined.

A transportation problem, in which there are m origins and n destinations and in which the total amount of resource available exactly balances the total amount of resource required, can be written in LP form as

$$\text{Minimize } z = \sum_{i=1}^{m} \sum_{j=1}^{n} c_{ij} x_{ij} \qquad (4.1)$$

subject to the constraints

$$\sum_{j=1}^{n} x_{ij} = s_i \quad \text{for } i = 1, 2, \ldots, m$$

$$\sum_{i=1}^{m} x_{ij} = d_j \quad \text{for } j = 1, 2, \ldots, n$$

and all

$$\sum_{i=1}^{m} s_i = \sum_{j=1}^{n} d_j$$

$$x_{ij} \geq 0$$

where
x_{ij} = amount transported from origin i to destination j
c_{ij} = cost of transporting 1 unit of resource from origin i to destination j
d_j = number of units required at destination d_j
s_i = number of units available at origin (source) i
z = total transportation cost
m = number of origins
n = number of destinations.

4.2.2 The Hardy Ready-Mix Concrete Company

The transportation problem may be illustrated by the following example. The Hardy Ready-Mix Concrete Company has received a contract to supply concrete for three bridges on the north Wessex motorway extension. The bridges are located at Mellstock, Blackmore Vale, and Casterbridge and it has been estimated that the following amounts of concrete will be required at the three sites:

Bridge site	Location	Weekly requirement (m^3 concrete)
A	Mellstock	50
B	Blackmore Vale	90
C	Casterbridge	60
	Total required	200

The Hardy Company has three ready-mix concrete plants located in the towns of Exonbury, Sandbourne, and Emminster. The chief dispatcher at Hardy has calculated the amounts of concrete which can be supplied by each plant:

Plant	Location	Amount available/week (m³ concrete)
I	Exonbury	70
II	Sandbourne	100
III	Emminster	30
	Total required	200

In this problem it is fortunate that the total amount of resource required is exactly equal to the total amount available. Such a problem is known as a *balanced transportation problem* and is very unlikely to occur in actual practice. However, it allows the basic ideas of the transportation method to be described. The unbalanced case in which supply and demand are unequal will be described in Section 4.2.9.

The transportation costs per m³ of concrete from each plant to each bridge have been calculated as follows:

	Cost per m³ of concrete		
From plant	To Bridge A	To Bridge B	To Bridge C
I	4	16	8
II	8	24	16
III	8	16	24

If the amount of concrete required at each bridge site and the amounts available at each plant are given, the problem becomes one of calculating how much concrete should be transported from each plant to each project in order to minimize the total transportation cost within the constraints imposed by the plant capacities and the project requirements. The problem is illustrated graphically in Figure 4.1.

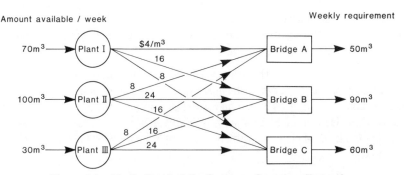

Figure 4.1 Hardy Ready-Mix Concrete Company Example

Table 4.1 The transportation matrix

(Plant) Origin	Destination (Bridge site)			m³ of concrete available
	A	B	C	
I	4	16	8	70
II	8	24	16	100
III	8	16	24	30
m³ of concrete required	50	90	60	200 / 200

4.2.3 The transportation matrix

The previous information may be displayed in a transportation matrix such as the one shown in Table 4.1.

In the transportation matrix, each row represents an origin and each column a destination. The cell at the intersection of each column and each row includes the transportation cost to link that origin–destination. The cost is shown in the left side of the cell. The remaining space in the cell will be used to allocate the amount of concrete transported between the given origin and destination.

Quantities available from the ready-mix plants are shown in the far right column. Site requirements are noted in the bottom line. These conditions are usually referred to as *rim requirements*. The transportation problem can therefore be summarized as the identification of the minimum cost (or maximum profit) solution which satisfies the rim requirements.

4.2.4 The initial feasible solution — the Northwest Corner Rule

Having established the transportation matrix, the next step, as shown in Figure 4.2, is to develop an initial feasible solution. This is achieved by allocating quantities so that

(i) the rim requirements are satisfied,
(ii) the number of allocations is one less than the number of origins plus the number of destinations i.e. $(m + n - 1)$.

The initial solution in the transportation method may be obtained by a systematic procedure known as the *northwest corner rule* which is summarized by the flow chart in Figure 4.3(a). The results of applying the rule to the problem described in Section 4.2.2 are shown in Table 4.2.

An explanation of each assignment made by the northwest corner rule is now presented:

CELL IA Starting in the upper left-hand corner (the northwest corner), the

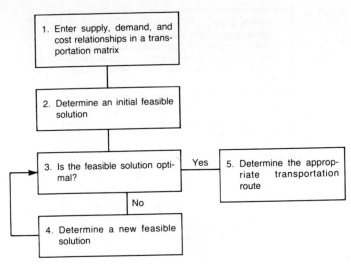

Figure 4.2 Flow chart for the Transportation Method

quantity of concrete available at plant I (in m³) is compared with the quantity required at bridge A (50 m³). The requirement for bridge A is met completely by plant I leaving 20 m³ extra concrete for the other bridges. The next step is to move to the right along the same row to the second column i.e. cell IB.

CELL IB The remaining 20 m³ available at plant I can now all be allocated to bridge B. This leaves bridge B short of 70 m³ of concrete. However, since all of the concrete from plant I has been used, the next step is to move down to the next row in the same column to cell IIB.

CELL IIB Bridge B still requires 70 m³ of concrete and this concrete can be supplied by plant II which leaves 30 m³ available to supply the remaining bridge. The next step is therefore to move along to the next column to cell IIC.

CELL IIC The 30 m³ of concrete remaining at plant II can now be allocated to bridge C leaving an outstanding 30 m³ still required at bridge C. Therefore, the final step is to move down to the last row to cell IIIC.

CELL IIIC The 30 m³ of concrete remaining at plant III is allocated to bridge C thus completing the rim requirements.

Thus the initial feasible solution can be summarized as:

Plant	Bridge	Quantity of concrete (m³)
I	A	50
I	B	20
II	B	70
II	C	30
III	C	30

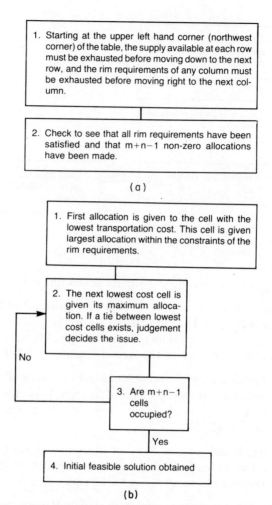

Figure 4.3 (a) Flow chart for the Northwest Corner Rule. (b) Minimum cost method for obtaining an initial basic feasible solution

Table 4.2 The initial basic feasible solution

Origin	Destination			m^3 of concrete available
	A	B	C	
I	4 50	16 20	8	70
II	8	24 70	16 30	100
III	8	16	24 30	30
m^3 of concrete required	50	90	60 200	200

The cells with finite values (e.g. cell IA has a value of 50) are known as *basic cells* and the remaining cells are non-basic cells. The allocation in *non-basic cells* is zero.

The cost of the solution can be obtained by multiplying the allocations in the basic cells by the associated unit transportation costs as shown below:

Basic cell	Quantity of concrete transported	×	Unit cost	=	Total cost
IA	50		4		200
IB	20		16		320
IIB	70		24		1680
IIC	30		16		480
IIIC	30		24		720
	Total transportation cost				$3400

In the initial feasible solution, the number of basic cells should be equal to one less than the number of origins and destinations. In the present case this condition is satisfied, but if a solution (initial or otherwise) does not obey this rule, it is said to be *degenerate*. A procedure for dealing with degeneracy will be discussed in Section 4.2.13. However, it is important to note that each solution should be tested for degeneracy.

4.2.5 An alternative method of obtaining an initial feasible solution — the minimum-cost method

Another method of obtaining an initial feasible solution, the minimum-cost method, is now briefly described. In this method as much allocation as possible is made to the cell with the minimum cost. Then as much allocation as possible is made to the cell with the next lowest cost, and so on. As before, in each allocation the rim requirements must be satisfied. The flow chart for this method is shown in Figure 4.3(b). In general this method will yield initial feasible solutions which are much closer to the optimal solution than solutions obtained by the northwest corner rule. This is to be expected since the northwest corner rule uses only the rim requirements and makes no use of the cost information in the original transportation matrix.

4.2.6 Testing a solution for possible improvement

Having obtained an initial solution, the next step is to ascertain whether this is the optimal solution. This can be achieved by examining the non-basic cells in the transportation matrix to determine if it is possible to make a shipment to one of them and also reduce the total transportation cost. Two procedures will be examined in Sections 4.2.7 and 4.2.12 respectively:

(i) The stepping-stone method.
(ii) The modified distribution (MODI) method.

The stepping-stone method has been replaced by the quicker and more convenient MODI method, but it will be presented here as a useful introduction to the MODI method.

4.2.7 The stepping-stone method

In the stepping-stone method, every cell which contains a non-zero allocation is known as a *stepping stone* and the cells containing zero allocation are known as *waters*. By placing one unit in any of the waters and adjusting other allocations to satisfy the rim requirements it is possible to determine the associated change in the overall transportation cost. This operation can be formalized in a simple way as shown in Figure 4.4. For example, if a quantity θ is sent down route IIA, then a similar quantity must be subtracted from IA in order to maintain the column constraint. Likewise θ must be added to IB and subtracted from IIB in order to maintain the balance.

Thus, $8\theta - 4\theta + 16\theta - 24\theta = -4\theta$ will be added to the total cost. If θ consists of 1 unit (1 m^3 of concrete in the present case) the total cost of the operation will be decreased by 4 units, and therefore the new solution will be better than the initial basic feasible solution. This change in the total cost is known as the *improvement index* and can be calculated for each water cell in turn. For example, the improvement index for cell IIIA can be obtained by the closed path shown in Table 4.3. Note that an improvement index is essentially the same as the simplex coefficient described in Chapter 3, with the simplification that the structural coefficients have values of ± 1. The improvement index for water cell IIIA is

$$C_{\text{IIA}} - C_{\text{IIIC}} + C_{\text{IIC}} - C_{\text{IIB}} + C_{\text{IB}} - C_{\text{IA}}$$
$$= 8 - 24 + 16 - 24 + 16 - 4 = -12.$$

The improvement index for water cell IC is

$$C_{\text{IC}} - C_{\text{IB}} + C_{\text{IIB}} - C_{\text{IIC}} = 8 - 16 + 24 - 16 = 0.$$

The improvement index for water cell IIIB is

$$C_{\text{IIIB}} - C_{\text{IIIC}} + C_{\text{IIC}} - C_{\text{IIB}} = 16 - 24 + 16 - 24 = -16.$$

The improvement index for each of the water cells is now summarized:

Water cell	Improvement index	Closed path
IIA	−4	IIA → IIB → IB → IA
IIIA	−12	IIIA → IIIC → IIC → IIB → IB → IA
IIIB	−16	IIIB → IIIC → IIC → IIB
IC	0	IC → IB → IIB → IIC

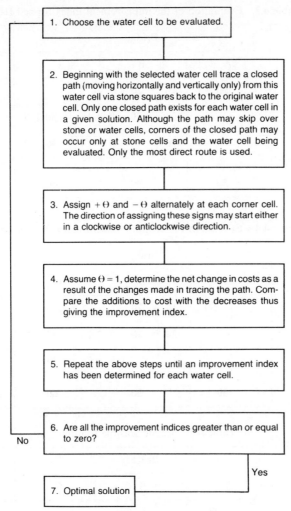

Figure 4.4 Stepping-stone method for testing a solution for possible improvement

Each negative improvement index represents the amount by which the total transportation costs could be reduced if 1 m³ of concrete were transported by that origin–destination. Given three routes with negative improvement indices, which route should be chosen? It is usual to choose the route with the most negative improvement index, in the present case IIIB. The next question to be considered is 'How much concrete can be re-routed via cell IIIB?' This can be ascertained by referring to the associated transportation matrix in Table 4.4. If θ is greater than 30 then the allocation IIIC becomes negative, which is not allowed. (This is equivalent to the minimum b_i/a_{ij} value of Section 3.7, step 6.) Consequently, 30 is the maximum amount of concrete which may be re-routed via cell IIIB. Thus, the second feasible solution can be written in matrix form as shown in Table 4.5.

Table 4.3 Evaluation of the improvement index for cell IIIA

Plant	Bridge A		Bridge B		Bridge C		Concrete available
I	4	50 − θ	16	20 + θ	8		70
II	8		24	70 − θ	16	30 + θ	100
III	8	+ θ	16		24	30 − θ	30
Concrete required	50		90		60		200 / 200

Table 4.4 Evaluation of the improvement index for cell IIIB

Plant	Bridge A		Bridge B		Bridge C		Concrete available
I	5	50	16	20	8		70
II	8		24	70 − θ	16	30 + θ	100
III	8		16	+ θ	24	30 − θ	30
Concrete	50		90		60		200 / 200

Table 4.5 Second feasible solution

Plant	Bridge A		Bridge B		Bridge C		Concrete available
I	4	50	16	20	8		70
II	8		24	40	16	60	100
III	8		16	30	24		30
Concrete required	50		90		60		200 / 200

It will be noted that in this second feasible solution the rim requirements are satisfied and there are $m + n - 1 = 5$ stone cells. The cost of the solution can be obtained as before by multiplying the allocation in the stone cells by the associated unit transportation costs:

Stone cell	Quantity of concrete transported	×	Unit cost	=	Total cost
IA	50		4		200
IB	20		16		320
IIB	40		24		960
IIIB	30		16		480
IIC	60		16		960
	Total transportation cost				$2920

It can be seen immediately that a substantial saving of $3400 − $2920 has been made in the second feasible solution. This solution must now be tested to see if it is possible to further reduce the total transportation cost. Thus, the improvement indices for each water cell are evaluated by exactly the same method as that used previously to obtain the second feasible solution. These improvement indices are as follows:

Water cell	Improvement index	Closed path
IIA	$8 - 24 + 16 - 4 = -4$	IIA → IIB → IB → IA
IIIA	$8 - 16 + 16 - 4 = 4$	IIIA → IIIB → IB → IA
IC	$8 - 16 + 24 - 16 = 0$	IC → IB → IIB → IIC
IIIC	$24 - 16 + 24 - 16 = 16$	IIIC → IIC → IIB → IIIB

Only route IIA may be chosen to re-route the concrete so that the total transportation cost can be reduced. The amount of concrete which can be re-routed via cell IIA can be determined by referring to the associated transportation matrix shown in Table 4.6.

If θ is greater than 40, then a negative allocation results and this is not allowed. Consequently, 40 is the maximum amount of concrete which may be re-routed

Table 4.6 Evaluation of the improvement index for cell IIA

Plant	Bridge			Concrete available
	A	B	C	
I	4 $50 - \theta$	16 $20 + \theta$	8	70
II	8 $+ \theta$	24 $40 - \theta$	16 50	100
III	8	16 30	24	30
Concrete required	50	90	60 200	200

Table 4.7 The third feasible solution

Plant	Bridge A	Bridge B	Bridge C	Concrete available
I	4 · 10	16 · 60	8	70
II	8 · 40	24	16 · 60	100
III	8	16 · 30	24	30
Concrete required	50	90	60	200 / 200

via cell IIA. Thus, the third feasible solution can be written in matrix form as shown in Table 4.7.

It will again be noted that in the third feasible solution the rim requirements are satisfied and there are $m + n - 1 = 5$ stone cells. The cost of the solution is:

Stone cells	Quantity of concrete transported	×	Unit cost	=	Total cost
IA	10		4		40
IIB	40		8		320
IB	60		16		960
IIIB	30		16		480
IIC	60		16		960
	Total transportation cost				$2760

It can be seen that a further saving of $2920 − $2760 has been made in the third feasible solution. As before, this solution must now be tested to see if any further reductions in the total transportation cost can be obtained. The improvement indices for the non-basic cells in Table 4.7 are:

Water cells	Improvement index	Closed path
IIB	4	IIB → IB → IA → IIA
IC	−4	IC → IA → IIA → IIC
IIIA	4	IIIA → IIIB → IB → IA
IIIC	12	IIIC → IIC → IIA → IA → IB → IIB

Route IC is chosen in order to re-route the concrete so that the total transportation cost can be reduced. The amount of concrete which can be re-routed is again determined by referring to the associated transportation matrix shown in Table 4.8.

Table 4.8 Evaluation of the improvement index for cell IC

Plant	Bridge A	Bridge B	Bridge C	Concrete available
I	4 $10 - \theta$	16 60	8 $+\theta$	70
II	8 $40 + \theta$	24	16 $60 - \theta$	100
III	8	16 30	24	30
Concrete required	50	90	60	200 / 200

Table 4.9 The fourth feasible solution

Plant	Bridge A	Bridge B	Bridge C	Concrete available
I	4	16 60	8 10	70
II	8 50	24	16 50	100
III	8	16 30	24	30
Concrete	50	90	60	200 / 200

If θ is greater than 10, then a negative allocation results and this is not allowed. Consequently, 10 is the maximum amount of concrete which may be re-routed via cell IC.

Thus the fourth feasible solution can be written in matrix form as shown in Table 4.9.

Rim requirements are satisfied and the number of basic cells in the fourth feasible solution is equal to $m + n - 1 = 5$. The cost of the solution is:

Stone cells	Quantity of concrete transported	×	Unit cost	=	Total cost
IIA	50		8		400
IB	60		16		960
IIIB	30		16		480
IC	10		8		80
IIC	50		16		800
Total transportation cost					$2720

A further saving of $2760 − $2720 has been made in the fourth feasible solution. To check whether this solution is optimal, the improvement indices for the water cells in Table 4.9 are as follows:

Water cells	Improvement index		Closed path
IA	4 − 8 + 16 − 8	= 4	IA → IIA → IIC → IC
IIIA	8 − 16 + 16 − 8 + 16 − 8	= 8	IIIA → IIIB → IB → IC → IIC → IIA
IIB	24 − 16 + 8 − 16	= 0	IIB → IIC → IC → IB
IIIC	24 − 8 + 16 − 16	= 16	IIIC → IC → IB → IIIB

Since all of the improvement indices are greater than or equal to zero, the fourth feasible solution is optimal. In other words, no further reduction in the total transportation cost can be obtained by re-routing concrete. The optimal solution can thus be summarized as:

Each week deliver
  50 m³ concrete from Sandbourne (Plant II) to Mellstock (bridge A)
  60 m³ concrete from Exonbury (Plant I) to Blackmore Vale (bridge B)
  30 m³ concrete from Emminster (Plant III) to Blackmore Vale (bridge B)
  10 m³ concrete from Exonbury (Plant I) to Casterbridge (bridge C)
  50 m³ concrete from Sandbourne (Plant II) to Casterbridge (bridge C)

The total transportation cost is equal to $2720.

4.2.8 Alternative optimal solutions

It will be noted that the improvement index for water cell IIB is equal to zero in the fourth and final solution to the transportation problem which has just been solved. A zero improvement index for a water cell implies that if this route were brought into the solution, the allocation would change but the total transportation cost would remain constant. Thus, an alternative optimal solution exists. To complete the solution to the transportation problem this other optimal solution should be found, as this solution may be more convenient than the optimal solution already found. The amount of concrete which can be re-routed via IIB can be determined by referring to the associated transportation matrix given in Table 4.10.

If θ is greater than 50, then a negative allocation results and this is not permissible. Consequently, 50 m³ is the maximum amount of concrete which may be re-routed via cell IIB. Thus the fifth possible solution can be written in matrix form as shown in Table 4.11.

Table 4.10 Evaluation of the improvement index for cell IIB

Plant	Bridge			Concrete available
	A	B	C	
I	4	16 $60-\theta$	8 $10+\theta$	70
II	8 50	24 $+\theta$	16 $50-\theta$	100
III	8	16 30	24	30
Concrete required	50	90	60	200 / 200

Table 4.11 Fifth possible solution

Plant	Bridge			Concrete available
	A	B	C	
I	4	16 10	8 60	70
II	8 50	24 50	16	100
III	8	16 30	24	30
Concrete required	50	90	60	200 / 200

The cost of this solution is:

Stone cell	Quantity of concrete transported	×	Unit cost	=	Total cost
IIA	50		8		400
IB	10		16		160
IIB	50		24		1200
IIIB	30		16		480
IC	60		8		480
	Total transportation cost				$2720

The total transportation cost is identical to that for the other optimal solution (the fourth feasible solution). It is important to check whether this solution can be improved upon by evaluating the improvement indices for the water cells.

Water cell	Improvement index	Closed path
IA	$4 - 8 + 24 - 16 = 4$	IA → IIA → IIB → IB
IIA	$8 - 16 + 24 - 8 = 8$	IIIA → IIIB → IIB → IIA
IIC	$16 - 8 + 16 - 24 = 0$	IIC → IC → IB → IIB
IIIC	$24 - 8 + 16 - 16 = 16$	IIIC → IC → IB → IIIB

As none of the improvement indices are less than zero, the fifth basic feasible solution is also optimal. This solution can be summarized as:

Each week deliver

50 m^3 concrete from Sandbourne (Plant II) to Mellstock (bridge A)
10 m^3 concrete from Exonbury (Plant I) to Blackmore Vale (bridge B)
50 m^3 concrete from Sandbourne (Plant II) to Blackmore Vale (bridge B)
30 m^3 concrete from Emminster (Plant III) to Blackmore Vale (bridge B)
60 m^3 concrete from Exonbury (Plant I) to Casterbridge (bridge C)

Alternative optimal solutions provide the decision maker with greater flexibility.

4.2.9 Unbalanced transportation problems

Unbalanced transporation problems in which supply and demand are unequal can be solved by introducing dummy origins or destinations. The optimal solution identifies the requirement which cannot be satisfied or the location of available supplies which remain unused.

A *dummy origin* is added to the transportation matrix when requirements are greater than supplies available. The cost of shipping from this origin to each destination is zero. The excess requirement is entered as a rim requirement for the dummy origin.

A *dummy destination* is included in the transportation matrix when supply is greater than demand. The cost of shipping from each origin to this destination is zero. The excess supply is entered as a rim requirement for the dummy destination.

Examples of dummy origins and destinations for the Hardy Ready-Mix Concrete Company example are shown in Figures 4.5(a) and 4.5(b) respectively.

4.2.10 Maximization

Transportation problems are usually posed as cost minimization problems. Sometimes, however, transportation profits rather than costs are associated with each route and the object is to maximize the total profit. To solve such problems, the profits in each cell are replaced by the amount by which the profit for that cell falls short of the largest profit in the transportation matrix. This gives a measure of the lost opportunity and the problem may then be treated as a minimization problem by the procedures outlined in the previous sections. Finally, the total

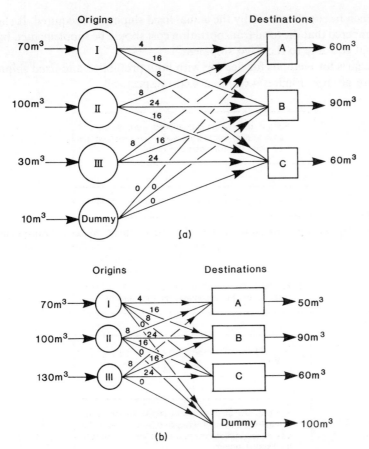

Figure 4.5 (a) Example of a dummy origin. (b) Example of a dummy destination

profit for the allocations is calculated by replacing the adjusted 'profit shortfalls' in the transportation matrix by the original profits.

4.2.11 Prohibited and fixed shipment routes

Sometimes it is not possible to transport commodities via a certain route in a transportation problem. To model such a situation it is necessary to introduce an associated fictitious cost M which has a very large value compared with the other costs, so that all shipment via the route is avoided.

Another possibility is that a fixed shipment S via a certain route is required for some reason. Problems of this nature are first transformed to prohibited shipment problems by subtracting from the associated rim requirements the fixed shipment S. The problem is then solved as a prohibited shipment problem as outlined above.

A fixed shipment of zero should result for the route in question. This allocation should then be supplemented by the actual fixed shipment S required. It should be remembered that the total transportation cost should be supplemented by the additional cost due to the fixed shipment.

Flowcharts for methods for dealing with both prohibited and fixed shipment routes are given in Figures 4.6(a) and 4.6(b) respectively.

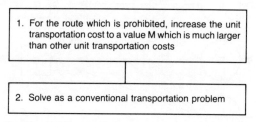

Figure 4.6 (a) Method for dealing with prohibited shipment routes in transportation problems

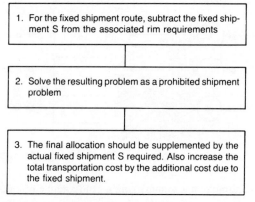

Figure 4.6 (b) Method for dealing with fixed shipment routes in transportation problems

4.2.12 The modified distribution (MODI) method

Compared with the stepping-stone method, the modified distribution (MODI) method is a more efficient method for solving transportation problems. The main difference is that it provides a quick and simple means for obtaining improvement indices for water cells. Figure 4.7 shows a flowchart for the MODI method which will now be used to solve the concrete distribution problem described in Section 4.2.2. It will be assumed that the initial feasible solution has been obtained by the northwest corner rule. The next step is to calculate the dual variables u_i and v_j for each row i and column j by use of the equation

$$u_i + v_j = c_{ij} \quad \text{(the cost at stone cell } ij\text{).} \tag{4.2}$$

As there are only $m + n - 1$ stone cells and $m + n$ dual variables it is necessary to arbitrarily set the first dual variable $u_1 = 0$ so that the other dual variables can be

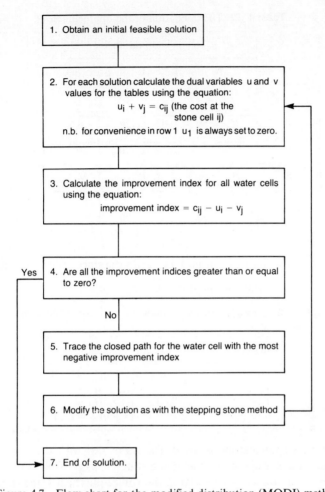

Figure 4.7 Flow chart for the modified distribution (MODI) method

calculated. In the problem given in Section 4.2.2 the following values for the dual variables can be easily calculated:

By definition $\quad u_I = 0$

From stone cell IA $\quad u_I + v_A = C_{IA}$ or $0 + v_A = 4 \quad \therefore v_A = 4$

From stone cell IB $\quad u_I + v_B = C_{IB}$ or $0 + v_B = 16 \quad \therefore v_B = 16$

From stone cell IIB $\quad u_{II} + v_B = C_{IIB}$ or $u_{II} + 16 = 24 \quad \therefore u_{II} = 8$

From stone cell IIC $\quad u_{II} + v_C = C_{IIC}$ or $8 + v_C = 16 \quad \therefore v_C = 8$

From stone cell IIIC $\quad u_{III} + v_C = 24$ or $u_{III} + 8 = 24 \quad \therefore u_{III} = 16.$

This result is written in the transportation matrix shown in Table 4.12. Note that, in general, $u_i + v_j \neq c_{ij}$ for the water cells.

Table 4.12 The evaluation of the dual variables

Origin u_i \ v_j	Destination A: 4		B: 16		C: 8		m³ concrete available
I 0	4	50	16	20	8		70
II 8	8		24	70	16	30	100
III 16	8		16		24	30	30
m³ concrete required		50		90		60	200 / 200

The improvement indices can now be easily obtained using the equation

$$\text{Improvement index} = c_{ij} - u_i - v_j. \tag{4.3}$$

Thus the improvement indices for all of the water cells are immediately available:

Water cell	Improvement index
IIA	$8 - (8 + 4)\ \ =\ -4$
IIIA	$8 - (16 + 4) =\ -12$
IIIB	$16 - (16 + 16) = -16$
IC	$8 - (8 + 0)\ \ =\ \ 0$

As with the stepping-stone method, the water cell with the most negative improvement index is chosen and as much concrete as possible is re-routed via this cell while simultaneously satisfying the rim requirements and maintaining non-negative allocations in all of the affected cells. The procedure at this stage is identical to that described earlier for the stepping-stone method shown in Table 4.4 and results in the new feasible solution shown in Table 4.13. The dual variables are again calculated using equation (4.2) for each stone cell.

The improvement indices can again be evaluated for the water cells:

Water cell	Improvement index
IIA	$8 - (8 + 4) = -4$
IIIA	$8 - (0 + 4) =\ \ 4$
IC	$8 - (8 + 0) =\ \ 0$
IIIC	$24 - (0 + 8) =\ 16$

The new feasible solution and the associated dual variables are shown in Table 4.14.

Table 4.13 The second feasible solution using the MODI method

Origin u_i \ v_j		A 4	B 16	C 8	m³ concrete available
I	0	4 | 50	16 | 20	8 |	70
II	8	8 |	24 | 40	16 | 60	100
III	0	8 |	16 | 30	24 |	30
m³ concrete required		50	90	60	200 / 200

Table 4.14 The third feasible solution using the MODI method

Origin u_i \ v_j		A 4	B 16	C 12	m³ concrete available
I	0	4 | 10	16 | 60	8 |	70
II	4	8 | 40	24 |	16 | 60	100
III	0	8 |	16 | 30	24 |	30
m³ concrete required		50	90	60	200 / 200

The associated improvement indices are:

Water cell	Improvement index
IIIA	$8 - (0 + 4) = 4$
IIB	$24 - (16 + 4) = 4$
IC	$8 - (0 + 12) = -4$
IIIC	$24 - (0 + 12) = 12$

The negative improvement index associated with cell IC means that an optimal solution has not yet been reached. Upon re-routing as much concrete as possible via cell IC the fourth feasible solution is obtained as shown in Table 4.15.

Table 4.15 The fourth feasible solution using the MODI method

Origin u_i \ v_j	A 0		B 16		C 8		m³ concrete available
I 0	4		16	60	8	10	70
II 8	8	50	24		16	50	100
III 0	8		16	30	24		30
m³ concrete required		50		90		60	200 / 200

The associated improvement indices are:

Water cell	Improvement index
IA	$4 - (0 + 0) = 4$
IIIA	$8 - (0 + 0) = 8$
IIB	$24 - (8 + 16) = 0$
IIIC	$24 - (0 + 8) = 16$

As there are no negative improvement indices, the solution given in Table 4.15 is the optimal solution and is identical with that given by the stepping-stone method in Section 4.2.7.

4.2.13 Degeneracy

Degeneracy occurs at any stage in the solution of a transportation problem when there are fewer than $m + n - 1$ stone cells. When the northwest corner rule is used to obtain an initial basic feasible solution, degeneracy occurs when the assignment to any cell (except the last cell (m, n)) satisfies both of the associated rim requirements. To overcome this problem it is usual to assign either of the adjacent cells (to the right or below) as a stone cell with zero allocation.

When degeneracy occurs in an iteration, it is because a tie exists between the stone cells of the entering stone cell's loop (i.e. the stone cells which have $-\theta$ added to them). The procedure for dealing with this difficulty is to assume that all but one of these cells remain as stone cells with zero value. Flowcharts for dealing with both types of degeneracy are shown in Figures 4.8 and 4.9.

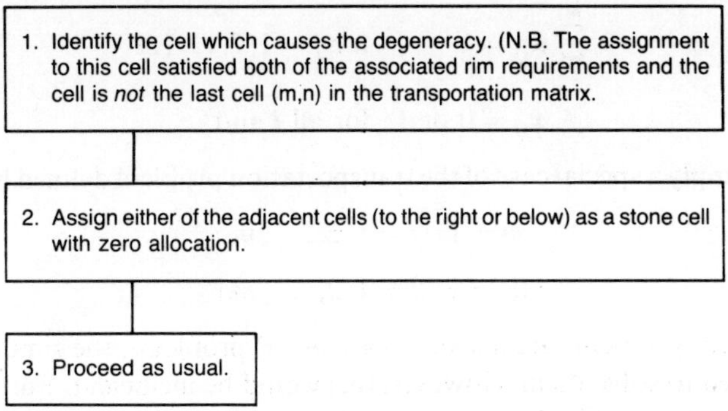

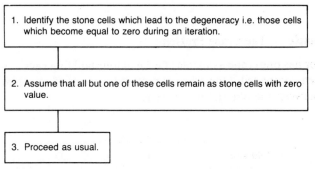

Figure 4.8 Flowchart for the procedure for dealing with degeneracy which occurs when using the northwest corner rule

1. Identify the stone cells which lead to the degeneracy i.e. those cells which become equal to zero during an iteration.

2. Assume that all but one of these cells remain as stone cells with zero value.

3. Proceed as usual.

Figure 4.9 Flowchart for the procedure for dealing with degeneracy which occurs during an iteration

4.3 THE ASSIGNMENT METHOD

4.3.1 Introduction

This section describes a method for solving a special kind of transportation problem known as an assignment problem. In an assignment problem only *one* unit of resource is available at each origin and only *one* unit of resource is required at each destination.

In an assignment problem there are $m = n$ origins and destinations and the total amount of resource available exactly balances the total amount of resource required. The problem can be written in LP form as

$$\text{Minimize (Maximize):} \quad z = \sum_{i=1}^{m} \sum_{j=1}^{m} c_{ij} x_{ij} \qquad (4.4)$$

subject to

$$\sum_{j=1}^{m} x_{ij} = 1 \qquad \text{for } i = 1, 2, \ldots, m$$

$$\sum_{i=1}^{m} x_{ij} = 1 \qquad \text{for } j = 1, 2, \ldots, m$$

$$x_{ij} = 0 \text{ or } 1 \quad \text{for all } i \text{ and } j.$$

(N.B. It is simply a special case of the transportation problem defined by (4.1) with

$$s_i = 1 \quad i = 1, 2, \ldots, m$$
$$d_j = 1 \quad j = 1, 2, \ldots, m.)$$

As assignment problems are a special case of LP problems, the simplex method could be used to solve them. However, this would be inefficient. Furthermore, if the transportation method was used to solve problems, degeneracy would occur at each iteration thus leading to an inefficient solution. A more efficient method for solving assignment problems known as the Hungarian method of assignment (or the assignment method) will now be illustrated by means of an example.

4.3.2 Example — The Crane Problem

A contractor has one crane available at each of four building sites at which work has just been completed. One crane is required at each of four new sites. The distances between the old and new sites is:

	New site			
Old site	A	B	C	D
I	21	24	19	40
II	30	33	37	32
III	42	36	35	33
IV	11	17	29	21

The problem is to make an assignment of the cranes from old sites to new sites which minimizes the total distance moved by the cranes.

A flowchart for the assignment method is shown in Figure 4.10 and this will now be used to solve the crane problem. The first step is to determine the so-called opportunity–cost table from the table of distances given above. By subtracting the lowest entry in each column of the table from all entries in that column the following table is obtained:

	A	B	C	D
I	10	7	0	19
II	19	16	18	11
III	31	19	16	12
IV	0	0	10	0

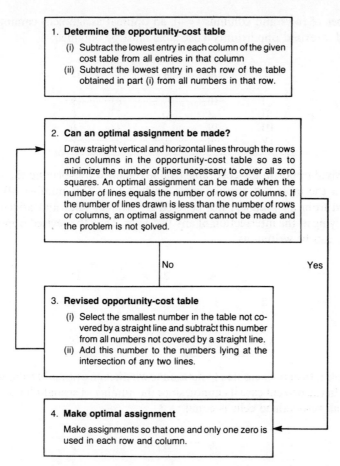

Figure 4.10 Flow chart for the assignment method

By subtracting the lowest entry in each row of the above table from all entries in that row the opportunity–cost table is obtained:

	A	B	C	D
I	10	7	0	19
II	8	5	7	0
III	19	7	4	0
IV	0	0	10	0

This table is now examined to determine whether an optimal assignment can be made. Straight horizontal and vertical lines are drawn through the rows and columns of the opportunity–cost table so as to minimize the number of lines necessary to cover all zero-valued cells. As the number of lines (= 3) is less than the number of rows and columns (=4), an optimal assignment cannot yet be made and a revised opportunity–cost table must be obtained:

	A	B	C	D
I	~~10~~	~~7~~	~~0~~	~~19~~
II	8	5	7	0
III	19	7	4	0
IV	~~0~~	~~0~~	~~10~~	~~0~~

The revised opportunity–cost table can be obtained by selecting the smallest number in the table not covered by a straight line. This number (4) is then subtracted from all numbers not covered by straight lines and added to the numbers lying at the intersection of any two lines. Thus the revised opportunity cost table can be written as

	A	B	C	D
I	10	7	0	23
II	4	1	3	0
III	15	3	0	0
IV	0	0	10	4

The table is then re-examined to determine whether an optimal assignment can be made. In the present case it cannot, since the number of straight lines required to cover all zero-valued cells is equal to 3.

	A	B	C	D
I	10	7	0	23
II	4	1	3	0
III	15	3	0	0
IV	~~0~~	~~0~~	~~10~~	~~4~~

A revised opportunity–cost table must now be obtained using the same procedure as before. The smallest number (1) in the table not covered by a straight line is subtracted for all numbers not covered by straight lines and added to the numbers lying at the intersection of any two lines. This leads to the following revised opportunity–cost table:

	A	B	C	D
I	9	6	0	23
II	3	0	3	0
III	14	2	0	0
IV	0	0	11	5

Using this table it is possible to make an optimal assignment since the minimum number of lines required to cover all zero-valued cells is equal to the number of rows or columns.
The assignments are made so that one and only one zero is used in each row and column.
Thus, the optimal assignment is obtained from the last opportunity–cost table.

	A	B	C	D
I	9	6	⓪	23
II	3	⓪	3	0
III	14	2	0	⓪
IV	⓪	0	11	5

Therefore,
The crane at old site   I goes to new site C a distance of 19 miles
The crane at old site II goes to new site B a distance of 33 miles
The crane at old site III goes to new site D a distance of 33 miles
The crane at old site IV goes to new site A a distance of 11 miles
Total distance travelled by all cranes = 96 miles.

4.4 INTEGER PROGRAMMING

4.4.1 Introduction

Integer programming [6-11] is required when decision variables (x_j) must assume integer values. For example, in deciding how many precast prestressed beams to produce, it seems sensible to express the optimal production mix in integer quantities. In this section, Gomory's cutting plane algorithm[6] for dealing with integer programming will be described and illustrated by an example.

An integer programming problem is basically a special LP problem which can be written as

$$\text{Minimize } z = \sum_{j=1}^{n} c_j x_j \quad (4.5)$$

subject to

$$\sum_{j=1}^{n} a_{ij} x_j (\geqslant, =, \leqslant) b_i \quad i = 1, 2, \ldots, m$$

x_j is a non-negative integer

in which c_j, a_{ij}, and b_j are known constants as defined in the previous chapter and x_j are the variables to be evaluated.

4.4.2 Example — Excavation Equipment

A typical integer programming problem will now be described. A contractor wants to hire some special excavation equipment for a large earthmoving job. Two types of suitable excavators are available and each requires the following manpower:

	Skilled operators	Unskilled assistants
Excavator 1	1	5
Excavator 2	2	4

The contractor can use no more than 6 skilled operators and 20 unskilled assistants. After taking rental and labour costs into consideration, the contractor finds that by using either type of excavator he can make a profit of $10/hr. How many units of each type should he hire to maximize his hourly profit? This problem is clearly an integer programming problem since for a realistic solution the number of units of each excavator must be an integer.

4.4.3 Graphical solution

If x_1 and x_2 are the respective number of units of excavators of types 1 and 2 to be hired, then the excavation equipment problem can be expressed in the following integer programming form:

$$\text{Maximize } z' = 10x_1 + 10x_2 \tag{4.6}$$

subject to

$$5x_1 + 4x_2 \leq 20$$

$$x_1 + 2x_2 \leq 6$$

x_1, x_2 are non-negative integers.

The graphical solution to the excavation equipment problem is shown in Figure 4.11. The overall optimal LP solution is a non-integer solution and is of no practical value.

Three alternative optimal integer solutions exist, each yielding an hourly profit of $40. From this graphical solution it can be seen that a straightforward application of the simplex method described in Chapter 3 would lead to the non-integer solution: $x_1 = 8/3$, $x_2 = 5/3$, $z = 130/3$. Rounding up to the nearest integer would give a result: $x_1 = 3$, $x_2 = 2$, $z = 50$ which is unfortunately infeasible; in other words there would not be enough operators and assistants to run the excavators. Truncating to the integer value below the solution results in a new solution: $x_1 = 2, x_2 = 1, z = 30$ which although feasible gives less profit than could be achieved without breaking any constraints. Consequently there is a need for an algorithm which is able to find optimum integer solutions. Such an

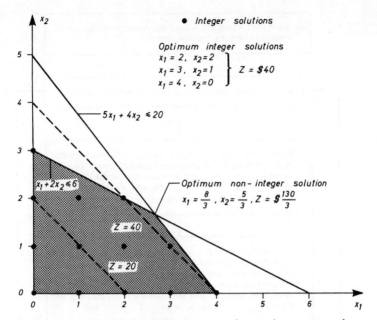

Figure 4.11 Graphical solution to excavation equipment example

algorithm is described in the next section and used to solve the excavation equipment problem.

4.4.4 Gomory's Cutting Plane method

Many algorithms for the solution of integer programming problems have been proposed. In this section Gomory's cutting plane method[6] is illustrated using the problem outlined in Section 4.4.3. This method gradually reduces the feasible region by sequentially introducing additional constraints

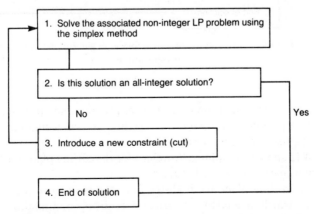

Figure 4.12 Simplified flow chart for cutting-plane method

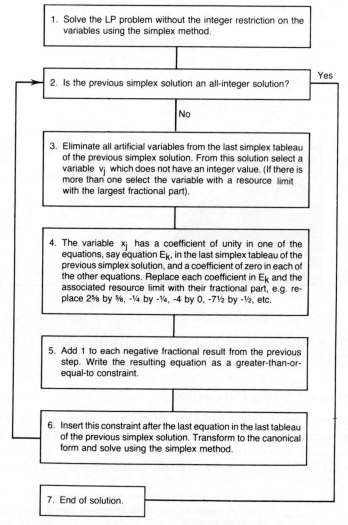

Figure 4.13   Detailed flow chart for Gomory's cutting-plane solution for integer programming

(cuts) until an optimal solution of the integer programming problem is obtained. The most important feature of this method is the technique adopted for constructing the new constraints.

Figure 4.12 shows a simplified flowchart for the cutting plane method. A more detailed flowchart is given in Figure 4.13.

Using the cutting plane method, the first step is to solve the associated LP problem in which the integer restriction is dropped. Equation (4.6) can be written in canonical form by changing to a minimization problem and adding slack variables x_3 and x_4:

$$\text{Minimize } z = -10x_1 - 10x_2 \tag{4.7}$$

$$5x_1 + 4x_2 + x_3 = 20$$

$$x_1 + 2x_2 + x_4 = 6$$

$$x_1, x_2 \geq 0.$$

The simplex method is now used to solve this problem. The simplex tableaux are:

	c_j			-10	-10	0	0	b_i
	$c^{(i)}$	Basis	b_i	v_1	v_2	v_3	v_4	$\overline{a_{ij}}$
I	0	x_3	20	5	4	1		5
	0	x_4	6	1	2		1	3
		z_j	0	0	0	0	0	
		Δz_j	0	0	-10	-10		
II	0	x_3	8		3	1	-2	$\frac{8}{3}$
	-10	x_2	3	1	$\frac{1}{2}$		$\frac{1}{2}$	6
		z_j	-30	-5			-5	
		Δz_j	-30	-5			5	
III	-10	x_1	$\frac{8}{3}$	1		$\frac{1}{3}$	$-\frac{2}{3}$	
	-10	x_2	$\frac{5}{3}$		1	$-\frac{1}{6}$	$\frac{5}{6}$	
		z_j	$-\frac{130}{3}$			$-\frac{5}{3}$	$-\frac{5}{3}$	
		Δz_j	$-\frac{130}{3}$			$\frac{5}{3}$	$\frac{5}{3}$	

Thus the solution is

$$z = -\tfrac{130}{3} \quad x_1 = \tfrac{8}{3} \quad x_2 = \tfrac{5}{3}. \tag{4.8}$$

As this is not an integer solution it is necessary to introduce a cut in order to reduce the feasible region thereby forcing an integer solution. Both variables have non-integer values at the current optimal solution. The next step is to select a non-integer-valued variable — it is usual to choose the variable with the largest fractional part. In this case, as both x_1 and x_2 have the same fractional part, x_1 is chosen arbitrarily. The new constraint is obtained in the following manner. The equation in the last tableau of the previous simplex tableaux in which x_1 has a unit coefficient can be written as

$$1x_1 + 0x_2 + \tfrac{1}{3}x_3 - \tfrac{2}{3}x_4 = \tfrac{8}{3}. \tag{4.9}$$

Consider only fractional parts. Upon adding 1 to coefficients with a negative fraction part, (4.9) can be re-written with a greater-than-or-equal-to constraint:

$$0x_1 + 0x_2 + \tfrac{1}{3}x_3 + \tfrac{1}{3}x_4 \geq \tfrac{2}{3}. \qquad (4.10)$$

This constraint is now added to the last set of equations in the previous simplex tableau:

$$\text{Minimize } z = -\tfrac{130}{3} + 0x_1 + 0x_2 + \tfrac{5}{3}x_3 + \tfrac{5}{3}x_4 \qquad (4.11)$$

$$x_1 + 0x_2 + \tfrac{1}{3}x_3 - \tfrac{2}{3}x_4 = \tfrac{8}{3}$$

$$0x_1 + x_2 - \tfrac{1}{6}x_3 + \tfrac{5}{6}x_4 = \tfrac{5}{3}$$

$$0x_1 + 0x_2 + \tfrac{1}{3}x_3 + \tfrac{1}{3}x_4 \geq \tfrac{2}{3}.$$

This problem can be transformed into its canonical form by adding an excess variable x_5 and an artificial variable x_6 to the added constraint. The problem can then be solved using the big M method as described in the previous chapter:

$$\text{Minimize } z = -\tfrac{130}{3} + 0x_1 + 0x_2 + \tfrac{5}{3}x_3 + \tfrac{5}{3}x_4 + 0x_5 + Mx_6 \qquad (4.12)$$

$$x_1 + 0x_2 + \tfrac{1}{3}x_3 - \tfrac{2}{3}x_4 + 0x_5 + 0x_6 = \tfrac{8}{3}$$

$$0x_1 + x_2 - \tfrac{1}{6}x_3 + \tfrac{5}{6}x_4 + 0x_5 + 0x_6 = \tfrac{5}{3}$$

$$0x_1 + 0x_2 + \tfrac{1}{3}x_3 + \tfrac{1}{3}x_4 - x_5 + x_6 = \tfrac{2}{3}$$

$$x_1, \ldots, x_6 \geq 0.$$

The last equation is used to eliminate x_6 from the first equation and the resulting set of equations can then be solved by the simplex method.

		c_j		0	0	$\tfrac{5}{3}-M/3$	$\tfrac{5}{3}-M/3$	M	0	$\dfrac{b_i}{a_{ij}}$
	$c^{(i)}$	Basis	b_i	v_1	v_2	v_3	v_4	v_5	v_6	
	0	x_1	$\tfrac{8}{3}$	1		$\tfrac{1}{3}$	$-\tfrac{2}{3}$			8
I	0	x_2	$\tfrac{5}{3}$		1	$-\tfrac{1}{6}$	$\tfrac{5}{6}$			-10
	0	x_6	$\tfrac{2}{3}$			$\tfrac{1}{3}$	$\tfrac{1}{3}$	-1	1	2
		z_j	$-\tfrac{130}{3}+2M/3$			0	0	0		
		Δz_j	0			$(-M/3+\tfrac{5}{3})$	$(-M/3+\tfrac{5}{3})$	M		
	0	x_1	2	1			-1	1	-1	-2
II	0	x_2	2		1		1	$-\tfrac{1}{2}$	$\tfrac{1}{2}$	2
	$\tfrac{5}{3}-M/3$	x_3	2			1	1	-3	3	2
		z_j	$-\tfrac{120}{3}$				$\tfrac{5}{3}-M/3$	$-5+M$	$(5-M)$	
		Δz_j	$\tfrac{10}{3}-2M/3$				0	5	$(M-5)$	

Thus the solution is

$$z = -40 \quad x_1 = 2 \quad x_2 = 2 \quad (x_3 = 2) \qquad (4.13)$$

which is the integer solution which was required.

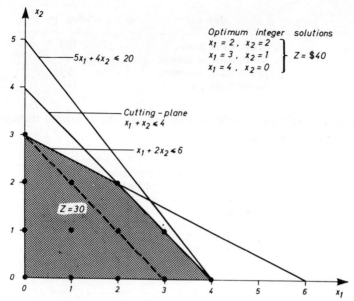

Figure 4.14 Optimal integer solution using Gomory's cutting plane method

(It should also be noted that an alternative optimal solution exists since a zero cost coefficient exists in tableau II.) This solution is expressed graphically in Figure 4.14. Note that in Figure 4.14 the cutting plane is expressed in terms of x_1 and x_2 using the first two constraints in (4.11).

4.5 EXERCISES

4.5.1 A concrete transit-mix company owns three plants with capacities and production costs as follows:

Plant number	Daily capacity (m^3)	Production cost ($/m^3$)
I	160	10
II	160	9
III	80	13

The company is under contract to supply concrete for a bridge construction and is scheduled to deliver to the various job sites the following quantities of concrete:

Job site	Amount (m³)
1. South bank pier	100
2. South bank abutment	40
3. North bank abutment	80
4. North bank pier	140

Based on distance, traffic, and site delays, the following transportation costs are estimated:

Job site number	Transportation cost ($/m³) Plant number		
	I	II	III
1	1	1	3
2	1	2	3
3	2	1	2
4	3	2	1

Schedule tomorrow's production to minimize total cost to the company.

4.5.2 A contractor has five locations A, B, C, D, and E on a road contract to which crushed stone is to be delivered. The stone, which is all of the same quality, is to be supplied from three quarries 1, 2, and 3. The table shows the relative costs per cubic metre of transporting stone from each source to each location, the quantity of stone which is required at each location, and the quantity which will be available at each quarry.

Treat this as a transportation problem and allocate quantities of stone from each quarry to each location for the least value of total transportation cost.

Quarry	Location					Quarry output (m³)
	A	B	C	D	E	
	Relative transportation costs					
1	9	10	12	11	10	120
2	12	8	6	3	7	255
3	10	9	4	13	6	150
	45	105	150	135	90	
	Quantity required at each location (m³)					

4.5.3 The Brunel Gravel Company has received a contract to supply gravel for three new construction projects, located in the towns of I, II, and III.

Construction engineers have estimated the required amounts of gravel which will be needed at three construction projects:

Project	Location	Weekly requirement
A	I	72
B	II	102
C	III	41
	Total	215

The Brunel Gravel Company has three gravel pits located in towns 1, 2, and 3. The gravel required for the construction projects can be supplied by these three plants. Brunel's chief dispatcher has calculated the amounts of gravel which can be supplied by each plant and the unit transportation costs:

Plant	Location	Amount available/week (truckloads)
W	1	56
X	2	82
Y	3	77
	Total available	215

	Cost per truckload		
From	To Project A	To Project B	To Project C
Plant W	$ 4	$ 8	$ 8
Plant X	$16	$24	$16
Plant Y	$ 8	$16	$24

Given the amounts required at each project and the amounts available at each plant, the Company's problem is to schedule shipments from each plant to each project in such a manner as to minimize the total transportation cost. Solve using the transportation method.

4.5.4 In the above problem, suppose that Plant W has a capacity of 76 truckloads per week rather than 56. The Company would be able to supply 235 truckloads per week. However, the project requirements remain the same. Solve this unbalanced transportation problem.

4.5.5 Figure 4.15 represents a railway network. The points labelled a, b, c, and d represent collieries and those labelled A, B, and C represent the coke ovens of a steelworks. The rail mileages from these points to the junctions of the network, and from one junction to another, are shown by the number on the lines. The collieries can supply the following amounts:

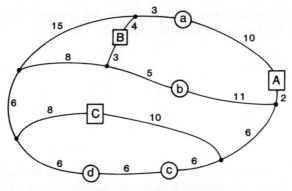

Figure 4.15 Railway network

a 3000 tons/month
b 2000 tons/month
c 8000 tons/month
d 3000 tons/month.

The requirements for coke ovens are:

A 2000 tons/month
B 4000 tons/month
C 6000 tons/month.

If the transportation cost is £3 per ton mile, find that distribution of coal from the collieries to coke ovens which minimizes the total transportation cost.

4.5.6 A company produces a special type of precast concrete component for use in prefabricated construction. The company has three factories, A, B, and C which supply five contractors (working on different sites) with the precast components. The production capacities of the factories and the demands of the contractors — assumed constant — and distribution costs are given below. How should the components be distributed in order to supply the contractors with their demands in the cheapest way?

Factory	Production capacity (000's per month)
A	5
B	10
C	10

Distribution costs per unit

Factory	Contractor				
	a	b	c	d	e
A	5	7	10	5	3
B	8	6	9	12	4
C	10	9	8	10	15
Contractor's requirements (000's)	3	3	10	5	4

4.5.7 Industries I_1, I_2, and I_3 require 80, 30, and 90 ($\times 10^5$) cu ft of water per day respectively to be supplied from three reservoirs R_1, R_2, and R_3. Reservoir R_1 can supply 100 ($\times 10^5$) cu ft per day. Reservoir R_2 can supply 25 ($\times 10^5$) cu ft per day. Reservoir R_3 can supply 75 ($\times 10^5$) cu ft per day. Given the following transportation cost matrix, show how to obtain an allocation of water to minimize the total transportation cost.

Transportation cost matrix

	I_1	I_2	I_3
R_1	5	10	2
R_2	3	7	5
R_3	6	8	4

4.5.8 An engineering firm produces a steel structural component for use in offshore oil production platforms. Three work centres are required to manufacture, assemble, and package the product. Four locations are available within the plant. The materials handling cost at each location for the work centres is given by the following cost matrix. Determine the location of work centres that minimizes the total materials handling costs.

	Location			
	1	2	3	4
Manufacturing	18	18	16	13
Assembly	16	11	×	15
Packaging	9	10	12	8

The symbol × implies that assembly cannot be performed in Location 3.

4.5.9 A civil engineering consulting firm has a backlog of four contracts. Work on these projects must be started immediately. Three project leaders are available for assignment to these contracts. Because of varying work experience of the project leaders, the profit to the consulting firm will vary based on the assignment:

	Contract			
Project leader	1	2	3	4
A	13	10	9	11
B	15	17	13	20
C	6	8	11	7

The unassigned contract can be assigned by sub-contracting the work to an outside consultant. Determine the assignment which optimizes the overall profit.

4.5.10 An aggregate producer has four identical mobile crushing–screening plants and four sources of raw material which he can use during the coming construction season. Given the following profit matrix, how many plants should be assigned to each site?

Profit matrix

No. of plants assigned	Raw-materials site			
	1	2	3	4
1	47	39	24	35
2	81	62	47	51
3	105	62	47	61
4	132	91	87	68

4.5.11 The following matrix gives the cost of using each of five different earth-movers to perform five separate jobs on a civil engineering site. Assign one job to each earth-mover in order that the total cost is optimized.

Earth-mover	Job				
	A	B	C	D	E
1	10	5	9	18	11
2	13	19	6	12	14
3	3	2	4	4	5
4	18	9	12	17	15
5	11	6	14	19	10

4.5.12 A construction company has to move four large cranes from old construction sites to new construction sites. The distance (in miles) between the old locations and the new are given in the following matrix:

Old construction sites	New construction sites			
	N_1	N_2	N_3	N_4
O_1	25	30	17	43
O_2	20	23	45	30
O_3	42	32	18	26
O_4	17	21	40	50

Any of the cranes work equally well on any of the new sites. Determine a plan for moving the cranes that will minimize the total distance involved in the move.

4.5.13 An engineering firm wishes to assign five of its personnel to the task of designing five projects. Given the following time estimates required by each designer to design a given project, find the assignments which minimize total time:

Engineer	Project				
	1	2	3	4	5
1	3	10	3	1	8
2	7	9	8	1	7
3	5	7	6	1	4
4	5	3	8	1	4
5	6	4	10	1	6

4.5.14 A contractor wants to hire some special excavation equipment for a large earthmoving job. Two types of suitable excavators are available and each requires the following manpower:

	Operators
Excavator 1	5
Excavator 2	2

The contractor can use no more than 17 operators. After taking rental and labour costs into consideration the contractor finds he can make a profit of $10 per hour on excavator 1 and $3 per hour on excavator 2. *Use Gomory's cutting plane method* to find how many units of each type he should hire to maximize his hourly profit. (N.B. Having worked out the first cut constraint, rework the simplex method from the beginning.) *Check your answer graphically.*

4.5.15 Large orders for four steel components to be used on North Sea oil rigs are to be assigned to four man–machine centres. Some machines are better suited to produce certain components and their operators are more proficient at producing some components then others. The costs to produce each component at each centre are:

Component	MMC1	MMC2	MMC3	MMC4
C1	12	9	11	13
C2	8	8	9	6
C3	14	16	21	13
C4	14	15	17	12

(i) Which component (C) should be assigned to each man–machine centre (MMC)?
(ii) Following a union agreement man–machine centre MMC2 cannot be used to produce component C1. How does this affect the assignment of components to machines?
(iii) Assume a new machine (MMC5) has been added to the facilities above. One old machine is to be phased out. The operator on the old machine will operate the new one if it can lead to an assignment which is less expensive than the assignment made in part (i). The estimated cost using the new machine is

Component	C1	C2	C3	C4
Production cost at MMC5	11	7	15	10

Ignoring the union agreement in (ii), should the new machine be used? If so, which component should it produce?

4.6 REFERENCES

1. Hitchcock, F. L. (1941). 'Distribution of a product from several sources to numerous localities.' *Journal of Mathematical Physics*, **29**.
2. Koopmans, T. C. (1949). 'Optimum utilization of the transportation system.' *Econometrica*, **17** (supplement).
3. Dantzig, G. B. (1951). 'Application of the simplex method to a transportation problem.' Chapter 22 of T. C. Koopmans (Ed.), *Activity Analysis of Production and Allocation*, Cowles Commission Monograph, 13. Wiley, New York.
4. Loomba, N. P., and Turban, E. (1974). *Applied Programming for Management*, Holt, Rinehart, and Winston, New York. Chapters 5 and 6.
5. Dantzig, G. B. (1963). *Linear Programming and Extension*, Princeton University Press, Princeton, New Jersey.
6. Gomory, R. E. (1967). 'All-integer programming algorithm.' In J. F. Muth and G. L. Thompson (Eds), *Industrial Scheduling*, Prentice-Hall, Englewood Cliffs, New Jersey. pp. 193–206.
7. Balinski, M. L. (1970). 'On recent developments in integer programming.' In H. W. Kuhn (Ed.), *Proceedings of the Princeton Symposium on Mathematical Programming*, Princeton University Press, Princeton, New Jersey.

8. Garfinkel, R. S., and Nemhauser, G. L. (1972). *Integer Programming*, Wiley, New York.
9. Greenberg, N. (1971). *Integer Programming*, Academic Press, New York.
10. Hadley, G. (1964). *Non-linear and Dynamic Programming*, Addison-Wesley, Reading, Mass.
11. Daellenbach, H. G., and George, J. A. (1978). *Introduction to Operations Research Techniques*, Allyn and Bacon, Boston, Mass.

Chapter 5
Non-Linear Programming

5.1 INTRODUCTION AND SCOPE

The formal definition of a non-linear programming problem is the determination of a set of design or decision variables $\mathbf{x} = (x_1, x_2, \ldots, x_n)^T$ that provides a solution for the problem

$$\text{Minimize } z(\mathbf{x}) \tag{5.1}$$

subject to

$$g_i(\mathbf{x}) \geq 0 \quad i = 1, 2, \ldots, m$$

in which some or all of the problem functions z or g_i are non-linear with respect to the variables.

It will be recalled that a maximization problem may be cast in this form by reversing the sign of the objective function z (see Section 1.6.1(b)) and in the remainder of this chapter minimization will be implied.

Within this broad classification of mathematical programming problems it is possible to identify several special cases dependent on the distribution and nature of the non-linearities within the problem functions. For example, an important group of problems concerns a non-linear objective function subject to strictly linear constraints. Then again, the ability of the objective function to be separated into distinct terms, each dependent on a single design variable, may affect the method of solution. The nature of the constraints, whether they be concave, convex, or linear, may also allow the development of a special problem-solving algorithm for a particular sub-class of problem. A wealth of literature exists concerned with these various aspects of non-linear programming (see Section 5.11 for a few examples) and it would be naive to assume that within the scope of the present work anything more than a very superficial treatment of the subject could be attempted.† Non-linear problems are sufficiently important, however, to prohibit the total omission of the subject from this book and it is hoped to give an overview of problem classification which will at least allow recognition of the major problem types and some of the better known solution techniques. In

---

† An excellent review of different non-linear programming techniques is given by Siddall (Chapter 7).[1]

addition, a few of the commoner algorithm types are developed in detail and presented with illustrative examples and documentation to enable the reader to obtain solutions to practical problems of modest size and complexity.

Optimization of a function of a single variable may be achieved relatively easily by tabulation or graphing. Depending on the problem, the design and cost calculation associated with each point on such a table or graph may be tedious, but the availability of calculators, especially of the programmable type, has substantially reduced this problem. When possible, there is really no substitute for being able to display graphically the dependence of a cost function on a design parameter. Not only is the identification of the optimal point confirmed by inspection, but the designer obtains valuable insight into the sensitivity of the cost to any changes in the design parameters. In cases where such an optimization must be carried out repeatedly for a range of parametric values, or as a component of a more complex design calculation, it may be helpful to apply an automatic routine (such as subroutine GOLDEN discussed later) which will locate a minimum point of an arbitrary cost function.

When two variables are involved, the method of tabulation or enumeration leads to the generation of a surface which may be depicted by graphing contours or iso-cost lines. Figure 5.1 illustrates such a situation in which the optimum cost value (minimum) is obtained at point A, from which the optimal values of the design variables x_1 and x_2 may be obtained.

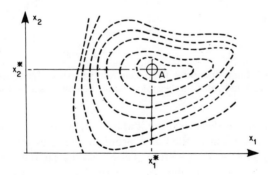

Figure 5.1   Iso-cost lines for an objective function of 2 variables

When more than two design variables are involved, the labour of tabulation or enumeration becomes too costly and other strategies must be devised. This chapter explores the various methods by which this may be done.

5.2 EFFECT OF NON-LINEARITIES ON THE SOLUTION

When the objective function and also the constraint functions of the problem of (5.1) are all linear in the design variables, the problem reduces to one of linear

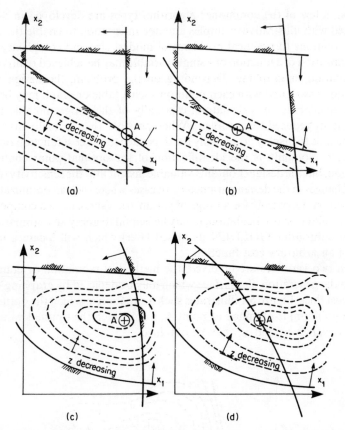

Figure 5.2 Effect of non-linearities. (The constraints illustrated here are all inequalities; the arrow indicates the feasible side of the constraint boundary)

programming which is described in Chapter 3. It is instructive to recall certain properties of the solution to the linear problem and to examine to what extent these are changed by the introduction of non-linearities in the problem functions.

These various changes are illustrated in the four diagrams of Figure 5.2, in which for simplicity it is assumed that the problem is two-dimensional (the design variables being x_1 and x_2) and the objective function may be represented by the contours of a surface in one quadrant of the x_1–x_2 space.

In Figure 5.2(a), both objective function and constraints are linear in x_1 and x_2. The linear constraints define a convex space. The linear objective function may be visualized as an inclined plane such that the iso-cost contours are all straight, parallel, and uniformly spaced. Even without the more detailed treatment of Chapter 3, it can be intuitively seen for this simple case that the minimum cost solution must lie on one of the vertices of the feasible space.

Figure 5.2(b) illustrates essentially the same problem in which one of the constraints is now non-linear. The existence of a linear objective function ensures that the solution remains on the exterior of the feasible space, but now there is the possibility that the solution may be found at any point along the edge of the space bounded by the constraints and not necessarily at a vertex.

The introduction of a non-linear objective function changes the character of the iso-cost lines. What was previously a plane surface may now exhibit curvature, saddle points, and turning values. In Figures 5.2(c) and (d), the same non-linear objective function surface is seen in relation to different sets of constraints. In Figure 5.2(c) the constraints are not active and the solution is found in the interior of the feasible space. In Figure 5.2(d) one of the constraints has become active and the solution occurs on the edge of the feasible space.

Intermediate cases may occur in which the objective function, although non-linear, may be shown to possess special properties which allow particular properties of the solution to be predicted. For example, it may be shown that if an objective function is composed of separable, concave, monotonic cost functions and is subject to linear constraints, then the solution must lie on one of the vertices of the feasible space. In general, however, no such helpful assumptions may be made, and the entire feasible space must be searched for the solution. Also, it is quite common to find a surface which exhibits not one but two or more 'minima' and the identification of local minima from the overall or global minimum is one of the more intractable aspects of non-linear programming.

5.3 OPTIMUM-SEEKING STRATEGIES

When the objective function and constraints take a simple, differentiable and easily manipulated form it may be possible to employ methods of calculus to locate and identify turning points. Some of these techniques are described in Chapter 2. In general, however, as the objective function and the associated constraints become more complex, the opportunity to use classic analytical techniques becomes rare and recourse must be made to numerical methods. Two general approaches will be discussed in some detail here but many special strategies and techniques exist.

Both methods require a starting point to be assumed — i.e. some initial trial values for the design variables which have to be optimized. They then attempt to improve on this solution by selecting a direction of search, through or towards the feasible space, represented by successive sets of values for the vector of design variables represented by $\mathbf{x} = (x_1, x_2, \ldots, x_n)^T$.

Gradient methods select the direction of search on the basis of evaluation of the objective function and also derivatives of the objective function at the current position.

Direct search methods rely solely on evaluation of the objective function at the current position together with experience gained from previous trial positions. No derivatives are involved.

In examining these techniques it is convenient to consider first *unconstrained* problems and then study methods to introduce the constraint relationships of *constrained* problems. Also it is usually helpful for both explanation and visualization to tackle problems involving functions of a *single variable* before extending the arguments and methods to functions involving *several variables*.

Accordingly, the layout of the chapter will follow the headings:

(i) Unconstrained function of a single variable.
(ii) Unconstrained function of several variables.
(iii) Constrained function of a single variable.
(iv) Constrained function of several variables.

5.4 UNCONSTRAINED FUNCTION OF A SINGLE VARIABLE

5.4.1 Gradient methods

Figure 5.3 illustrates a function $z(x)$ of the single design variable x. The function is assumed to be unimodal — i.e. the function possesses only a single turning value in the interval of x within which interest is concentrated. Moreover, it is assumed for this case that the turning value is a *minimum* and occurs at a value

$$x = x^*.$$

Also shown in Figure 5.3 is the first derivative of the function, $z'(x)$, from which it is observed that $z'(x) = 0$ at $x = x^*$, or

$$\frac{dz(x)}{dx} = 0 \quad \text{at} \quad x = x^*. \tag{5.2}$$

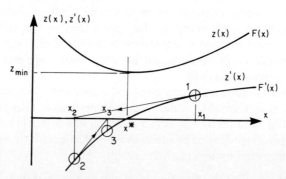

Figure 5.3 Minimizing a function of a single variable by the Newton–Raphson method

This gives an implicit definition of the point x^* which may be obtained by a trial and error procedure, or an iterative technique such as the Newton–Raphson method (see Section 2.8). With reference to Figure 5.3, the process starts by assuming an initial starting point 1 from which successive improved estimates 2, 3, etc., are obtained by the relation

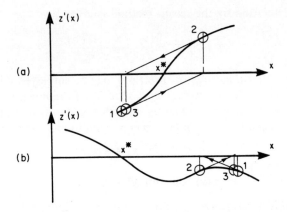

Figure 5.4   Examples of non-convergence using the Newton–Raphson method

$$x_{r+1} = x_r - \frac{z'(x_r)}{z''(x_r)}. \qquad (5.3)$$

Unfortunately, as illustrated in Figure 5.4, the success of the Newton–Raphson method is very dependent on the shape of the function; the process may converge to the wrong root, or it may converge only very slowly, or not converge at all, depending on the initial guess x_1 and on the possible existence of turning values or points of contraflexure in the function. Figure 5.4(a) shows an objective function $z(x)$ which results in a curve of $z'(x)$ which exhibits a point of contraflexure near the solution. The process may cycle indefinitely around the solution unless steps are taken to detect and break the oscillation. A similar situation is shown in Figure 5.4(b) resulting from a turning value in the $z'(x)$ curve. Conditions such as these can rarely be foreseen by an examination of the objective function.

It will be observed that the process of determining a turning value in $z(x)$ reduces to the location of roots of the equation

$$z'(x) = 0$$

for which the Newton–Raphson method is but one of several available techniques.

The *secant method* (or method of false-position) is obtained by replacing the second derivative $z''(x)$ by the secant slope defined by two points on the $z'(x)$ curve. These points are, in general, the current position and the preceding position, and it is necessary to provide two initial guesses in order to start the iteration.

Thus, if

$$z''(x_r) \simeq \frac{z'(x_r) - z'(x_{r-1})}{x_r - x_{r-1}} \qquad (5.4)$$

the improved iteration by the secant method is given by

$$x_{r+1} = x_r - \frac{(x_r - x_{r-1})z'(x_r)}{z'(x_r) - z'(x_{r-1})}.$$

The *method of interval halving* (or bisection) may be employed if it is known in advance that x^* lies in the interval $x_a \leqslant x^* \leqslant x_b$. If the function $z(x)$ is indeed unimodal this means that $z'(x)$ changes sign once, and only once, in the interval $(x_b - x_a)$. Also if the turning value of $z(x)$ is a minimum it can be stated that

$$z'(x) \leqslant 0 \quad \text{for } x_a \leqslant x \leqslant x^*$$

and

$$z'(x) > 0 \quad \text{for } x^* < x \leqslant x_b.$$

The algorithm is described by the flow diagram of Figure 5.5.

The methods described all represent some form of root-solving technique

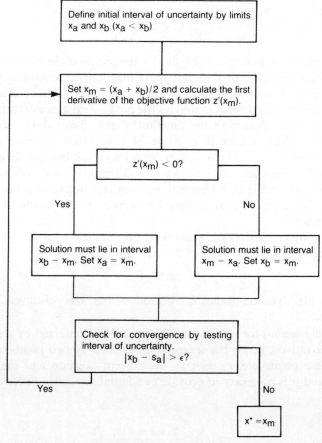

Figure 5.5 Flow chart of interval-halving algorithm

applied to the first derivative of the objective function. The derivative may be obtained analytically or numerically depending on the nature and complexity of the cost functions. Appendix A contains documentation and listing of a FORTRAN subroutine called BISECT which employs the interval-halving procedure described here. The routine contains a preliminary coarse search procedure to define the initial interval of uncertainty and also a test to distinguish between a true root and a function discontinuity. Although the routine may be used for normal root-solving problems, it is applied in the following example to a problem of cost minimization.

5.4.2 Optimization of a sedimentation process

Example

The cost of processing an industrial waste by means of sedimentation is known to depend on the retention time t in hours, adopted for the design of the plant. Three cost components are identified, which are dependent on the flow capacity Q:

(i) A fixed charge of $3Q$ cost units.
(ii) Plant cost defined as $0.8Q^2 t^{3.25}$.
(iii) A tax or penalty cost dependent on the quality of the effluent and which can be described by the exponential term $14Q\,e^{-t}$.

The objective function may thus be represented as the sum of these three terms and the aim is to find a value of t for which the cost is a minimum, when the value of Q is (say) 5 m^3/s. That is,

$$\underset{t^*}{\text{Minimize}}\; z = 20t^{3.25} + 70\,e^{-t} + 15 \qquad (5.5)$$

subject to

$$t > 0.$$

The constraint is rather self-evident and the problem may be treated as one of unconstrained minimization of a function of a single variable.

For fixed cost coefficients the problem is probably most easily solved by tabulation and graphing; Figure 5.6 shows the nature of the objective function and the fact that it is indeed unimodal with a minimum near $t = 0.75$ hr. However, if it is intended to test the sensitivity of this result to changes in the flow rate Q, or if the result of the optimization is to be incorporated in a more complex design calculation, it may be desirable to use one of the algorithms described in the previous section.

Solution by Newton–Raphson method The objective function in this case is quite simple and the first and second derivatives may be obtained directly by differentiation. Thus

$$z = 20t^{3.25} + 70\,e^{-t} + 15 \qquad (5.6)$$

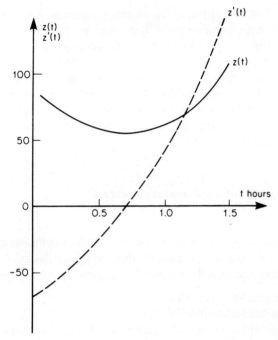

Figure 5.6 Sedimentation process — objective function $z(t)$ and derivative $z'(t)$

$$z' = 65t^{2.25} - 70\,e^{-t} \tag{5.7}$$

$$z'' = 146.25t^{1.25} + 70\,e^{-t}. \tag{5.8}$$

The iteration scheme obtained by the Newton–Raphson method is

$$t_{r+1} = t_r - z'(t_r)/z''(t_r).$$

The following sequence of values is generated by starting with $t_1 = 1.0$:

n	t_n	$z'(t_n)$	$z''(t_n)$	Δt_n	t_{n+1}
1	1.000	39.25	172.00	−0.228	0.772
2	0.772	3.94	138.15	−0.029	0.743
3	0.743	0.06	134.22	−0.0004	0.7429

In this example the function is well conditioned and converges rapidly for any initial approximation greater than the root. For very small positive initial values, the first iteration yields a value close to unity since

$$z' \simeq -z'' \quad \text{for } t > 0 \simeq 0.$$

Negative values of t are of course both practically and mathematically infeasible.

Solution by interval halving Determination of the root of the equation formed by setting the first derivative of the objective function equal to zero may be carried out by the method of interval halving without the need for second derivatives z''. This involves finding the solution of the equation

$$z' = 65t^{2.25} - 70\,e^{-t} = 0 \tag{5.7'}$$

assuming an initial interval of uncertainty given by $0 \leqslant t \leqslant 2.0$. The sequence of iteration is tabulated, convergence being assumed when the interval reduces to less than 0.01.

t_A	t_B	$(t_B - t_A)$	$t_m = (t_A + t_B)/2$	$z'(t_m)$	Remarks
0.000	2.000	2.000	1.000	39.25	$t_B = t_m$
0.000	1.000	1.000	0.500	−28.79	$t_A = t_m$
0.500	1.000	0.500	0.750	0.96	$t_B = t_m$
0.500	0.750	0.250	0.625	−14.89	$t_A = t_m$
0.625	0.750	0.125	0.688	− 7.22	$t_A = t_m$
0.688	0.750	0.063	0.719	− 3.20	$t_A = t_m$
0.719	0.750	0.031	0.734	− 1.14	$t_A = t_m$
0.734	0.750	0.016	0.742	− 0.09	$t_A = t_m$

Thus $t^* \simeq (0.742 + 0.750)/2 = 0.746$.

Note that the relative efficiency of the Newton–Raphson method is not proportional to the number of iterations but rather to the number of function evaluations.

Solution by the secant method The secant or false position method is again a technique for determining the root of the equation $z'(t) = 0$. Trial points at $t_1 = 0.0$ and $t_2 = 1.0$ are used yielding the following sequence of iterations. Note that the method does not require the initial trial values to be chosen such that $t_0 \leqslant t^* \leqslant t_1$.

n	t_n	$z'(t_n)$	$(t_n - t_{n-1})$	$(t_{n+1} - t_n)$
0	0.000	−70.00		
1	1.000	39.25	1.000	−0.359
2	0.641	−13.01	−0.359	0.089
3	0.730	− 1.69	0.089	0.013
4	0.744	0.09	0.013	−0.001
5	0.743	0.00	−0.001	

In terms of the number of function evaluations the secant method is comparable with the Newton–Raphson method, but only because the objective function can be differentiated twice analytically. In the case of an objective function which cannot be differentiated by calculus, the derivatives must be determined by finite differences. In general this involves two evaluations of the objective function for the first derivative and three for second derivatives, etc. Thus for each iteration the Newton–Raphson method requires at least three function evaluations. The secant and interval-halving method require only two evaluations per iteration, but in this example convergence requires twice as many iterations. The routine BISECT described in Appendix A may also be used to solve the same problem, as illustrated below.

Solution by routine BISECT The user must provide a suitable driving program and also a subroutine to generate values of the function the root of which is being sought. In the coding illustrated in Figure 5.7, it is assumed that the objective function may not be differentiated so that the first derivative must be approximated by a forward finite difference quotient. The interval used to generate the derivative is taken to be of the same order of magnitude as the acceptable interval of uncertainty at convergence.

It will be noted that routine BISECT employs a coarse search procedure to detect an initial interval of uncertainty within which the sign of the function changes. The starting point and direction of this coarse search must be specified in the calling statement. The program is written so that the coefficients of the objective function may be easily changed.

The results obtained are shown in Figure 5.7. It will be noted that during the coarse search the function (i.e. dz/dt) changes sign between $t = 0.6$ and 1.4 which becomes the interval of uncertainty for the fine search.

5.4.3 Direct search methods

All direct search procedures are characterized by the fact that no evaluation of derivatives is involved, the movement towards the solution being guided solely by evaluations of the objective function at the current and previous positions. Considering a function of a single variable, Figure 5.8 illustrates a unimodal function $z(x)$ with a minimum within an initial interval of uncertainty defined by $x_a \leqslant x^* \leqslant x_b$. Two additional points are selected arbitrarily at x_1 and x_2 such that

$$x_a < x_1 < x_2 < x_b.$$

Function evaluations are then obtained at the two new points and on the basis of this information one or other of the outer segments of the range may be discarded by the following argument:

(i) If $z(x_1) < z(x_2)$ then x^* must lie in the range $x_a < x^* < x_2$. Thus the segment x_2 to x_b may be eliminated from the search by setting $x_b = x_2$.
(ii) If $z(x_1) > z(x_2)$ then $x_1 < x^* < x_b$ and the segment x_a to x_1 may be discarded by setting $x_a = x_1$.

When the interval $|x_b - x_a|$ is acceptably small the process may be assumed to have converged.

```
      PROGRAM TST(OUTPUT,TAPE6=OUTPUT)
C     PROGRAM TO COMPUTE OPTIMUM SEDIMENTATION TIME
      EXTERNAL DERIV
      COMMON /COEFS/C1,C2,C3
      Q=5.0
      C1=0.8*Q*Q
      C2=14.0*Q
      C3=3.0*Q
      TSTART=0.0
      DT=0.2
      EPS=0.001
      CALL BISECT(DERIV,TSTART,DT,EPS,1,SOLN)
C     GET COST FOR OPTIMUM TIME
      CALL COST(SOLN,CST)
      WRITE(6,10)SOLN,CST
   10 FORMAT(6H AT T=,F8.4,10H MIN COST=,F10.3)
      STOP
      END
C
      SUBROUTINE COST(T,CST)
      COMMON /COEFS/C1,C2,C3
      CST=C1*T**3.25 + C2*EXP(-T) + C3
      RETURN
      END

C
      SUBROUTINE DERIV(X,DFDX)
      EPS=0.001
      CALL COST(X,FX0)
      CALL COST(X+EPS,FX1)
      DFDX=(FX1-FX0)/EPS
      RETURN
      END
```

```
          IN BISECT      0.0000      -70.0000
          IN BISECT      0.2000      -55.5724
          IN BISECT      0.6000      -17.8222
          IN BISECT      1.4000      121.3185
          IN BISECT      1.0000       39.3345
          IN BISECT      0.8000        7.9609
          IN BISECT      0.7000       -5.5638
          IN BISECT      0.7500        1.0272
          IN BISECT      0.7250       -2.3105
          IN BISECT      0.7375       -0.6523
          IN BISECT      0.7437        0.1848
          AT T =  0.7437  MIN COST =  55.9143
```

Figure 5.7   FORTRAN code for solution of the sedimentation problem

The effectiveness of such a search procedure may be measured either by the number of function evaluations required to reduce the interval of uncertainty to an acceptable size, or by the maximum interval of uncertainty which can remain following a prescribed number of trial evaluations. Clearly such a measure of effectiveness will depend on the strategy or policy adopted for positioning the trial points within the current interval of uncertainty.

One very efficient method involves the use of the Fibonacci constant or Golden

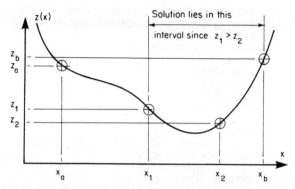

Figure 5.8 Direct search minimization of a function of a single variable

Section. The Fibonacci series is generated by starting with 1 and defining succeeding terms by the rule

$$n_{r+1} = n_r + n_{r-1}.$$

Thus the series is uniquely defined as

$$1, 1, 2, 3, 5, 8, 13, 21, 34, 55, 89 \ldots$$

$$\ldots 75\,025, 121\,393, 196\,418 \ldots.$$

The Fibonacci constant is obtained from two adjacent terms in the series and for higher-order terms in the series approaches a constant value defined as

$$F = \frac{n_r}{n_{r+1}} = \frac{n_{r+1}}{n_r} - 1 = 0.618033989.$$

The rather fascinating property that $F^2 = 1 - F$ provides (among many other aesthetic and scientific facts) a means of positioning the trial points so that although only one new point need be added at each iteration, the proportions of the subdivided interval of uncertainty remain constant. Figure 5.9 illustrates the process with respect to an interval of 1.0. Function evaluations are made at $x_1 = 0$ and $x_2 = 1.0$. Two additional trial points at x_3 and x_4 are located symmetrically within the interval $0.0 \leqslant x \leqslant 1.0$ such that

$$(x_2 - x_3) = (x_4 - x_1) = 0.618(x_2 - x_1)$$

and the function is evaluated at these new points.

Now by the same reasoning as presented before, since $z(x_3) < z(x_4)$ the segment from x_4 to x_2 may be discarded, as indicated in iteration 1 (see Figure 5.9).

The interval of uncertainty is now reduced to 0.618 and contains three function evaluations at x_1, x_3, and x_4. A fourth point is added at x_5 such that

$$(x_4 - x_5) = (x_3 - x_1) = 0.618(x_4 - x_1).$$

Note that the three segments of the interval (x_4-x_1) remain in the same proportion (0.382:0.236:0.382).

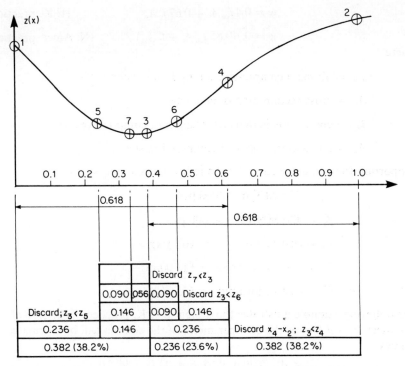

Figure 5.9 Direct search using the Fibonacci constant (Method of Golden Section)

A second iteration results in segment $(x_5 - x_1)$ being discarded, reducing the interval of uncertainty to 0.382 or $(0.618)^2$. A third and fourth iteration are completed, each by the addition of only a single point and each reducing the interval of uncertainty by a factor of 0.618. It can easily be shown that in n iterations a total of $(3 + n)$ function evaluations are required and the initial interval of uncertainty is reduced by a factor of $(0.618)^n$. In the example of Figure 5.9 after four iterations only seven points have been evaluated on the objective function and the interval of uncertainty is reduced to 0.146 or $(0.618)^4$.

An algorithm based on the above argument is presented in subroutine GOLDEN, which is described in Appendix A. The following example illustrates the use of this routine.

5.4.4 Design of a short concrete column

Example

A short, square-sectioned, reinforced concrete column is to be designed for minimum cost. The axial collapse load is to be 3.2 MN and is given by the following expression:

$$w = 0.4 f_{cu} A_c + 0.67 f_y A_{sc} \qquad \text{(U.K. practice)}$$
$$w = 0.7(0.85 f_{cu} A_c + f_y A_{sc}) \qquad \text{(N. Amer. practice)}$$
where

f_{cu} = ultimate compressive strength in concrete

A_c = cross-section area of concrete

f_y = characteristic (tensile) strength of compressive steel

A_{sc} = cross-section area of compression steel.

Proportion the column section for the following cost and stress data:

$f_{cu} = 37.5$ MN/m^2 (4000 psi)

$f_y = 420$ MN/m^2 (60 000 psi)

C_s = cost of steel = £0.15/kg

C_c = cost of concrete = £11.00/m^3

C_f = cost of forming = £5.00/m^2

The following example illustrates the design in U.K. practice. If the side of the cross-section is d metres the cost per metre of the column can be calculated as follows:

$$z = A_{sc}.7830 \times 0.15 + A_c.11.0 + 4d.5.0$$

where the unit weight of steel is taken as 7830 kg/m^3. Strictly speaking this objective function is to be minimized subject to certain constraints, i.e.

$$15.0 A_c + 281.4 A_{sc} = 3.2$$

and

$$A_c + A_{sc} - d^2 = 0.0.$$

Also

$$A_c, A_{sc}, \text{ and } d \geqslant 0.0.$$

Because of the simple nature of these constraints it is a simple matter to incorporate them in the objective function by substitution, thus eliminating two of the three variables.

For example, selecting A_c as the main design variable we obtain by substitution

$$A_{sc} = (3.2 - 15.0 A_c)/281.4$$

and

$$d = (A_{sc} + A_c)^{1/2}.$$

These substitutions can be carried out in the coding of a cost subroutine, including any necessary checks to ensure that the variables are non-negative. Typical coding for the main program is given in Figure 5.10.

The cost subroutine can be written in a very general way since all relevant data are transferred in through the labelled common block. Notice that this method also allows calculated values of other design parameters (e.g. side and steel area)

```
C   MAIN PROGRAM FOR R.C. COLUMN DESIGN
    EXTERNAL COST
C   SET UP COMMON BLOCK FOR USE BY ROUTINE COST
    COMMON /DATA/ASC,D,CS,CC,CF,FCU,FY,W
C   DEFINE COST, STRESS AND LOAD DATA
    CS=0.15
    CC=11.0
    CF=5.0
    FCU=37.5
    FY=420.0
    W=3.2
C   SET INITIAL, INCREMENTAL VALUES FOR AC
    ACMIN=0.1
    DAC=0.01
    CALL GOLDEN(COST,ACMIN,DAC,0.001,1,AC,CMIN)
C   CONVERT ASC TO SQ.MM
    ASC=ASC*1.0E6
    WRITE(6,10)CMIN
10  FORMAT(16H MINIMUM COST/M=,F12.2)
    WRITE(6,20)D,AC,ASC
20  FORMAT(7H SIDE =,F10.3,2H M,/
   +       7H CONC =,F10.3,3H M2,/
           7H STEEL=,F10.3,4H MM2)
    STOP
    END

    SUBROUTINE COST(AC,CST)
C   FINDS COST PER UNIT LENGTH FOR A SHORT R.C. COLUMN.
C   DUPLICATE COMMON BLOCK STATEMENT IN CALLING PROGRAM
    COMMON /DATA/ASC,D,CS,CC,DF,FCU,FY,W
C   CALCULATE REQUIRED STEEL AREA
    ASC=(W-0.4*FCU*AC)/(0.67*FY)
C   CHECK THAT ASC NON-NEGATIVE
    IF(ASC.LT.0.0) ASC=0.0
C   CALCULATE SIDE
    D=SQRT(AC+ASC)
C   COMPUTE COST PER UNIT LENGTH
    CST = 7830.0*ASC*CS+AC*CE+4.0*D*CF
    RETURN
    END

    IN GOLDEN              0.1000      14.7082
    IN GOLDEN              0.1100      14.4766
    IN GOLDEN              0.1262      14.0783
    IN GOLDEN              0.1524      13.3823
    IN GOLDEN              0.1947      12.1551
    IN GOLDEN              0.2633      13.1577
    IN GOLDEN              0.2209      11.8300
    IN GOLDEN              0.2371      12.3461
    IN GOLDEN              0.2047      11.8505
    IN GOLDEN              0.2271      12.0285
    IN GOLDEN              0.2085      11.7328
    IN GOLDEN              0.2071      11.7778
    IN GOLDEN              0.2185      11.7536
    IN GOLDEN              0.2080      11.7500
    MINIMUM COST/M = 11.73
    SIDE = 0.456 M
    CONC = 0.209 M2
    STEEL= 285.160 MM2
```

Figure 5.10 Main FORTRAN program, subroutine and typical results for column design example (Section 5.4.4) using routine GOLDEN

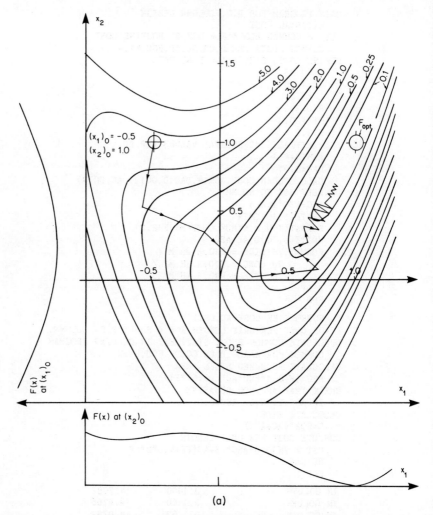

Figure 5.11 (a) Objective function of equation 5.9 (fixed-step gradient method shown).

to be transferred back to the calling program for inclusion in the solution. The routine is listed in Figure 5.10.

The subroutine GOLDEN is added at this point in the program deck or alternatively may reside in a source or semi-compiled library of subroutines for inclusion either during compilation or loading.

The results obtained are shown in Figure 5.10 and show five increments in the coarse search, followed by a further eight iterations converging to the solution.

For this example it is sufficient to incorporate in the main program a number of statements defining the load, stress, and cost data. Should one or more of these

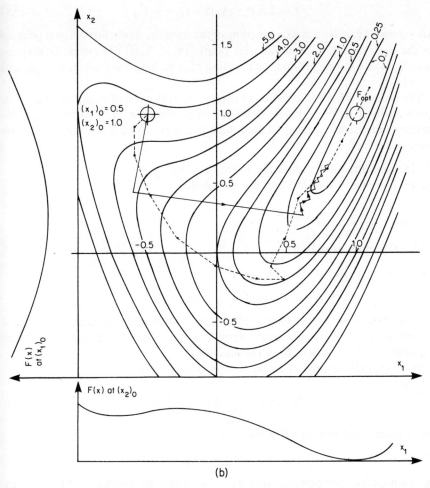

Figure 5.11 (b) Objective function of equation 5.9 (variable-step gradient method and Hooke–Jeeves pattern search shown)

require to be changed repetitively, the input might be more conveniently made by means of a 'read' or 'input' statement for use in either batch or time-sharing mode.

5.5 UNCONSTRAINED FUNCTION OF SEVERAL VARIABLES — GRADIENT METHODS

In this and the following section, attention is directed at the optimization of a function of several variables, free from constraints. The present section is concerned with methods involving the calculation of gradients or derivatives.

It is helpful to consider initially two-dimensional problems which can be easily visualized and depicted. Figure 5.11(a) illustrates a surface generated by the function

$$z = 2.5(x_1^2 - x_2)^2 + (1 - x_1)^2. \tag{5.9}$$

Also shown against each axis are plots of the function along lines drawn parallel to the axis through an initial trial point $[(x_1)_0, (x_2)_0]$. These plots may be visualized as sections produced by cutting planes along the lines $x_1 = (x_1)_0$ and $x_2 = (x_2)_0$.

The process of finding a minimum involves the selection of an initial trial point, followed by a series of moves aimed at improving the approximation to the solution. The strategy used to determine the move to be made comprises two distinct decisions:

(i) In what direction should the move be made?
(ii) By how much should the point be moved?

5.5.1 Direction of search

Let the coordinates of the new approximation be given by

$$[(x_1)_1, (x_2)_1] = [(x_1)_0 + \Delta x_1, (x_2)_0 + \Delta x_2].$$

It is clear from the gradients of Figure 5.11(a) that the same amount of movement in the two coordinate directions will result in different degrees of improvement (i.e. reduction) of the function z.

In general the change in z may be approximated by

$$\Delta z \simeq dz = \frac{\partial z}{\partial x_1} \Delta x_1 + \frac{\partial z}{\partial x_2} \Delta x_2. \tag{5.10}$$

For a given total movement defined by

$$h = (\Delta x_1^2 + \Delta x_2^2)^{1/2}$$

the orthogonal components Δx_1 and Δx_2 must be proportioned so as to maximize the change in z. This situation may be visualized by approximating the surface z in the vicinity of the point $[(x_1)_0, (x_2)_0]$ by a plane the contours of which will be straight, parallel, and equally spaced as in Figure 5.12.

The total change in z can be expressed as

$$dz = \frac{\partial z}{\partial x_1} \Delta x_1 + \frac{\partial z}{\partial x_2} (h^2 - \Delta x_1^2)^{1/2} \tag{5.11}$$

which will have an extreme value when $d(dz)/d(\Delta x_1) = 0$ or

$$\frac{\partial z}{\partial x_1} - \frac{\partial z}{\partial x_2} \frac{2\Delta x_1}{2(h^2 - \Delta x_1^2)^{1/2}} = 0 \tag{5.12}$$

therefore

$$\frac{\Delta x_1}{\Delta x_2} = \frac{(\partial z/\partial x_1)}{(\partial z/\partial x_2)}.$$

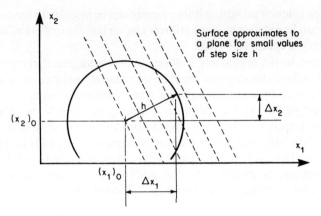

Figure 5.12 Defining the direction of search

Thus the search direction which will result in the largest change in z is the line of greatest slope.†

5.5.2 Fixed step size

The simplest strategy to determine the amount by which the point should be moved is to adopt a constant, arbitrary step size h. The incremental change in each variable can then be calculated as a function of the gradients. Thus

$$\frac{\partial z}{\partial x_1}\Delta x_1 + \frac{\partial z}{\partial x_2}\Delta x_2 = 0$$

or

$$\frac{\Delta x_1}{\Delta x_2} = -\frac{(\partial z/\partial x_2)}{(\partial z/\partial x_1)}.$$

The direction of the dip, or line of greatest slope, is given by the negative reciprocal of the contour direction. Hence, with reference to Figure 5.12

$$\frac{\Delta x_1}{\Delta x_2} = \frac{(\partial z/\partial x_1)}{(\partial z/\partial x_2)}$$

$$\Delta x_1 = \pm \frac{\partial z}{\partial x_1}.h \bigg/ \left[\left(\frac{\partial z}{\partial x_1}\right)^2 + \left(\frac{\partial z}{\partial x_2}\right)^2\right]^{1/2}$$

$$\Delta x_2 = \pm \frac{\partial z}{\partial x_2}.h \bigg/ \left[\left(\frac{\partial z}{\partial x_1}\right)^2 + \left(\frac{\partial z}{\partial x_2}\right)^2\right]^{1/2}$$

(5.13)

in which the positive and negative signs apply respectively to maximization ('hill-climbing') and minimization of the objective function. As illustrated in Figure 5.11(a), the path taken by the search responds to changes in the direction of greatest slope if the step size h is not too large. Once the search has reached the

† The direction of a contour may be defined by the condition that $dz = 0$, i.e. from equation (5.10).

vicinity of the solution no further improvement can be found with the current step size and h must be reduced by some arbitrary factor (say 0.1 to 0.5) until the region of uncertainty is acceptably small.

The process works reasonably well on surfaces with gentle curvature but in steep-sided valleys the step size may be prematurely reduced in size and the iteration proceeds by a large number of small oscillatory moves.

If sharp-edged valleys are encountered during the search, the method may terminate prematurely without locating the true minimum, since from points on either side of the valley ridge the direction of steepest slope yields no prospect of improvement. This fault can be particularly serious when the surface of the objective function is modified by the addition of penalty terms to ensure feasibility in the presence of constraints (see Section 5.9).

5.5.3 Variable step size

An alternative strategy is to maintain a given direction of search until no further improvement in the function can be found. For a known initial (or current) position, the gradients $(\partial z/\partial x_1)$ and $(\partial z/\partial x_2)$ are known. The coordinates of the next position may now be defined as

$$(x_1)_1 = (x_1)_0 - (\partial z/\partial x_1)\rho \tag{5.14}$$

$$(x_2)_1 = (x_2)_0 - (\partial z/\partial x_2)\rho$$

where

$$\rho = h \bigg/ \left[\left(\frac{\partial z}{\partial x_1}\right)^2 + \left(\frac{\partial z}{\partial x_2}\right)^2 \right]^{1/2} \tag{5.15}$$

and minimization is assumed. The value of the function z_1 at the new location is now dependent on the single variable ρ and the optimum value of ρ is implicitly defined by the condition $(\partial z/\partial \rho) = 0$. For this optimization of ρ to be carried out analytically the objective function must be simple and easily differentiated so that a solution to the problem might be more easily obtained by classical methods of calculus. For more practical problems, ρ must be optimized by a one-dimensional search technique such as the one used in subroutine GOLDEN.

The FORTRAN code illustrated in Figure 5.13(a) shows how this might be done for the objective function of equation (5.9).

The results obtained by means of this program are also shown in Figure 5.11(b). One significant feature of the strategy is that successive search directions must be mutually at right angles. It is analogous to the progress of a skier who maintains his or her initial direction until all sustaining gradient is lost, at which point the skis are re-aligned down the direction of greatest slope. Considering the number of function evaluations required to optimize the step size, the strategy may prove rather inefficient and depending on the shape of the surface the technique may prove equally hazardous for both numerical optimization and skiing.

The results obtained are shown in Figure 5.13(b).

```
      DIMENSION X(2),DFDX(2)
      COMMON /GLDEN/ X,DFDX
      COMMON /DRIVS/ NX,EPS
      COMMON /NCALLS/ NCNT
      EXTERNAL FUN2
      NCNT=0
      NX=2
      EPS=0.001
C   INITIALIZE VARIABLES
      X(1)=-0.5
      X(2)= 1.0
      NPRINT=1
    5 CONTINUE
C   COMPUTE DERIVATIVES TO DEFINE DIRECTION OF
C   SEARCH FOR USE IF ROUTINE FUN2
      CALL DERIVS(X,DFDX)
C   USE ROUTINE GOLDEN TO OPTIMIZE RHO
      CALL GOLDEN(FUN2,0.0,0.1,0.005,0,RHO,RMIN)
C   UPDATE VALUES OF VARIABLES X()
C   AND EVALUATE OBJECTIVE FUNCTION
      DO 20 IX=1,2
         X(IX)=X(IX) - RHO*DFDX(IX)
   20 CONTINUE
      CALL COST(X,FNEW)
      IF(NPRINT.GT.0) WRITE(6,30)X,FNEW
   30 FORMAT(3X,5F11.4)
      SUMSQ=0.0
      DO 40 IX=1,2
         SUMSQ=SUMSQ + DFDX(IX)*DFDX(IX)
   40 CONTINUE
C   GET STEP SIZE TO CHECK CONVERGENCE
      H=RHO*SQRT(SUMSQ)
      IF(ABS(H).GT.EPS) GOTO 5
      WRITE(6,50)NCNT,X,FNEW
   50 FORMAT("    SOLN. IN",I5," COST CALLS",/,5F10.3)
      STOP
      END
C
C
      SUBROUTINE COST(X,F)
C ******************************************************
C   THIS IS A USER DEFINED ROUTINE TO COMPUTE THE
C   VALUE OF AN OBJECTIVE FUNCTION AS A FUNCTION OF
C   A VECTOR OF ARGUMENTS.
C      X    = ARRAY CONTAINING CURRENT VALUES OF
C             ARGUMENTS
C      F    = COMPUTED OBJECTIVE FUNCTION
C ******************************************************
      DIMENSION X(1)
      COMMON /NCALLS/ NCNT
      X1=X(1)
      X2=X(2)
      F1=X1*X1 - X2
      F=2.5*F1*F1 + (1.0-X1)*(1.0-X1)
```

Figure 5.13 (a) Minimization of (5.9) by variable-step gradient method

```
      NCNT=NCNT+1
      RETURN
      END
C
C
      SUBROUTINE FUN2(RHO,ANS)
C *****************************************************
C  THIS ROUTINE EVALUATES THE OBJECTIVE FUNCTION
C  COST(X,F) AT A TRIAL POINT A DISTANCE RHO FROM
C  THE CURRENT POSITION ALONG THE DIRECTION OF
C  GREATEST SLOPE.  THE ROUTINE IS USED BY GOLDEN
C  TO OPTIMIZE THE STEP SIZE RHO.
C    RHO   = TRIAL VALUE OF STEP SIZE
C    ANS   = COMPUTED VALUE OF COST AT TRIAL PT.
C *****************************************************
      DIMENSION X(2),DFDX(2),X2(2)
      COMMON /GLDEN/ X,DFDX
      DO 10 IX=1,2
        X2(IX)=X(IX) - RHO*DFDX(IX)
   10 CONTINUE
      CALL COST(X2,F)
      ANS=F
      RETURN
      END
C
C
      SUBROUTINE DERIVS(X,DFDX)
C *****************************************************
C  THE ROUTINE CALCULATES DERIVATIVES OF A USER
C  DEFINED OBJECTIVE FUNCTION COST(X,F) WITH RESPECT
C  TO EACH OF THE ARGUMENTS X().  THE FORWARD DIFF-
C  ERENCE EPS AND THE NUMBER OF VARIABLES NX MUST BE
C  PASSED THROUGH LABELLED COMMON BLOCK /DERIVS/
C    X     = ARRAY OF CURRENT VALUES OF ARGUMENTS
C    DFDX  = ARRAY CONTAINING COMPUTED VALUES OF
C            DERIVATIVES.
C  USES LABELLED COMMON BLOCK:-
C    COMMON /DERIVS/NX,EPS
C *****************************************************
      DIMENSIONX(1),DFDX(1)
      COMMON /DRIVS/ NX,EPS
      CALL COST(X,CSTO)
      DO 10 IX=1,NX
        X(IX)=X(IX) + EPS
        CALL COST(X,CST1)
        X(IX)=X(IX) - EPS
        DFDX(IX) = (CST1-CSTO)/EPS
   10 CONTINUE
      RETURN
      END
```

Figure 5.13(a) — continued

```
         -.6104      .4476     2.6075
          .6632      .2732      .1829
          .6194      .3563      .1467
          .6919      .3729      .1229
          .6768      .4380      .1055
          .7291      .4481      .0908
          .7196      .5017      .0793
          .7634      .5095      .0694
          .7532      .5492      .0617
          .7882      .5580      .0549
          .7797      .5905      .0493
          .8095      .5989      .0442
          .8012      .6268      .0401
          .8281      .6341      .0362
          .8189      .6597      .0331
          .8454      .6649      .0301
          .8391      .6779      .0276
          .9288      .7991      .0152
          .9113     ·.8112      .0088
          .9100      .8264      .0081
          .9168      .8267      .0074
          .9989     1.0097      .0004
         1.0031     1.0073      .0000
         1.0025     1.0065      .0000
         1.0024     1.0047      .0000
         1.0021     1.0047      .0000
    SOLN. IN   384 COST CALLS
       1.002      1.005       .000
    END OF PROGRAM
```

Figure 5.13(b)   Result of minimization of (5.9) by variable step method

5.5.4 Extension to n variables

Development of either of the foregoing gradient search methods for multivariable problems presents little difficulty. The constant step size technique possesses the advantage of simplicity and in some cases may even be more efficient, and is therefore considered here.

The objective function may be expressed as

$$z(\mathbf{x}) \quad \text{where} \quad \mathbf{x} = (x_1, x_2, \ldots, x_n)^T.$$

An initial position is selected at x_0 for which a vector of gradients may be calculated. Assuming minimization, the new position of the search is then defined as

$$\mathbf{x}_{r+1} = \mathbf{x}_r - h\mathbf{d}_r \tag{5.16}$$

where

$$d_i = \frac{\partial z}{\partial x_i} \bigg/ \left[\sum_{i=1}^{n} \left(\frac{\partial z}{\partial x_i}\right)^2\right]^{1/2} \tag{5.17}$$

and h is the specified stepsize.

This procedure is continued until no further improvement can be found with the current step size, at which point the step size h is reduced by a fractional factor,

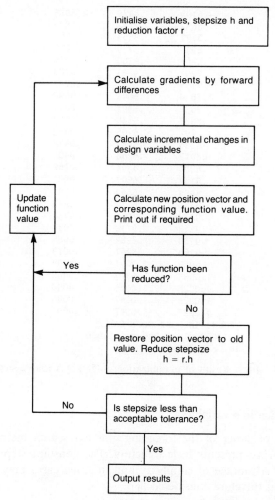

Figure 5.14 Flow diagram of gradient method (fixed step-size)

typically in the range 0.1 to 0.5. A further search is initiated and the whole process repeated until the step size is smaller than the specified acceptable interval of uncertainty. The algorithm is summarized in the flow diagram of Figure 5.14.

5.6 UNCONSTRAINED FUNCTION OF SEVERAL VARIABLES — DIRECT SEARCH METHODS

As in the case of single-variable functions, direct search techniques rely on the current position together with information from previous positions to plan strategies for future moves. No calculation of gradients is involved. A great many direct search algorithms have been developed and in this section it will be possible to give only a brief classification and description with detailed consideration given to one specific technique.

A broad classification of methods might be as follows:

(i) Tabulation —exhaustive trials of all possible solutions.
(ii) Random search —variables selected at random in an effort to shrink the zone of uncertainty.
(iii) Directed vector —strategy adopted to select directions for search.

These are considered briefly, following which a particular directed vector search algorithm is presented.

5.6.1 Tabulation

The simplest but computationally most expensive procedure is to tabulate all possible combinations of variable values. The range of uncertainty for each variable is defined as

$$(\mathbf{x}_{low}) \leqslant \mathbf{x} \leqslant (\mathbf{x}_{high})$$

where $\mathbf{x} = (x_1, x_2, \ldots, x_n)^T$.

If the kth variable has the range $(x_{high})_k - (x_{low})_k$ divided into r_k equal intervals then the objective function $z(\mathbf{x})$ must be evaluated at $(r_k + 1)$ points. Extending this to n variables the total number of function evaluations is given by

$$\prod_{i=1}^{n} (r_i + 1).$$

To gain some impression of the magnitude of such a task, a problem involving 10 variables, each of which is subdivided in 9 intervals (i.e. $r = 9$), requires 10^{10} function evaluations. If each calculation requires 0.01 s of machine time, tabulation will take over 3 years! For problems with only 2 or 3 variables, however, the method may be competitive with other search techniques especially if the function is of complex or multimodal shape (i.e. more than one turning value).

5.6.2 Random search

It is assumed once again that upper and lower limits may be specified for each of the variables. Random numbers uniformly distributed in the interval

$$0 < p_i \leqslant 1$$

are generated for each variable x_i. A trial point is then defined by the position vector

$$x_i = (x_i)_{low} - p_i[(x_i)_{high} - (x_i)_{low}] \tag{5.18}$$

The method is time-consuming but may be useful for escaping from a local optimum or in searching for a function with several false optima. After (say) 100 trials, the range may be shrunk around the minimum and the process repeated. Such a search is sometimes used following a gradient or directed vector search to provide some guidance as to whether or not the solution is indeed global.

5.6.3 Direction vector methods

These are techniques in which a set or sequence of directions is selected according to some strategy and a search initiated along these directions. Usually provision is made for varying the direction during the course of the search.

One method of *alternate variable search* involves holding constant $(n - 1)$ of the n variables, the objective function being minimized with respect to the single variable. Thus the search is parallel to each coordinate axis in turn, changing direction each time a minimum is located. The method results in a zig-zag, oscillating search and may be very slow if one of the principal axes of the objective function is not approximately parallel to one of the axes, as shown in Figure 5.15.

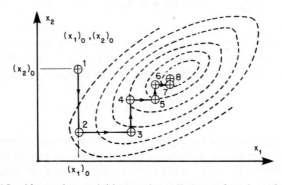

Figure 5.15   Alternating-variable search applied to a function of two variables

5.7 PATTERN SEARCH

One popular directed vector method is the *pattern search* algorithm developed by Hooke and Jeeves (1957).[1] The technique is designed to define a search direction by establishing a series of position vectors $\mathbf{b}_r$, called base points, which tend to follow a principal axis. Each move involves an extrapolation of any success achieved by moving from the previous base point to the current one, followed by an exploration of the surface around the extrapolated point to provide corrections for local conditions. The algorithm thus employs a double strategy and the rth iteration comprises

(i) a *pattern move* from base point $\mathbf{b}_{r-1}$ to a new trial centre $\mathbf{c}_r$,
(ii) a *local exploration* around the trial centre $\mathbf{c}_r$ as a result of which a new basepoint $\mathbf{b}_r$ is established.

These position vectors may be defined as

$$\mathbf{b}_r = [(b_1)_r, (b_2)_r, \ldots, (b_n)_r]^T$$
$$\mathbf{c}_r = [(c_1)_r, (c_2)_r, \ldots, (c_n)_r]^T$$

in which

$(b_i)_r$ = the value of the ith variable at the rth basepoint

$(c_i)_r$ = the value of the ith variable at the rth trial centre.

The sequence of operations is described in detail in the flow diagram of Figure 5.16 and later illustrated by reference to a two-dimensional problem.

With reference to Figure 5.17, the pattern search algorithm may be easily visualized with respect to a two-dimensional problem. The initial position is located at point 1 which becomes the first basepoint $\mathbf{b}_0$. Since no previous basepoint and therefore no pattern exists, the pattern move or extrapolation results in zero change and $\mathbf{c}_1$ is also at point 1. The local exploration about this trial centre involves function evaluations at points 2, 3, and 4 resulting in a new basepoint $\mathbf{b}_1$ at point 4. The second iteration commences with a finite pattern move to trial centre $\mathbf{c}_2$ (point 5) and after an exploration at points 6, 7, and 8 the basepoint $\mathbf{b}_2$ is established (point 8).

The third iteration proceeds in the same way leading to basepoint $\mathbf{b}_3$ (point 11), but it is found that $z(\mathbf{b}_3) \not< z(\mathbf{b}_2)$ so that $\mathbf{b}_3$ is abandoned and relocated at point 8, the pattern is lost (since now $\mathbf{b}_2 = \mathbf{b}_3$), and a new direction has to be sought.

Iteration four has a zero pattern move and terminates in basepoint $\mathbf{b}_4$ (point 13).

Similarly the fifth and sixth iterations lead to basepoints $\mathbf{b}_5$ (point 15) and $\mathbf{b}_6$ (point 22).

Iteration seven commences with a pattern move to a trial centre $\mathbf{c}_7$ (point 23), but local exploration around $\mathbf{c}_7$ results in basepoint $\mathbf{b}_7$ being located at point 25 which is coincident (in this case) with point 22 and basepoint $\mathbf{b}_6$. At this stage, the search can proceed no further with the present size of increment. The procedure from this point would involve a reduction of step size (assuming that the convergence criterion $\delta x \leq \delta x_{min}$ had not yet been met) and a recommencement of the search from basepoint $\mathbf{b}_7$ (point 25). These subsequent moves are not shown on Figure 5.17 but may be visualized as a repetition of the procedure from the initial basepoint $\mathbf{b}_0$.

5.7.1 Implementation of Pattern Search algorithm

The Hooke and Jeeves pattern search strategy is implemented in a FORTRAN subroutine HJMIN, the listing and documentation for which is contained in Appendix A. As with other routines mentioned in this chapter, provision is made for the name of the objective function routine to appear in the argument list. In application, it is necessary for the user to provide a calling or main program and a subroutine function. The main program carries out the following functions:

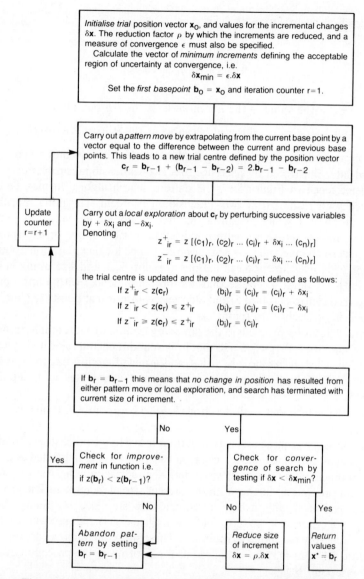

Figure 5.16 Flow diagram of the Hooke and Jeeves algorithm

(i) Set up storage for variables and increment sizes.
(ii) Cause variables and increment sizes to be initialized and assign values to the increment reduction factor, the convergence criterion, and the maximum number of function evaluations to be allowed.
(iii) Call HJMIN.
(iv) Output results as required.

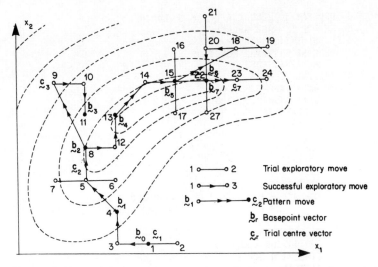

Figure 5.17 Hooke and Jeeves pattern search applied to a function of two variables

The objective function routine may have any name, as long as this appears as an EXTERNAL statement in the calling program, but must have only a single input parameter consisting of a vector of design variables. Any additional data required by the objective function routine must be either built-in to the body of the routine or transferred from the calling program through common block (preferably labelled). The following examples illustrate the application of the method to simple problems. In practice, of course, the objective function may involve quite complex design calculations involving many ad hoc subroutines.

5.7.2 Example

Find the minimum of the following two-dimensional function by means of the HJMIN routine, making provision for the coefficients to be varied easily in successive runs:

$$z(\mathbf{x}) = 2.5(x_1^2 - x_2)^2 + (1 - x_1)^2.$$

The FORTRAN code of Figure 5.18 should be self-explanatory. The results of the analysis are shown plotted in Figure 5.11(b).

5.7.3 Optimum design of a reinforced concrete tank

Example

A reinforced concrete water tank is to be designed to hold 1500 m^3. The tank is to be rectangular, open-topped and sunk flush with existing ground level (see Figure 5.19). Freeboard of 0.3 m should be allowed.

```
C   PROGRAM TO ILLUSTRATE USE OF HJMIN
C   SET UP STORAGE ARRAYS
      DIMENSION VAR(2),DVAR(2)
C   DECLARE EXTERNAL FUNCTION
      EXTERNAL COST
C   INITIALISE DATA
      DATA VAR/-0.5,1.0/
      DATA DVAR/0.01,0.01/
C   SET PARAMETERS FOR HJMIN2
      RHO=0.5
      EPS=0.01
      NPRINT=1
      NMAX=200
      CALL HJMIN(VAR,DVAR,2,ANS,RHO,EPS,COST,NPRINT,NMAX)
      WRITE(6,30)ANS,NMAX,VAR
30    FORMAT(9H MINIMUM=,F12.3,9H FOUND IN,I5,
     +       12H EVALUATIONS,/,15H OPT. POLICY IS,2F14.4)
      STOP
      END
      SUBROUTINE COST(X,FX)
      DIMENSION X(2)
      X1=X(1)
      X2=X(2)
      F1 = X1*X1-X2
      FX = 2.5*F1*F1 + (1.0-X1)*(1.0-X1)
      RETURN
      END
```

Figure 5.18   FORTRAN code for minimization of (5.9) by routine HJMIN

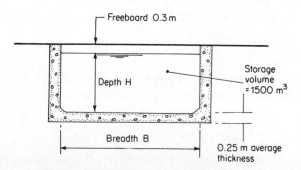

Figure 5.19   Reinforced concrete water tank

The floor is priced for an average thickness of 0.25 m and the wall thickness is estimated from the resistance moment equation

$$BM = 0.4 \times 10^6 . bd^2 \text{ m-N}.$$

Cost of construction is based on the following rates:

Item	Unit	Rate
Reinforced concrete	m³	18.00
Forming unexposed surfaces	m²	2.70
Forming vertical internal surfaces	m²	4.50
Excavation and disposal	m³	1.25

Let the inside dimensions of the tank be B, L, and $(H + 0.3)$ m (i.e. water depth is H m). Since the wall thickness can be designed for a specified water depth H, all costs can be calculated from these three variables. The problem may then be stated as

$$\underset{B^*,H^*,L^*}{\text{Minimize}} \; z = z(B, L, H)$$

subject to

$$BLH = 1500.$$

Due to the simple nature of the constraint, any one variable may be expressed easily in terms of the other two, so that the problem may be reduced to an unconstrained objective function in only two variables.

The steps in the cost calculation may be summarized as follows:

(i) Select or obtain values for B and H.
(ii) Calculate $L = 1500/(BH)$.
(iii) Calculate bending moment for wall for the case of water of depth H with no outside backfill. (This will usually be the severest test of an open-topped water-retaining structure since earth pressure moments are usually smaller and need not be designed for no cracking.) i.e. $BM = \gamma H^3/6$ where γ = unit weight of water.
(iv) Design wall thickness from

$$d = (BM/400\,000b)^{1/2}.$$

(v) Obtain cost of concrete, formwork, and excavation.

Coding for the solution of this problem may be organized as shown in Figure 5.20.

The calling statement is designed to suppress printout of intermediate basepoints prior to the final solution and the results obtained are illustrated in Figure 5.20. As might be expected, the plan shape of the optimized tank is nearly square since for any depth this minimizes the perimeter. Clearly if this fact were used in the design, the problem would reduce to an optimization of a function of a single variable, the water depth, with the other two dimensions given by

$$B = L = (1500/H)^{1/2}.$$

5.8 CONSTRAINED FUNCTION OF A SINGLE VARIABLE

In most practical problems of optimization the model will include constraints. These may be either *equality* or *inequality* constraints.

```
      C   COST MINIMIZATION OF WATER TANK
      C   SET UP STORAGE ARRAYS, COMMON BLOCK FOR USE
      C   BY COST ROUTINE AND DECLARE EXTERNAL ROUTINE
          DIMENSION VAR(2),DVAR(2)
          COMMON /DESIGN/ CRC,CFORMO,CFORMI,CEXC,XL,D
          EXTERNAL COST
      C   INITIALISE DATA
          DATA DVAR/10.0,2.0/
          DATA DUAR/0.25,0.1/
          DATA RHO,EPS,NMAX/0.25,0.01,200/
      C   SPECIFY COSTS
          CRC=18.0
          CFORMO=2.20
          CFORMI=4.50
          CEXC=1.25
          CALL HJMIN(VAR,2,DVAR,ANS,RHO,EPS,COST,0,NMAX)
          WRITE(6,10)NMAX,ANS
       10 FORMAT("     AFTER",I4," COST EVALNS. MIN.COST=",F12.2)
          WRITE(6,20)VAR(1),XL,VAR(2),D
       20 FORMAT(10X,"BREADTH =",F10.3,/,
         +       10X,"LENGTH  =",F10.3,/,
         +       10X,"DEPTH   =",F10.3,/,
         +       10X,"WALL TH.=",F10.3)
          STOP
          END
      C   OBJECTIVE FUNCTION SUBROUTINE
          SUBROUTINE COST(VAR,CST)
      C******************************************************
      C   ROUTINE EVALUATES TOTAL COST OF WATER TANK BASED
      C   ON TRIAL VALUES OF BREADTH AND DEPTH.
      C   VOLUME OF STORAGE =1500.0 CUB. METRES
      C   VAR(1)     = INSIDE BREADTH
      C   VAR(2)     = INSIDE DEPTH INCL. FREEBRD.
      C   CST        = COMPUTED COST
      C   REQUIRES LABELLED COMMON BLOCK /DESIGN/ I.E.
      C      COMMON /DESIGN/CRC,CFORMO,CFORMI,CEXC,XL,D
      C******************************************************
          DIMENSION VAR(2)
          COMMON /DESIGN/ CRC,CFORMO,CFORMI,CEXC,XL,D
          B=VAR(1)
          H=VAR(2)
          XL=1500.0/(B*H)
      C   NOTE: XL TRANSFERRED BACK TO CALLING PROGRAM
      C   VIA COMMON BLOCK
          BM=1000.0*9.81*H*H*H/6.0
          D=SQRT(BM/400000.0)
          HTOT = H+0.25+0.30
          CONC=(0.25*B*XL) + 2.0*(B+XL+D)*HTOT*D
          FORMO=2.0*(B+XL+2.0*D)*HTOT
          FORMI=2.0*(B+XL)*(H+0.30)
          EXC=(B+2.0*D)*(XL+2.0*D)*HTOT
          CST=CRC*CONC + CFORMO*FORMO + CFORMI*FORMI + CEXC*EXC
          RETURN
          END

              AFTER 103 COST EVALNS. MIN.COST = 8343.45
                     BREADTH =    24.875
                     LENGTH  =    24.903
                     DEPTH   =     2.422
                     WALL TH.=      .241
          END OF PROGRAM
```

Figure 5.20   FORTRAN code for the water tank problem (Section 5.7.3)

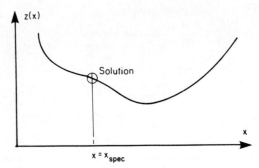

Figure 5.21 Equality constraint on function of a single variable

If the objective function $z(x)$ depends on only a single variable x then clearly the imposition of an *equality* constraint dictates a unique solution, with no opportunity to modify the objective function. Figure 5.21 illustrates the situation. It should be noted that when the constraint is of the general form

$$h(x) = 0$$

this may always be transformed into one or more simple equality constraints of the form

$$x = (x_{spec})_i \quad i = 1, 2, \ldots$$

where $(x_{spec})_i$ represent the real roots of the function $h(x)$. If more than one real root exists then each must be tested to obtain the constrained value of x for which $z(x)$ is a minimum. For example,

$$\underset{x^*}{\text{Minimize}} \; z = x^2 + 2x + 4$$

subject to

$$x^2 = 4.$$

Thus

$$x = +2, -2$$

$$z(+2) = 12; \quad z(-2) = 4; \quad x^* = -2 \quad \text{and} \quad z_{min} = 4.$$

The existence of *inequality* constraints on the other hand need not reduce the feasible region to one or more points. As with equality constraints, inequality constraints of any arbitrary functional form applied to an objective function of a single variable may always be reduced to one or more simple constraints of the form

$$x \geqslant x_{min}$$

or

$$x \leqslant x_{max}.$$

As illustrated in Figure 5.22, the constraints may or may not be active, i.e. the optimum solution may or may not be affected by the presence of the constraints.

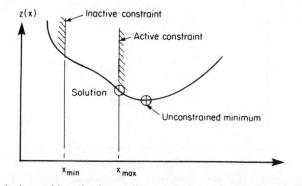

Figure 5.22   Active and inactive inequality constraints on a function of a single variable

Two strategies may be employed in the solution of such problems:

(i) At each iteration of the design variable x, check if any of the inequality constraints are violated. If so, set the variable equal to the stipulation. Using this procedure, if two successive values of the variable differ by less than an acceptable amount, the search has converged.

The algorithm is illustrated in the flowchart of Figure 5.23.

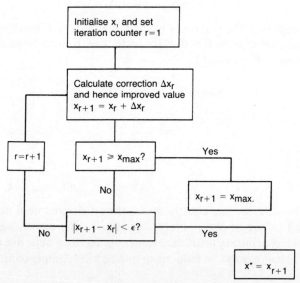

Figure 5.23   Flow diagram illustrating incorporation of a simple inequality constraint (function of a single variable)

(ii) A more general approach involves a modification of the objective function by the addition of penalty terms the values of which are proportional to the degree by which the constraints are violated. For minimization the penalty terms are positive, thus forcing the solution away from infeasible regions. Thus the following model would be transformed from a constrained to an unconstrained problem by creating an artificial objective function:

$$\underset{x^*}{\text{Minimize}}\ z(x)$$

subject to

$$x \leqslant x_{\max}$$
$$x \geqslant x_{\min}.$$

Alternatively,

$$\underset{x^*}{\text{Minimize}}\ z_{\text{mod}} = z(x) + 10^6(x - x_{\max})\delta_1 + 10^6(x_{\min} - x)\delta_2 \quad (5.19)$$

where

$$\delta_1 = 1 \quad \text{for } (x - x_{\max}) > 0$$
$$= 0 \quad \text{for } (x - x_{\max}) \leqslant 0$$
$$\delta_2 = 1 \quad \text{for } (x_{\min} - x) > 0$$
$$= 0 \quad \text{for } (x_{\min} - x) \leqslant 0.$$

The terms δ_1 and δ_2 are included simply to express mathematically the fact that the terms $(x - x_{\max})$ and $(x_{\min} - x)$ are included only when positive. In practice it is likely that such a strategy would be used only in a computer-coded solution in which the necessary logic can be easily incorporated in the objective function subroutine, e.g.

```
SUBROUTINE FUN(X,Z)
Z=..................        (suitable coding)
PEN1=1000.0*(X-XMAX)
PEN2=1000.0*(XMIN-X)
IF(PEN1.LT.0.0)PEN1=0.0
IF(PEN2.LT.0.0)PEN2=0.0
Z=Z+PEN1+PEN2
RETURN
END
```

The penalty term strategy has the effect of erecting very steep ramps on top of the unconstrained objective function as illustrated in Figure 5.24.

5.9 CONSTRAINED FUNCTION OF SEVERAL VARIABLES

When a function of several variables is to be optimized subject to constraints, the model takes the following general form:

$$\underset{x^*}{\text{Minimize}}\ z = z(\mathbf{x}) \quad \mathbf{x} = (x_1, x_2, \ldots, x_n)^{\text{T}}$$

subject to

$$g_i(\mathbf{x}) = 0 \quad i = 1, 2, \ldots, r$$
$$g_j(\mathbf{x}) \leqslant 0 \quad j = r+1, r+2, \ldots, r+s.$$

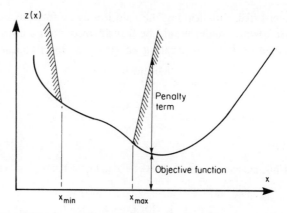

Figure 5.24   Inequality constraints simulated by penalty terms

The type of constraint may be either equality or inequality. The functional form $g(\cdot)$ of the constraint may be linear or non-linear, explicit or implicit, and may involve any number of the n design variables. It should be noted that inequality constraints may be written either as 'greater than' or 'less than' constraints simply by reversing the sign of the function $g(\cdot)$ and it is good practice to adopt a particular format as standard.

5.9.1   Equality constraints

In general, when *equality constraints* are encountered the objective function and the functional form of the constraint should be carefully examined to check if the constraint may be incorporated into the objective function by substitution for one of the design variables. This has the advantage of reducing the number of design variables by one for each equality constraint so treated. This device is simple if the form of the constraint is such that one of the variables may be expressed explicitly in terms of other variables and/or known numerical constants. The example of Section 5.7.3 is typical. For the constraint

$$BLH = 1500$$

any one of the three design variables can be expressed in terms of the other two.

In cases where the functional form of the constraint $g(\mathbf{x})$ is more complex and is not amenable to manipulation into an explicit form, it may be computationally advantageous to solve for one of the design variables in terms of the others by some iterative procedure. The case study of Section 12.5 provides a typical illustration of such an approach.

Frequently, however, the equality constraint is too complex to allow substitution in this way and it is necessary to adopt the device of the penalty term. As for functions of a single variable, this involves creating an artificial objective function by the addition of terms which have an arbitrarily high value when the constraint is violated and zero value when the constraint is satisfied. The sign of this penalty term will be positive for problems of minimization and vice versa. To reduce the effect of the discontinuity on the objective function 'surface' it is usual to make the magnitude of the penalty term proportional in some way to the extent to which the constraint is violated. Indeed, there may be cases where the form of penalty term is designed to provide a gentle transition between the original surface of the objective function within the feasible region and the steep gradients arising from the penalty term. The following examples illustrate the procedure.

Example

The functions of this example are extremely simple so that visualization of the problem is made easier. As a result, direct substitution would be possible but is not employed.

$$\underset{x_1^*, x_2^*}{\text{Minimize}}\ z = (x_1 - 10)^2 + (x_2 - 5)^2$$

subject to

$$x_1 + 2x_2 = 14.$$

Figure 5.25 shows the contours of the objective function which form a series of concentric circles around the point $\mathbf{x} = (10, 5)^T$ with a value of z corresponding to the square of the radius.

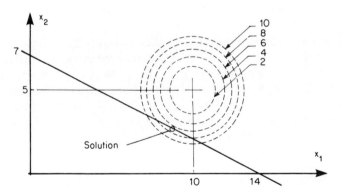

Figure 5.25 Spherical objective function subject to a single equality constraint

The constraint appears on Figure 5.25 as a straight line passing through the points (0, 7) and (14, 0) and since the solution must satisfy this equality constraint, the feasible space becomes the set of points forming that line. Clearly the solution must be at that point on the line which is tangent to one of the circles, and could be found for this simple problem by coordinate geometry. The solution may be visualized as the lowest point on the section produced by a cutting plane along the line $x_1 + 2x_2 = 14$.

To apply the device of penalty terms the objective function is modified to the following form:

$$\underset{x_1^*, x_2^*}{\text{Minimize}}\ z_{\text{art}} = (x_1 - 10)^2 + (x_2 - 5)^2 + 10^2|(x_1 + 2x_2 - 14)|$$

where $|\cdot|$ denotes the absolute value. Thus the penalty term will have a non-zero value for any point not satisfying the constraint. The constant multiplier of 10^2 is somewhat arbitrary but certain problems may be sensitive to the scaling of the penalty term in relation to the objective function. Figure 5.26 shows the resulting surface of the now unconstrained function z_{art}. It will be appreciated that the

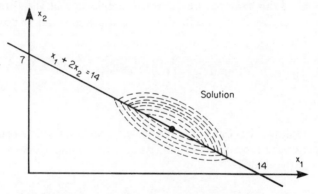

Figure 5.26   Contours of artificial objective function

adoption of too large a multiplying constant on the penalty term may mask to a great extent the influence of the real objective function making a search procedure inaccurate or even impossible. As a general guide, the penalty term should increase the objective function by only one or two orders of magnitude in the vicinity of the solution. Even with this precaution it is good practice to evaluate the final optimum value of the objective function by means of the real objective, unmodified by penalty terms.

5.9.2   Inequality constraints

The treatment of inequality constraints is similar to the previous example with the difference that the feasible region is usually not reduced to a line segment and therefore a portion of the real objective function 'surface' is unaffected by the penalty terms. The situation is best illustrated by a modified form of the previous problem.

Example

$$\underset{x_1^*, x_2^*}{\text{Minimize}}\ z = (x_1 - 10)^2 + (x_2 - 5)^2$$

subject to
$$x_1 + 2x_2 \geq 14.$$

The constraint may be re-written as
$$14 - x_1 - 2x_2 \leq 0$$

and the modified objective function becomes
$$z_{art} = (x_1 - 10)^2 + (x_2 - 5)^2 + \delta_1 . 10^2 (14 - x_1 - 2x_2)$$
where
$$\delta_1 = 1 \quad \text{for } (14 - x_1 - 2x_2) > 0$$
$$= 0 \quad \text{for } (14 - x_1 - 2x_2) \leq 0.$$

As mentioned in Section 5.8, the inclusion of the term δ_1 is a mathematical formality and in practice the same result is achieved by a simple conditional test in the coding of the objective function subroutine. Figure 5.27(a) shows the contours of the artificial objective function and it is obvious that the constraint is inactive and the solution $x^* = (10, 5)$ is identical to that for the unconstrained function.

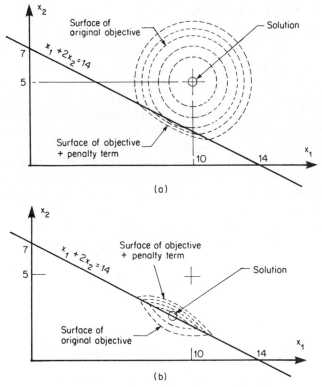

Figure 5.27 (a) Inactive inequality constraint. (b) Active inequality constraint

As a second example, consider now the effect of reversing the sign of the constraint, so that the problem becomes

$$\text{Minimize}_{x_1^*, x_2^*} z = (x_1 - 10)^2 + (x_2 - 5)^2$$

subject to

$$x_1 + 2x_2 \leqslant 14$$

or

$$x_1 + 2x_2 - 14 \leqslant 0.$$

Thus, the equivalent unconstrained function becomes

$$z_{\text{art}} = (x_1 - 10)^2 + (x_2 - 5)^2 + \delta_1 . 10^2 (x_1 + 2x_2 - 14)$$

where

$$\delta_1 = 1 \quad \text{for } (x_1 + 2x_2 - 14) > 0$$
$$= 0 \quad \text{for } (x_1 + 2x_2 - 14) \leqslant 0.$$

The corresponding surface is shown in Figure 5.27(b), from which it is seen that the solution is the same as in the problem of Section 5.9.1 with the equality constraint.

5.9.3 A transitional penalty term

From the illustrations of the previous two sections it will be readily appreciated that the intersection of the surface of a real objective function with the steep ramp resulting from a penalty term will frequently produce a very sharp-edged, steep-sided valley. Although special algorithms have been developed to negotiate such surfaces—notably by Bandler[7]—the search strategy used in algorithms similar to Hooke and Jeeves may well be frustrated by such discontinuities. The hazards can be reduced to some extent by careful weighting of the penalty terms to ensure that the influence of the real objective function is dominated but not completely lost. A further refinement is possible by designing a penalty term which is not linear in terms of the extent of violation, but provides a gentle, curved transition at the edge of the feasible region, something like a cove at the junction between floor and wall.

Many alternatives are possible and the following approach is suggested for purposes of illustration. Consider the problem represented by the following model:

$$\text{Minimize}_{x^*} z = z(\mathbf{x}) \tag{5.20}$$

subject to

$$g_i(\mathbf{x}) \leqslant 0 \quad i = 1, 2, \ldots, s.$$

This may be converted to the unconstrained objective function:

$$\text{Minimize}_{x^*} z_{\text{art}} = z(\mathbf{x}) + \sum_{i=1}^{s} \delta c_i (e^{g_i(\mathbf{x})} - 1) \tag{5.21}$$

where
$$\delta = 1 \quad \text{for } g_i(\mathbf{x}) > 0$$
$$\text{or} \quad \delta = 0 \quad \text{otherwise}$$
and
$$c_i = \text{a weighting factor for the } i\text{th constraint.}$$

By this device, the penalty term has zero value at $g_i(x) = 0$, and disappears also for $g_i(x) < 0$, but the magnitude of the penalty increases in a smooth exponential transition as the extent of constraint violation increases.

For equality constraints the form of the penalty term becomes

$$\sum_{i=1}^{r} c_i(e^{|g_i(\mathbf{x})|} - 1) \quad (5.22)$$

where $|\cdot|$ implies the absolute value.

With this arrangement there is the possibility that the solution obtained by a search procedure may violate a constraint by a very small amount. However, the chances of the search being prematurely terminated in a sharp valley are considerably reduced and the relative benefits and disadvantages must be weighed by the engineer in each case. As mentioned previously (Section 5.9.1) it is important to obtain the final evaluation of the objective function with the unmodified expression — i.e. without penalty terms — to avoid small residual penalties artificially increasing the true optimum value.

A more rigorous approach employing non-linear penalty terms has been suggested by Fiacco and McCormick[5] in which penalty terms are added both inside and outside the feasible space. Thus with reference to (5.20) an artificial unconstrained objective function is formed as follows:

$$\underset{\mathbf{x}^*}{\text{Minimize}} \; z_{\text{art}} = z(\mathbf{x}) - k \sum_{i=1}^{s} \frac{1}{g_i(\mathbf{x})} \quad (5.23)$$

in which the value of k is positive and approaches zero. Inside the feasible region $g(x) < 0$ and the penalty is large and positive. As $k \to 0$ the minimum is swept closer and closer to the constraint. Outside the feasible region $g(x) > 0$ and the penalty is large and negative, thus forcing the solution towards the constraint. Good explanations of the methods are given by Aoki[4] and by Daellenbach and George.[2]

5.10 EXERCISES

5.10.1 A two-bar cantilever is to be designed to support the vertical load W ($=500$ kN) at a distance L ($=3$ m) from the wall as shown in Figure 5.28. The members AB and BC are to be tubular in section, of diameter D and wall thickness T. The working stresses in the two members are given by

$$f(\text{tension}) \quad = 125 \text{ N/mm}^2$$
$$f(\text{compression}) = 125(1 - 0.01L/D) \text{ N/mm}^2.$$

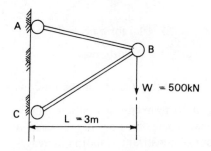

Figure 5.28 Two-bar cantilever

The vertical deflection of point B should be not greater than 15 mm assuming a modulus of elasticity of 200 kN/mm². Determine a design which will minimize the total weight of material assuming D and T to be continuous.

5.10.2 A roof member is to be formed as a folded plate with the cross-section shown in Figure 5.29. The member is to support a superimposed load of

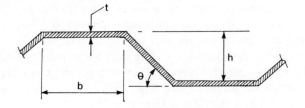

Figure 5.29 Folded plate roof member

3500 N/m² over a simply supported span of 4 m. The maximum allowable working stress is 100 N/mm² and the mid-span deflection due to superimposed load only is to be not greater than 1/180 of the span. The properties of one complete 'wave' of the member may be approximated by the following expressions:

$$A = 2bt + 2(h - t)(t/\sin \theta)$$
$$I = 2bt(h/2)^2 + (t/\sin \theta)(h - t)^3/6.$$

Assuming an elastic modulus of 200 kN/mm² design the dimensions for minimum weight.

5.10.3 Figure 5.30 shows the cross-section of a mass concrete wall to retain a height of $H = 3$ m of soil. For simplicity the pressure distribution on both front and rear faces may be assumed to be linear and given by:

$$p_{\text{active}} = 0.3\gamma h$$
$$p_{\text{passive}} = 1.8\gamma h$$

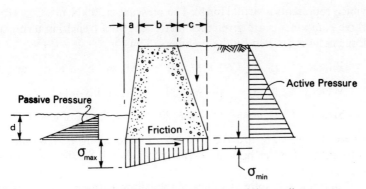

Figure 5.30   Mass concrete retaining wall

where γ is the specific weight of the soil (14 kN/m³). Friction on the rear face may be ignored but on the base a coefficient of 0.35 should be used. The wall is to be designed for the following criteria:

$$\sigma_{max} \leqslant 100 \text{ kPa}$$

$$\sigma_{min} \geqslant 0$$

Factor of safety against sliding = 2.

The cost of construction may be estimated on the basis of concrete and formwork using the following rates:

Concrete	$120/m³
Front face forms	$ 30/m²
Rear face forms	$ 15/m²

Proportion the dimensions a, b, c, and d for minimum cost (weight of concrete = 23 kN/m³ approximately).

5.10.4   Figure 5.31 shows the arrangement of sheet piling, waling beam, and struts to brace an excavation. The pressure distribution (greatly simplified for this analysis) is assumed to be uniform from the surface down to a point A

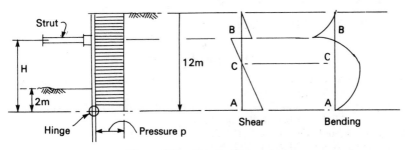

Figure 5.31   Sheet piling system

which represents a virtual hinge. The pressure is 30 kN/m². Costs for the three components are given (per metre length of trench) in terms of the force actions resisted by the members. Thus:

Piling: C_p ($/m) = 60 + M_p/40   (M_p in kN-m/m of trench)

Wale: C_w ($/m) = 4 + M_w/20   (M_w in kN-m)

Struts: C_s ($/m) = 124/S + F/20   (F in kN/m of trench)

where M_p and M_w are the moments in the piling and waling beam respectively, F is the thrust per metre of trench, and S is the strut spacing in metres.

Determine the optimal values of the height H (m) and strut spacing S (m) and hence the minimum cost per metre of trench. (Adapted from Stark and Nicholls.[3])

5.10.5 A catchment has been found to have an instantaneous unit response function as defined below, the values being at equal time intervals of 1 hour.

$$u(\ ) = \begin{matrix} 0.0 & 0.0 & 0.0464 & 0.0976 & 0.1585 \\ 0.1464 & 0.1196 & 0.0988 & 0.0773 & 0.0572 \\ 0.0428 & 0.0379 & 0.0304 & 0.0237 & 0.0173 \\ 0.0132 & 0.0106 & 0.0079 & 0.0059 & 0.0036 \\ 0.0025 & 0.0018 & 0.0009. & & \end{matrix}$$

It is intended to model this as the response of a cascade of 4 linear reservoirs and a linear channel in series. If the lag times of the reservoirs are respectively K_1, K_2, K_3, and K_4 and the channel lag is K_0 the response can be expressed as follows:

$$u(t_1) = 0 \quad \text{for } t_1 = t - K_0 < 0$$

$$u(t_1) = \sum_{j=1}^{n} \left[K_j^{n-1} e^{-t_1/K_j} \Big/ \prod_{\substack{i=1 \\ i \neq j}}^{n} (K_j - K_i) \right] \quad \text{for } t_1 = t - K_0 > 0$$

in which n = 4 for this case.

Determine the optimal lag values which best represent the observed response (Smith and Kimmett[8]).

5.10.6 Solve the problem of Section 1.4 by direct search.

5.11 REFERENCES

1. Siddall, J. N. (1972). *Analytical Decision-Making in Engineering Design*, Prentice-Hall, Englewood Cliffs, New Jersey.
2. Daellenbach, H. G., and George, J. A. (1978). *Introduction to Operations Research Techniques*, Allyn and Bacon, Boston, Mass.
3. Stark, R. M., and Nicholls, R. L. (1972). *Mathematical Foundations for Design: Civil Engineering Systems*, McGraw-Hill, New York.

4. Aoki, Masanao (1971). *Introduction to Optimization Techniques: Fundamentals and Applications of Nonlinear Programming*, McMillan, New York.
5. Fiacco, A. V., and McCormick, G. P. (1968). *Nonlinear Programming: Sequential Unconstrained Minimization Techniques*, Wiley, New York.
6. Wolfe, P. (1967). 'Methods of nonlinear programming.' In J. Abadie (Ed.), *Nonlinear Programming*, North-Holland, Amsterdam.
7. Bandler, J. W. *et al.* (1972). 'Minimax optimization of networks by grazor search.' *IEEE Trans. on Microwave Theory and Tech.*, **MTT-20**, No. 9, 596–604.
8. Smith, A. A., and Kimmett, D. (1978). 'Watershed modelling by system identification.' *CSCE Speciality Conference, Toronto, May 18–19, 1978*.

Chapter 6
Dynamic Programming

6.1 INTRODUCTION

Various mathematical programming models for civil engineering systems have been described in Chapters 3, 4, and 5. Such methods are mainly suited for single-stage decision problems. Many civil engineering systems, however, involve multistage decision problems in which the outcome of a decision at one stage affects the subsequent decision for the next stage and so on. Consider the following examples of multistage problems which are encountered in certain civil engineering systems:

(i) In the design of a trunk sewer, the selection of a steep gradient in one reach will result in cost reduction for that particular reach but may incur a cost penalty in downstream reaches as a result of the lowered invert level of the pipe.
(ii) The selection of floor, column, and foundation elements for a simple structure must involve not only the cost minimization of each component but also the effect of increased self-weight on 'downstream' elements. For example, reinforced concrete columns may be the cheapest type of column support, but savings may be offset by the added cost of foundations to support the increased transmitted load.
(iii) If several activities (for example, irrigation schemes) compete for a limited resource (for example, money or water) then each activity may be considered in turn using an arbitrarily chosen sequence. The decision regarding the amount of resource to be used in a given activity will affect the remaining amount available for activities considered after it. A problem of this type is described later in detail together with a useful computer program for its solution.
(iv) In selecting a path through a network (in order to minimize or maximize time, cost, distance, etc.) the choice of direction at any particular node in the network will affect only the path through the remaining nodes. A pipeline network problem is described in detail in Section 6.2. It should be noted that alternative methods for dealing with network problems can be found in Chapters 4, 7 and 8.

Techniques for solving multistage decision problems usually involve the breaking down of the system into stages, the formulation of recursive equations based on the serial relationships between the stages, and the sequential solution of these equations.

Dynamic Programming (DP), which is a general approach to the solution of multistage decision problems, will be described in this chapter with the aid of a number of examples and a small computer program. DP was developed originally by Bellman[1] and fellow workers at the Rand Corporation in 1957 and it has since found many applications in a wide variety of fields including those of direct interest to civil engineers.

6.2 A PIPELINE NETWORK PROBLEM

To demonstrate the DP approach, the simplified pipeline network problem shown in Figure 6.1 is considered. Town A requires a new source of water from river E. The terrain between the town and the river is such that a variety of alternative routes are available, each involving different levels of expenditure. It has been decided that there are four possible locations for the intake of water from river E (at points E1–E4) and 12 possible intermediate locations for the three pumping stations (B1–4, C1–4, and D1–4) through which the pipeline must pass as shown in Figure 6.1. The numbers attached to each pipeline link represent the construction cost and the number in each box is the cost of the pumping station. The problem is thus one of finding the pipeline route linking river E with town A resulting in minimum total cost.

In this small example it is simple (if tedious) to find the cheapest route by evaluating all possible routes. This procedure, known as enumeration, becomes almost impossible to apply for larger problems. A more efficient solution for such problems can be obtained using DP.

6.3 SOLUTION OF THE PIPELINE NETWORK PROBLEM

The pipeline network can be considered as a multistage process in which each of the four stages involves the choice of a direction (e.g. northeast, east, or southeast) for the pipeline route, from any one of a number of possible starting points (i.e. a pumping station or, in the case of the first section, the origin at town A).

Consider now a sub-problem comprising, for example, the continuation of the pipeline from station B2; the problem can again be regarded as a multistage process which has the same kind of structure as the original problem, i.e. each stage involves the choice of a pumping station site and the associated pipeline section. In this case, only three stages remain to be completed. An identical observation may be made not only for all other sites of the first pumping station, but also for the second, third, and fourth, except that fewer stages are left to be completed.

Assume that by some means, as yet unspecified, the minimum cost is known for completing the pipeline to the river E from each of the four alternative sites B1,

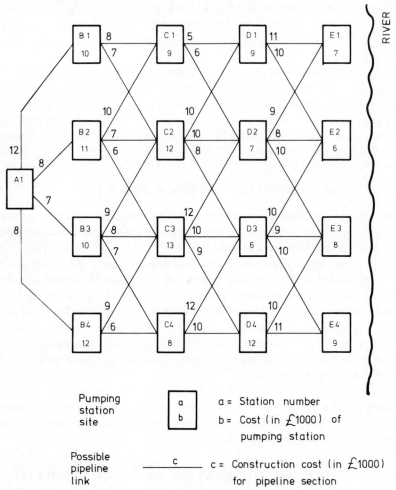

Figure 6.1 A pipeline network problem

B2, B3, and B4. The costs are (say) 54 from site B1, 57 from site B2, 59 from site B3, and 60 from site B4. See the table below for details:

Origin	Route	Destination	Cost of pipeline	Minimum completion cost from B	Total cost
A1	N	B1	12	54	66
A1	NE	B2	8	57	65
A1	SE	B3	7	59	66
A1	S	B4	8	60	68

It is now a simple matter to find the minimum cost from town A.

The lowest cost is obtained by the section from A1 to B2 then by taking the optimal route from there. It would therefore be possible to find the best route from the origin at town A to river E, if the optimal route to station E from each of the sites for station B was known. However, the problem of completing the pipeline from each site for station B is also a multistage process. It would thus be possible to adopt the same technique as before if the minimum cost of completing the pipeline from each of the four sites for station C was known.

This leads to the fundamental idea behind DP. Starting with only one stage to be considered, the minimum cost of completing the pipeline from each site for pumping station D is found. Then working in reverse order, stage by stage, the minimum cost of completing the pipeline from each site of the preceding station is evaluated until the origin at town A is reached and the overall minimum cost solution is found.

The solution for the whole network is summarized in Figure 6.2. First, the cost is noted for each of four possible sites for station E at the river. These costs are recorded in the large boxes representing sites E1–E4.

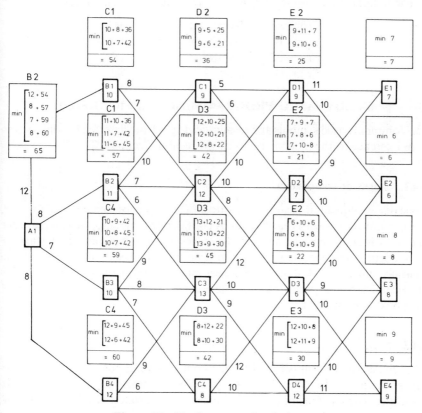

Figure 6.2 Pipeline network solution

Next, the best decision (i.e. route) for each site for station D is evaluated. For example, from D1 the following choices are available:

Origin	Route	Destination	Station D1	Cost of pipeline	Completion from E	Total
D1	E	E1	9	11	7	27
D1	SE	E2	9	10	6	25

The optimal decision at site D1 is a pipeline routed southeast (SE) to site E2 at a cost of 25 monetary units.* These calculations are shown in the box associated with site D1. The label E2 above this box signifies the best destination for the pipeline to take next.

The three other sites for station D are considered next; the reader can verify the calculations shown in Figure 6.2.

As the optimal route from each possible site for station D is now known, the optimal decision for each site of station C can be obtained in a similar manner. In this way it is possible to eventually evaluate the minimum cost for town A1.

The minimum cost route from A1 can now easily be found; it takes the form

From A1	Northeast	to	B2
From B2	Northeast	to	C1
From C1	Southeast	to	D2
From D2	East	to	E2.

6.4 DYNAMIC PROGRAMMING TERMINOLOGY

The commonly used symbols and terminology in DP are described in this section with reference to the pipeline network problem in Section 6.2.

6.4.1 Stages or components

Consider a sequential system consisting of *n stages* or *components*. In DP, optimization is frequently carried out by considering these stages in reverse order and so it is convenient to number the stages in the same manner. A stage or component corresponds to a point in time or to a specific milestone on the solution route. At each stage the engineer must decide on, or adopt, a policy for proceeding along a specific route to the next stage. Typically, stage i is denoted by a box as shown in Figure 6.3.

6.4.2 State

At each stage, a system is found to be in one of several possible conditions or *states* which are described by state variables. In other words, each stage is associated with either a finite or an infinite number of states. The decision of how

*£1000 = 1 monetary unit.

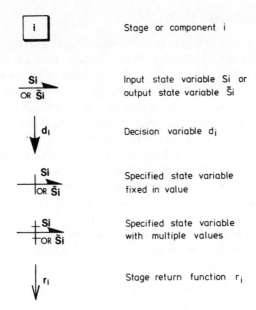

Figure 6.3   Notation used in dynamic programming

to move from one stage to another may frequently be a decision of how to transform the system from one state to another. On other occasions, the state variable may represent an important effect of the decision made in the previous stage or activity, the result of which is to impose some penalty or constraint on 'downstream' or subsequent activities (e.g. in the illustrative problems of Section 6.1 the depth of the sewer pipe, or the self-weight of the structure, may be seen as state variables). The optimal solution to the DP problem is a choice of the best transformations that take the engineer from the initial state to the final state.

In the pipeline network problem there are four possible states at the commencement of stage 1: i.e. states coinciding with station sites D1, D2, D3, and D4.

6.4.3 State variables

State variables carry information from stage to stage and are denoted by a half arrow as shown in Figure 6.3. Consequently, a state variable is an output from one stage and an input to another. The *input state variable* to stage i is denoted by S_i. The *output state variable*, $\bar{S}_i$, from the same stage, is related to the input state variable S_i and to the decision variable d_i through the *state transformation function* T_i by the expression

$$\bar{S}_i = T_i(S_i, d_i). \tag{6.1}$$

6.4.4 Decision variables

The *decision variable* d_i at stage i is symbolized by a full arrow. This is an input variable which supplies information to the system or describes a course of action which should be taken during the activity represented by the present stage.

Thus for a given state at stage i indicated by input state variable S_i and for a given value of the decision variable d_i at stage i, the state transformation function $T_i(S_i, d_i)$ gives the output state variables $\bar{S}_i$ which in turn indicates the new state reached by the solution at stage $i - 1$. T_i may be a mathematical function or matrix or, more simply, a set of rules governing the transformation from one state to another state at adjacent stages as in the pipeline network problem. Thus, for example, the input state variable S_1 at stage 1 in the pipeline network problem can take the values D1, D2, D3, or D4 and the output state variable $\bar{S}_1$ can take the values E1, E2, E3, or E4.

If the decision d_1 may take one of the three values Northeast (NE), East (E), or Southeast (SE), the state transformation function $T_1(S_1, d_1)$ may be represented by the following matrix:

	d_1		
S_1	NE	E	SE
D1	—	E1	E2
D2	E1	E2	E3
D3	E2	E3	E4
D4	E3	E4	—

6.4.5 Stage return function

A *stage return function* r_i at stage i, which is represented by an open arrow as shown in Figure 6.3, measures the effectiveness of the decision d_i for any specified value of the input state variable S_i and is given as

$$R_i = r_i(S_i, d_i). \tag{6.2}$$

This function R_i may be available as either:

(i) A mathematical function or set of functions which can be presented graphically as r_i against d_i for a set of state variables S_i.
(ii) A set of discrete values in the form of a matrix.

The return functions in the pipeline network problem, which are given discretely because of the nature of the problem, are shown in Figure 6.2. For example, the return (in this case a cost) for stage 2 (station C to station D) may be represented by the following matrix:

		d_2	
S_2	NE	E	SE
C1	—	14	15
C2	22	22	20
C3	25	23	22
C4	20	18	—

Note that each return (e.g. 14) comprises a component which is the cost of occupying a particular value of S_2 (e.g. 9 at station C1) and a further cost as a consequence of the decision d_2 (e.g. link cost of 5).

The total return is given as the sum of the stage functions

$$R = \sum_{i=1}^{n} r_i. \qquad (6.3)$$

Whether the function r_i is available in form (i) or (ii) it is usually tabulated for discrete values of S_i and d_i. Intermediate values can then be obtained by recourse to the original mathematical function, if it exists, or by linear interpolation.

6.4.6 Policy

A set of sequential decision rules covering each stage of the system is called a *policy*. For example in the pipeline network problem a typical — though non-optimal — policy might be represented by the decisions SE, NE, SE, E.

6.4.7 Optimal policy

A policy which optimizes the total return (e.g. cost) of the entire process (e.g. pipeline network) is called an *optimal policy*. The optimal policy in the pipeline network problem is (NE, NE, SE, E).

6.4.8 Specified state variables

The value of one or more state variables may frequently be specified. The specification of a variable is denoted by a vertical with one or more cross lines as shown in Figure 6.3. The input state variable S_4 at stage 4 in the pipeline network problem is specified as A1 since the pipeline must finish at town A. This is known as an initial-value problem.

6.5 THE PRINCIPLE OF OPTIMALITY

Bellman's Principle of Optimality states that: 'An optimal policy (set of decisions) has the property that whatever the initial state and decisions are, the remaining decisions must constitute an optimal policy with regard to the state resulting from the first decision.' Dynamic programming is based on this principle.

It is instructive to re-work the pipeline network problem using the standard notation and thereby develop the recurrence equations of dynamic programming.

6.5.1 A re-examination of the pipeline network problem

The desired objective is to minimize an n-stage objective function which is given by the sum of the individual stage return functions. Such a problem can be written as follows:

$$\operatorname*{Minimize}_{d} z = [r_n(S_n, d_n) + r_{n-1}(S_{n-1}, d_{n-1}) + \ldots + r_1(S_1, d_1)]. \quad (6.4)$$

Figure 6.4 shows a diagrammatic representation of an n-stage pipeline network problem using the notation described in Section 6.4. It can be seen that the stages are numbered in reverse order to simplify notational problems in dealing with the end stage first.

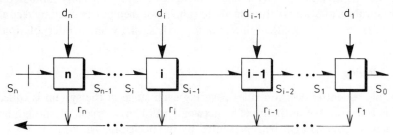

Figure 6.4 An n-stage minimization problem

Two important characteristics of the DP problem should be emphasized at this point:

(i) The general objective function should be capable of decomposition into n separable return functions each of which depends on a single decision variable. Moreover, each decision variable must appear in one and only one return function.
(ii) The problem should be capable of being described in a purely serial or sequential way. No 'feed-back' is allowable, whereby a decision at some stage may affect the outcome of another stage which has previously been considered.

The analysis begins with the sub-optimization of stage 1 as shown in step 1 of Figure 6.5. This involves the solution of the problem

$$\min_{d_1} [r_1(S_1, d_1)] = f_1^*(S_1). \quad (6.5)$$

The best value of the design variable, denoted as d_1^*, and the value of the minimum itself, denoted as f_1^*, depend on the condition of the input state variable

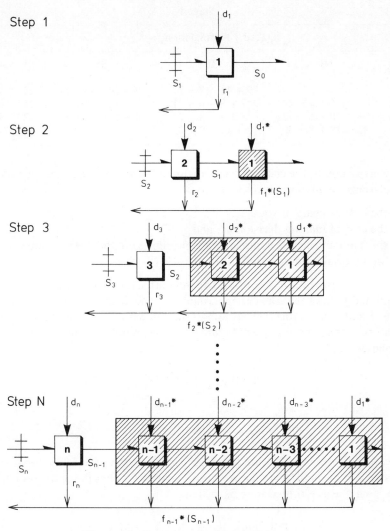

Figure 6.5 Sequence of sub-optimizations

that this last stage receives from stage 2 (i.e. on S_1). This dependence can be denoted symbolically by $d_1^*(S_1)$ and $f_1^*(S_1)$. Since the particular value of S_1 which occurs in the overall optimal policy cannot be known at this step in the solution, this initial sub-optimization problem is solved for the range of possible values of S_1 and the results tabulated.

For example, in the 4-stage problem discussed in Section 6.3, S_1 can take the values of D1, D2, D3, or D4. The decision d_1 can take the values (i.e. the compass directions) NE, E, or SE. For the 4 values of S_1 the optimal values are given by (6.5) and these results together with the corresponding optimal decision are tabulated as follows:

Stage 1

S_1	\multicolumn{3}{c}{$r_1(S_1, d_1) + f_0^*(S_0)$ for $d_1 =$}	f_1^*	d_1^*
	NE	E	SE		
D1	—	9 + 11 + 7 = 27	9 + 10 + 6 = 25	25	SE
D2	7 + 9 + 7 = 24	7 + 8 + 6 = 21	7 + 10 + 8 = 25	21	E
D3	6 + 10 + 6 = 22	6 + 9 + 8 = 23	6 + 10 + 9 = 25	22	NE
D4	12 + 10 + 8 = 30	12 + 11 + 9 = 32	—	30	NE

By inspection of the cost figures in Figure 6.1 it will be seen that each of the calculations involves three components:

(i) the cost of being at state S_1,
(ii) the cost of making decision d_1, and
(iii) the minimum cost of completing the pipeline from the downstream state S_0 which results from S_1 and d_1, i.e. $S_0 = T_1(S_1, d_1)$.

The next step in the DP solution is to combine the last two stages. The principle of optimality states that the design variables d_1 and d_2 must be adjusted so that stages 1 and 2 taken together are sub-optimized with respect to the input state variables S_2 as shown in step 2 of Figure 6.5. This sub-optimization can be written as follows:

$$\min_{d_2, d_1} [r_2(S_2, d_2) + r_1(S_1, d_1)] = f_2^*(S_2)$$

or (6.6)

$$\min_{d_2} \left[r_2(S_2, d_2) + \min_{d_1} \{r_1(S_1, d_1)\} \right] = f_2^*(S_2).$$

However, the optimal solution for stage 1 for various values of the input state variable S_1 is available from the table of f_1^* against S_1. Thus the optimal solution $f_1^*(S_1)$ may be substituted in (6.5) so that

$$\min_{d_2} [r_2(S_2, d_2) + f_1^*(S_1)] = f_2^*(S_2). \quad (6.7)$$

By making use of the state transformation function

$$S_{i-1} = T_i(S_i, d_i) \quad (6.8)$$

(6.7) can be written entirely in terms of S_2 and d_2 as

$$\min_{d_2} [r_2(S_2, d_2) + f_1^*(T_2(S_2, d_2))] = f_2^*(S_2). \quad (6.9)$$

For the 4 values of S_2 the 4 corresponding optimal values can be tabulated as follows, noting that:

$$S_1 = T_2(S_2, d_2). \quad (6.10)$$

Stage 2

S_2	NE	E	SE	f_2^*	d_2^*
		$r_2(S_2, d_2) + f_1^*(S_1)$ for $d_2 =$			
C1	—	9 + 5 + 25 = 39	9 + 6 + 21 = 36	36	SE
C2	12 + 10 + 25 = 47	12 + 10 + 21 = 43	12 + 8 + 22 = 42	42	SE
C3	13 + 12 + 21 = 46	13 + 10 + 22 = 45	13 + 9 + 30 = 51	45	E
C4	8 + 12 + 22 = 42	8 + 10 + 30 = 48	—	42	NE

It should perhaps be emphasized here that the number of variables that need to be considered during minimization has been reduced from 2 (i.e. d_1 and d_2) to 1(i.e. d_2) and reference to stage 1 is now through the table of f_1^* against S_1 rather than through a separate minimization problem.

The next step is to consider the last three stages in combination. The principle of optimality states that the design variables d_1, d_2, and d_3 must be adjusted so that stages 1, 2, and 3 taken together are sub-optimal with respect to the input state variable S_3 as shown in step 3 of Figure 6.5.

This sub-problem can be written as

$$\min_{d_3, d_2, d_1} [r_3(S_3, d_3) + r_2(S_2, d_2) + r_1(S_1, d_1)] = f_3^*(S_3) \quad (6.11)$$

or

$$\min_{d_3} \left[r_3(S_3, d_3) + \min_{d_2, d_1} \{r_2(S_2, d_2) + r_1(S_1, d_1)\} \right] = f_3^*(S_3)$$

or finally

$$\min_{d_3} [r_3(S_3, d_3) + f_2^*(S_2)] = f_3^*(S_3).$$

For the 4 values of S_3 the corresponding optimal values can be tabulated as follows:

Stage 3

S_3	NE	E	SE	f_3^*	d_3^*
		$r_3(S_3, d_3) + f_2^*(S_2)$ for $d_3 =$			
B1	—	10 + 8 + 36 = 54	10 + 7 + 42 = 59	54	E
B2	11 + 10 + 36 = 57	11 + 7 + 42 = 60	11 + 6 + 45 = 62	57	NE
B3	10 + 9 + 42 = 61	10 + 8 + 45 = 63	10 + 7 + 42 = 59	59	SE
B4	12 + 9 + 45 = 66	12 + 6 + 42 = 60	—	60	E

The final stage in the solution is to consider all four stages taken together. This sub-problem can be written as

$$\min_{d_4, d_3, d_2, d_1} [r_4(S_4, d_4) + r_3(S_3, d_3) + r_2(S_2, d_2) + r_1(S_1, d_1)] = f_4^*(S_4) \quad (6.12)$$

or
$$\min_{d_4} [r_4(S_4, d_4) + f_3^*(S_3)] = f_4^*(S_4).$$

The process may be visualized as the last step of Figure 6.5 in which $n = 4$. For the single value of S_4 the corresponding optimal value can be tabulated as:

Stage 4

	$r_4(S_4, d_4) + f_3^*(S_3)$ for $d_4 =$					
S_4	N	NE	SE	S	f_4^*	d_4^*
A1	12 + 54 = 66	8 + 57 = 65	7 + 59 = 66	8 + 60 = 68	65	NE

Having assembled the four tables describing the optimal performance at each of the four stages, it is possible to identify the minimum cost and also the optimum sequence of decisions which lead to this result. Although the least-cost path was identified in Section 6.3, it is instructive to note how the information contained in the tables is used to 'unroll' the optimum policy. The following steps describe the process:

(i) From the table for stage 4, only one input state (A1) is possible and the minimum cost is immediately given as 65 monetary units.

(ii) The optimum decision in stage 4 is a move in the NE direction. The state transformation function (as defined in Section 6.4.5) is rather self-evident in this case but may be formally described as

$$S_3^* = T_4(S_4^*, d_4^*)$$

where the asterisks denote that the values are associated with the optimal policy. Here,

$$S_3^* = T_4(A1, NE) = B2.$$

(iii) From the table for stage 3, the optimal decision for an input state variable of B2 is NE. This in turn leads to

$$S_2^* = T_3(S_3^* = B2, d_3^* = NE) = C1.$$

(iv) Repeating the process, using the tables for stages 2 and 1, the remaining optimal decision and also the optimal terminal state is found, e.g.

$$S_1^* = T_2(S_2^* = C1, d_2^* = SE) = D2$$
$$S_0^* = T_1(S_1^* = D2, d_1^* = E) = E2.$$

6.5.2 Generalization of the process

The development of the recurrence relationships may now be summarized. Suppose that this sub-optimization sequence has been carried on to include $i - 1$

of the end stages and the next step is to sub-optimize the i end steps as indicated by the equation

$$\min_{d_i, d_{i-1}, \ldots, d_1} [r_i(S_i, d_i) + r_{i-1}(S_{i-1}, d_{i-1}) + \ldots + r_1(S_1, d_1)] = f_i^*(S_i). \quad (6.13)$$

Since by this time the $i-1$ end stages have already been sub-optimized, the function

$$\min_{d_{i-1}, \ldots, d_1} [(r_i(S_{i-1}, d_{i-1}) + \ldots + r_1(S_1, d_1)] = f_{i-1}^*(S_{i-1}) \quad (6.14)$$

is available and hence may be used to reduce the dimensionality of the ith-stage sub-optimization (i.e. the numbers of decision variables) to

$$\min_{d_i} [r_i(S_i, d_i) + f_{i-1}^*(S_{i-1})] = f_i^*(S_i) \quad (6.15)$$

or again by making use of the state transformation function

$$\min_{d_i} [r_i(S_i, d_i) + f_{i-1}^*\{T_i(S_i, d_i)\}] = f_i^*(S_i). \quad (6.16)$$

Once again it should be noted that an i-dimensional optimization problem has been reduced to a 1-dimensional optimization problem by means of a series of solutions to the downstream† sub-optimization problem. The sequential sub-optimization strategy of dynamic programming is shown in Figure 6.5 in which it should be noted that the design variables are analysed one at a time rather than simultaneously.

In DP solutions the following points should be considered:

(i) Any optimization technique may be used to solve the recursive equation at a given stage.
(ii) Generally, the solution of equations of the type given in (6.13) requires tabular-type computations in which the optimal decision variables, and the optimal return functions, are in discrete form.
(iii) The use of tabular-type computations is obligatory when either the state variables are inherently discrete quantities or when the analytical representation of the return function is impractical. For the latter case, the state variables must be artificially discretized at each stage.
(iv) In formulating problems for solution by DP it is advisable to follow a standard procedure. Some elements of this have been introduced in Section 6.3, in particular the definition of stage, state, etc. However, the pipeline network problem was deliberately chosen for its simplicity and many features associated with it could be regarded as obvious whereas in other problems these same features require definition. Figure 6.6 shows a flowchart indicating a disciplined approach to formulating DP problems.

† The term refers to stages $i-1, \ldots, 1$ occurring after stage i; see Figure 6.5.

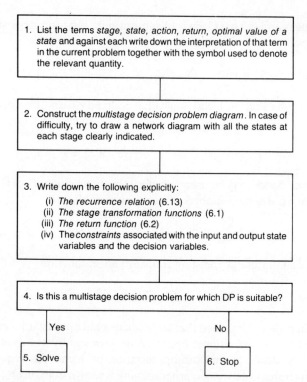

Figure 6.6 Dynamic Programming problem definition flow chart

Table 6.1 Specification of the pipeline network problem

Stage	A pipeline section from one pumping station site to an adjacent site (i = number of stages remaining)	i
State	A pumping station site	S_i
Decision	Taking a particular pipeline route from a pumping station site	d_i
Return	Cost of constructing a pipeline section and the pumping station at the upstream end of the pipeline section defined by S_i	$r_i(S_i, d_i)$
Optimal value of a state	Cost of constructing the pipeline network from the corresponding pumping station to the pumping stations at the river E under an optimal plan	$f_i^*(S_i)$

Table 6.1 shows a summary of the DP terms used in the pipeline network problem.

No standard form of DP model exists and consequently it is impossible to have a completely general computational tool such as the simplex method in LP which

1. **PHASE I** — Compute tables of optimal performance for each activity. Set stage counter $i = 1$.
2. For ith stage tabulate all feasible values of input state variable $S_i = S_{ij}$ $(j = 1, 2, 3 \ldots)$
3. Select input state variable S_{ij}
4. For selected S_{ij} tabulate all feasible values of decision variable $d_i = d_{ik}$ $(k = 1, 2, 3 \ldots)$
5. Select devision variable d_{ik}
6. Calculate return function $r_i(S_{ij}, d_{ik})$
7. Find resulting downstream state S_{i-1} from state transformation function $S_{i-1} = T_i(S_{ij}, d_{ik})$
8. Obtain minimum cost for downstream stages $f^*_{i-1}(S_{i-1})$
9. Combine costs to get $f_i(S_{ij}, d_{ik}) = r_i(S_{ij}, d_{ik}) + f^*_{i-1}(S_{i-1})$
10. Repeat steps 5 to 9 to obtain vector of costs $f_i(S_{ij})$ for all feasible decisions d_{ik} $(k = 1, 2, \ldots)$
11. Select $f_i^*(S_{ij}) = \min [f_i(S_{ij})]$ and note corresponding optimal decisions d^*_{ij}
12. Repeat Steps 3 to 11 to obtain vectors of f_i^* and d_i^* as functions of S_{ij} $(j = 1, 2, 3 \ldots)$
13. Repeat Steps 2 to 12 for stages $i, i+1, i+2, \ldots, n$.
14. **PHASE II** — 'Un-roll' Optimal Policy. Obtain optimal system cost as $\min f_n^*$
15. Note optimal decision d_n^* for n^{th} stage and corresponding input state variable S_n^*
16. Compute optimal downstream state variable $S^*_{n-1} = T_n(S_n^*, d_n^*)$
17. Look up tables for next downstream stage $n - 1$ to get optimal decision d^*_{n-1} corresponding to input state variable S^*_{n-1}
18. Repeat Steps 16 to 17 until all n stages have been examined and all n optimal decisions identified. Optimal policy is $d_n^*, d^*_{n-1}, d^*_{n-2}, \ldots, d_2^*, d_1^*$.

Figure 6.7 A two-pass algorithm for dynamic programming problems

was described in Chapter 3. However, all of the essential steps are represented in the problem of Section 6.5.1 and may be summarized in the two-pass algorithm described in Figure 6.7. In the remaining sections of this chapter some typical DP problems will be described with the aid of worked examples. In the next section, a general formulation of one-dimensional allocation problems is presented and then two typical DP problems will be considered.

6.6 ALLOCATION PROCESSES

6.6.1 General formulation

An allocation problem is an example of a single-period deterministic multiactivity process that can be transformed by DP into a multistage decision process with a finite number of stages. Only one-dimensional allocation problems will be considered in which a certain limited quantity Q of an economic resource (e.g. labour, land, machines, or water) is to be allocated among some potential recipients. The resource, which is used in the production of certain products or services, can be used in two or more alternative ways. Each possible use is called an *activity* and each single activity, where the resource is used, yields a *return*. The problem is how to allocate the resource to the alternative activities (or to users) in such a way that the total return is maximized. By the nature of the problem it is less likely that an allocation problem will be associated with the minimization of cost, hence the departure in this section from the standard practice of expressing problems as minimization models. It is, of course, a simple matter to employ the equivalence

$$\min z \simeq \max(-z)$$

to change the sense of the problem if desired.

In the solution of allocation problems by DP the following assumptions are made:

(i) Returns from different allocations can be compared. In other words they can be measured in a common monetary unit such as the pound or the dollar.
(ii) The return from any allocation is independent of the allocations to other activities.
(iii) The total return that can be obtained is the sum of the individual returns.

The most general mathematical formulation of the one-dimensional allocation problem involves the maximization of an objective function (total return) given as

$$\text{Maximize } R = r_n(S_n, d_n) + \ldots + r_2(S_2, d_2) + r_1(S_1, d_1) \tag{6.17}$$

subject to the constraint that the sum of the resources allocated to each activity must not exceed the available resource:

$$\sum_{i=1}^{n} d_i \leq Q(=S_n) \tag{6.18}$$

where

$Q = S_n$ is the total amount of resources
d_i is the quantity of the resource assigned to the ith activity or stage
$r_i(S_i, d_i)$ is the return of the ith activity
n is the number of possible activities or stages.

If the objective function is linear, then the allocation problem reduces to a linear programming problem and may be solved by the simplex method which was outlined in Chapter 3. However, in the more general case, where the objective function can take any form, dynamic programming may be used. First the problem must be converted to a multistage decision problem as follows:

(i) The first allocation goes to the nth activity or stage.
(ii) Then an allocation is made to activity $(n - 1)$.
(iii) An allocation is then made to activity $(n - 2)$.
(iv) The allocation proceeds in this manner until an allocation is made to activity 1.

It must be emphasized that the optimal policy will be the same irrespective of the order in which the activities are considered.

As described in Section 6.4 the allocation problem can be solved using DP by developing a sequence of recurrence relations.

Let $f_n^*(S_n)$ be the optimal return from an allocation of $S_n = Q$ to the n activities. Assuming $r_i(0, d_i) = 0$ for all i, which is usually the case, it follows that $f_n^*(0) = 0$. It is also assumed that

$$f_1(S_1) = r_1(S_1, d_1). \tag{6.19}$$

Let d_n be the allocation made to the nth activity where $0 \leq d_n \leq Q$. The remaining quantity, in other words the input state variable S_{n-1} to the $(n - 1)$th activity, is obtained from the state transformation function

$$S_{n-1} = T_n(S_n, d_n)$$

or

$$S_{n-1} = S_n - d_n \tag{6.20}$$
$$= Q - d_n.$$

The quantity S_{n-1} will be used in the $(n - 1)$ remaining activities. If this quantity S_{n-1} has already been allocated to the $(n - 1)$ activities in the optimal way and the allocation yielded a return of $f_{n-1}^*(Q - d_n)$, then by definition the return from the allocation of d_n to the nth activity is $r_n(Q, d_n)$. Thus the total return from allocating Q to all n activities is

$$R = r_n(Q, d_n) + f_{n-1}^*(Q - d_n)$$

or

$$R = r_n(S_n, d_n) + f_{n-1}^*(S_n - d_n). \tag{6.21}$$

Usually there are several choices for d_n in the nth activity and obviously the optimal one is that which maximizes R; that is

$$f_n^*(S_n) = \max_{0 \leq d_n \leq Q} [r_n(S_n, d_n) + f_{n-1}^*(S_n - d_n)]. \tag{6.22}$$

Thus the allocation problem defined by (6.14) has been replaced by the problem defined by (6.22). Before this problem can be solved, however, it is necessary to obtain $f_{n-1}^*(S_n - d_n)$ which can be obtained from the equation

$$f_{n-1}^*(S_{n-1}) = f_{n-1}^*(S_n - d_n) = \max_{0 \leq d_{n-1} \leq S_{n-1}} [r_{n-1}(S_{n-1}, d_{n-1}) + f_{n-2}^*(S_{n-2})] \tag{6.23}$$

where d_{n-1} is the amount allocated to the $(n - 1)$th activity. The amount of resource available for activity S_{n-2} is obtained by the stage transformation function

$$S_{n-2} = T_{n-1}(S_{n-1}, d_{n-1})$$
$$= S_{n-1} - d_{n-1}. \quad (6.24)$$

Note that in (6.21), in order to obtain $f^*_{n-1}(S_{n-1})$ the value $f^*_{n-2}(S_{n-2})$ must be found. The process continues backwards until the second stage is reached. At the second stage the optimal result of the first stage $f^*_1(S_1)$ is used and $f^*_1(S_1)$ is given from (6.19). Thus the entire process can be solved since $f^*_1(S_1)$ determines $f^*_2(S_2)$, $f^*_2(S_2)$ determines $f^*_3(S_3)$, and so on.

6.6.2 A resource allocation problem

Three irrigation projects are competing for capital of $3 million. The allocation is in discrete units of one million and the return functions are given as follows:

For project 1 $r_1(d_1) = \frac{2}{3}d_1^2$

project 2 $r_2(d_2) = 2d_2$

project 3 $r_3(d_3) = \dfrac{6}{(3)^{1/2}}(d_3)^{1/2}$

where d_i ($i = 1, 2, 3$) is the number of units allocated to each project.

Thus the profit or return functions may be tabulated as:

d_i	r_1	r_2	r_3
0	0	0	0
1	0.67	2.00	3.46
2	2.67	4.00	4.90
3	6.00	6.00	6.00

The transformation functions $T_i(S_i, d_i)$ are simple statements of continuity

$$S_0 = S_1 - d_1 \quad \text{or} \quad S_1 = d_1 + S_0 \quad (6.25)$$
$$S_1 = S_2 - d_2 \quad \text{or} \quad S_2 = d_2 + S_1$$
$$S_2 = S_3 - d_3 \quad \text{or} \quad S_3 = d_3 + S_2$$

where S_i and the other variables are as defined in Table 6.2.

In this problem there is also the constraint that the sum of the units of capital allocated to each activity must not exceed the available capital. Thus

$$d_1 + d_2 + d_3 \leqslant 3. \quad (6.26)$$

The sub-optimization of stage 1 is considered initially and this involves the solution of the maximization problem

Table 6.2 Specification of resource allocation problem

Stage	Allocation of capital to an irrigation project (i = number of projects remaining)	i
State	S_i units of capital remain to be allocated and i projects remain to be considered	S_i
Decision	Allocate d_i units of capital to project i	d_i
Return	Profit from allocation of d_i units of capital to irrigation project i	$r_i(S_i, d_i)$
Optimal value of a state	Total profit when state S_i is the starting state at stage i and an optimal plan is followed	$f_i^*(S_i)$

$$\max_{d_1} [r_1(S_1, d_1)] = f_1^*(S_1). \qquad (6.27)$$

The decision variable d_1 can take up to 4 values (0, 1, 2, or 3) depending on S_1. For the 4 values of S_1 (0, 1, 2, or 3) the 4 corresponding optimal values are given in tabular form as

Stage 1

	$R_1(S_1, d_1)$ for $d_1 =$					
S_1	0	1	2	3	$f_1^*(S_1)$	d_1^*
0	0.00	—	—	—	0.0	0
1	0.00	0.67	—	—	0.67	1
2	0.00	0.67	2.67	—	2.67	2
3	0.00	0.67	2.67	6.00	6.00	3

The end-value condition that $S_0 \geqslant 0$ implies the constraint that

$$S_1 - d_1 \geqslant 0$$

or $\qquad (6.28)$

$$d_1 \leqslant S_1.$$

Next the sub-optimization of stages 1 and 2 is performed and this involves the solution of the problem

$$\max_{d_2} [r_2(S_2, d_2) + f_1^*(S_1)] \qquad (6.29)$$

subject to

$$S_1 = S_2 - d_2.$$

The results are again tabulated as follows; note that for $S_2 = 3$ there is a tie for the optimal decision $d_2^* = 0$ or 3:

Stage 2

S_2	\multicolumn{4}{c}{$r_2(S_2, d_2) + f_1^*(S_1)$ for $d_2 =$}	$f_2^*(d_2)$	d_2^*
	0	1	2	3		
0	0.00 + 0.00	—	—	—	0.00	0
1	0.00 + 0.67	2.00 + 0.00	—	—	2.00	1
2	0.00 + 2.67	2.00 + 0.67	4.00 + 0.00	—	4.00	2
3	0.00 + 6.00	2.00 + 2.67	4.00 + 0.67	6.00 + 0.00	6.00	0, 3

Finally the three stages are considered together and this involves the solution of the problem

$$\max_{d_3} [r_3(S_3, d_3) + f_2^*(S_2)] = f_3^*(S_3) \qquad (6.30)$$

subject to

$$S_2 = S_3 - d_3.$$

Again the results are presented in tabular form:

Stage 3

S_3	\multicolumn{4}{c}{$r_3(S_3, d_3) + f_2^*(S_2)$ for $d_3 =$}	$f_3^*(d_3)$	d_3^*
	0	1	2	3		
0	0.00 + 0.00	—	—	—	0.00	0
1	0.00 + 2.00	3.46 + 0.00	—	—	3.46	1
2	0.00 + 4.00	3.46 + 2.00	4.90 + 0.00	—	5.46	1
3	0.00 + 0.00	3.46 + 4.00	4.90 + 2.00	6.00 + 0.00	7.46	1

Having completed the first pass through the three activities, and generated the working tables, the maximum return and corresponding optimal policy can be identified by a second pass through the stages in the opposite direction. Thus:

Maximum return $f_3^*(S_3) = 7.46$ million.

From stage 3 $\quad S_3^* = 3, \quad d_3^* = 1, \quad$ and $\quad S_2^* = S_3^* - d_3^* = 2$

stage 2 $\quad S_2^* = 2, \quad d_2^* = 2, \quad$ and $\quad S_1^* = S_2^* - d_2^* = 0$

stage 1 $\quad S_1^* = 0, \quad d_1^* = 0, \quad$ and $\quad S_0^* = S_1^* - d_1^* = 0.$

Thus the optimal policy is $d_1^* = 0$, $d_2^* = 2$, and $d_3^* = 1$.

In performing the 'second pass', it will be noted that the only information which must be recalled are the values of d_i^* as a function of S_i.

6.6.3 A waste-water treatment example

Three factories located on a river discharge waste water with varying degrees of biological oxygen demand (BOD). Each plant can provide alternative levels of treatment (primary, secondary, and tertiary) for varying costs, and with varying degrees of reduction in BOD. The following table shows the details of costs and reduced BOD:

Highest degree of treatment	BOD ppm x_{ik} Factory			Installation cost ($1000 units) Factory		
	$i = 1$	$i = 2$	$i = 3$	$i = 1$	$i = 2$	$i = 3$
$j =$ none	1.2	0.8	1.6	0	0	0
$j = 1$ primary	0.8	0.4	1.2	70	60	90
$j = 2$ secondary	0.4	0.0	0.8	120	100	140
$j = 3$ tertiary	0.0	0.0	0.0	170	130	220

If the total BOD should not exceed 1.6 parts per million (ppm), then the problem is one of finding the policy which minimizes the total cost of treatment. The specification of the variables in this problem is shown in Table 6.3.

Table 6.3 Specification of waste-water treatment problem

Stage	Factory ($i =$ number of factories remaining)	i
State	The amount (ppm) of BOD which exists before treatment in the remaining i factories	S_i
Decision	The highest level of water treatment (none, primary, secondary, tertiary)	d_i
Return	Cost of water treatment at factory	$r_i(S_i, d_i)$
Optimal value of a state	Total cost when state S_i is the starting state at stage i and an optimal plan is followed	$f_i^*(S_i)$

(Adapted from Daellenbach and George[7])

An important difference between this problem and the allocation example of Section 6.6.2 is in the more complex form of the state transformation functions. For any selected value of the input state variable, each trial decision must be tested for feasibility to ensure that the downstream state does not exceed the maximum allowable BOD. Thus, step 4 in the algorithm of Figure 6.7 requires more than a simple test of continuity (e.g. $S_{i-1} = S_i - d_{ik}$) and involves the use of the BOD load x_{ik}, i.e.

$$S_{i-1} = S_i + x_{ik}(d_k).$$

The sub-optimization of stage 1 is considered first and this involves the minimization problem

$$\text{Minimize}_{d_1} [r_1(S_1, d_1)] = f_1^*(S_1) \qquad (6.31)$$

subject to
$$S_0 = S_1 + x_{1k} \leqslant 1.6 \quad k = 0, 1, 2, 3.$$

There are 5 possible values of the input state variable S_1 and the optimal policies for these values are tabulated as:

Stage 1

S_1	\multicolumn{4}{c}{$r_1(S_1, d_1)$ for $d_1 =$}	$f_1^*(S_1)$	d_1^*
	0	1	2	3		
	1.0	0.8	0.4	0		
0.0	0	70	120	170	0	0
0.4	0	70	120	170	0	0
0.8	—	70	120	170	70	1
1.2	—	—	120	170	120	2
1.6	—	—	—	170	170	3

Next the sub-optimization of stages 1 and 2 is considered:

$$\underset{d_2}{\text{Minimize}} \; [r_2(S_2, d_2) + f_1^*(S_1)] = f_2^*(S_2)$$

subject to
$$S_1 = S_2 + x_{2k} \leqslant 1.6 \quad k = 0, 1, 2, 3.$$

The optimal policies for the five possible values of the input state variable S_2 are:

Stage 2

S_2	\multicolumn{4}{c}{$r_2(S_2, d_2) + f_2^*(S_2)$ for $d_2 =$}	$f_2^*(S_2)$	d_2^*
	0	1	2	3		
	0.8	0.4	0	0		
0.0	0 + 70	60 + 0	100 + 0	130 + 0	60	1
0.4	0 + 120	60 + 90	100 + 0	130 + 0	100	2
0.8	0 + 170	60 + 120	100 + 70	130 + 70	170	0, 2
1.2	—	60 + 170	100 + 120	130 + 120	220	2
1.6	—	—	100 + 170	130 + 170	270	2

Finally the optimization of stages 1, 2, and 3 is considered:

$$\underset{d_3}{\text{Minimize}} \; [r_3(S_3, d_3) + f_2^*(S_2)] = f_3^*(S_3) \qquad (6.32)$$

subject to
$$S_2 = S_3 + x_{3k} \leqslant 1.6 \quad k = 0, 1, 2, 3.$$

Stage 3

S_3	$r_3(S_3, d_3) + f_3^*(S_3)$ for $d_3 =$				$f_3^*(S_3)$	d_3^*
	0 1.0	1 1.2	2 .8	3 0		
0.0	0 + 270	90 + 220	140 + 170	220 + 60	270	0
0.4	—	90 + 270	140 + 220	220 + 100	320	3
0.8	—	—	140 + 270	220 + 170	390	3
1.2	—	—	—	220 + 220	440	3
1.6	—	—	—	220 + 270	490	3

The optimal policy can thus be summarized as

Minimum cost $270 000

$$S_3^* = 0.0, \quad d_3 = 0 \quad x_{3j} = 1.6$$
$$S_2 = 1.6, \quad d_2^* = 2 \quad x_{2j} = 0.0$$
$$S_1 = 1.6, \quad d_1^* = 3.$$

In other words, there should be no treatment at factory 3, both primary and secondary treatment at factory 2, and primary, secondary, and tertiary treatment only at factory 1.

6.7 COMPUTER SOLUTION

6.7.1 Scope of the routines

For obvious reasons the illustrative problems in this chapter have been relatively simple and of extremely modest size. Application of dynamic programming to problems of practical size is facilitated by computer routines since in general the method is recursive. As mentioned in Section 6.5.2, however, dynamic programming is more an approach to problem-solving than a specific technique and a completely general algorithm is almost impossible to implement in a computer program.

A general class of problem of the resource allocation type can be handled conveniently, and may even be applied to other problem types with a little ingenuity. The chief restrictions which apply to the routines illustrated here may be summarized as follows:

(i) In decomposing the problem into stages, the return or cost of each stage is assumed to be combined by addition (cf. in estimating probability of failure by DP the individual stage reliabilities may require to be combined as a product).
(ii) Each stage involves a decision with respect to a single variable.
(iii) The state and decision variables are assumed to be discrete rather than

continuous and a single array of values is used to define (each of) the state variables and decisions at every stage.
(iv) It is necessary for the user to write a main program and also three special purpose auxiliary routines to describe the return function, the state transformation function, and also to check the feasibility of selected state and decision variables.

The main routines are DYNAM and DYNSOL, and both are listed in Appendix A. Routine DYNAM performs the first pass through the system and generates the tables of optimum values and decisions for each stage as a function of the input state variable. Routine DYNSOL operates on these tables to identify the maximum or minimum return and 'unroll' the corresponding optimal policy. Together, the two routines generally follow the algorithm of Figure 6.7.

6.7.2 Specification of the auxiliary routines

(a) Routine FEASBL is intended to check the feasibility of currently selected values of the input state variable and the decision variable. Prior to the call of FEASBL in the main routine DYNAM a penalty term is set to a large positive or negative value (depending on whether the object is minimization or maximization respectively). Routine FEASBL should reset the penalty term to zero whenever a feasible condition is encountered. The coding of FEASBL may include a call of routine TRANSF to compute the downstream state variable.
(b) Routine TRANSF describes the state transformation function and computes the subscript JDS such that the array element STATE(JDS) is the downstream state corresponding to the currently selected stage, input (i.e. upstream) state, and decision. The routine may employ a simple utility subroutine to find the array subscript, the corresponding element of which is equal to a specified value (routine JSTATE).
(c) Routine RETFUN calculates the cost or return of a particular stage or activity corresponding to currently selected values for the input state and the decision variables.

To maintain as much flexibility as possible all of the above routines contain as input parameters the arrays of allowable state values and decision variables. The following input parameters are used in all three routines:

STATE —An array of allowable state values
D —An array of allowable decision variables
NSTAGE —The number of stages or activities
NSTATE —The number of state values
ND —The number of decision variables
I —The currently selected stage
J —The currently selected input state is STATE(J)
K —The currently selected decision is D(K).

The following output variables are used.

In FEASBL:
PEN —A penalty term set to zero if currently selected values represent a feasible condition; otherwise PEN is left unchanged.

In TRANSF:
IDS —The element of the array STATE(IDS) which describes the output (downstream) state.

In RETFUN:
RTN —The return from the current stage I under input state STATE(J) and decision D(K).

The following sections show the application of the routines to the problems solved previously. The illustrative coding together with the listings contained in Appendix A demonstrates the method of use.

6.7.3 The Resource Allocation Problem

The coding and results are shown in Figure 6.8. The main program includes the necessary dimension statements, an external statement to identify the auxiliary routines, and statements defining the system parameters. Two calling statements of DYNAM and DYNSOL complete the main program.

Routines FEASBL and TRANSF are very simple due to the elementary nature of the state transformation function in this problem. Routine RETFUN contains special purpose arrays and corresponding DATA statements to define the return functions. Linear interpolation is employed although in this special case the return functions could be described in algebraic terms.

The output (which may be suppressed if desired) contains not only the optimal result and policy but also the tables generated by routine DYNAM. These may be of use in checking the correctness of the solution.

6.7.4 The Waste-Water Treatment Problem

The main program for this problem differs from the previous case only in the dimensions of the problem, the sign of the parameter SIGN, and the values of state and decision variables (see Figure 6.9). The state transformation function is described in matrix form by the two-dimensional array BOD in routine TRANSF. The output state level JDS is set to zero if the allowable BOD level is exceeded. Otherwise, the appropriate value of JDS is computed by the utility routine JSTATE. Routine TRANSF thus forms the basis for routine FEASBL.

The return function is quite simple since costs are tabulated for every stage and decision, thus eliminating the need for interpolation.

```
C     MAIN PROGRAM TO SOLVE THE RESOURCE (I.E. CAPITAL)
C     ALLOCATION PROBLEM BY DYNAMIC PROGRAMMING.
C     REFER TO LISTING OF  "DYNAM" FOR DEFINITIONS.
      DIMENSION D(7),DOPT(3,7),FOPT(3,7)
      DIMENSION STATE(7),DSOL(3)
      EXTERNAL TRANSF,RETFUN,FEASBL
      NACT=3
      NSTATE=7
      ND=7
C     FOR MAXIMIZATION...
      SIGN=+1.0
C     BOTH DECISION AND STATE REFER TO AMOUNT OF CAPITAL
C     USED OR REMAINING.
      DATA D    /0.00,0.50,1.00,1.50,2.00,2.50,3.00/
      DATA STATE/0.00,0.50,1.00,1.50,2.00,2.50,3.00/
      CALL DYNAM(STATE,D,NACT,NSTATE,ND,SIGN,DOPT,FOPT,
     +           FEASBL,TRANSF,RETFUN)
      NPRINT=1
      CALL DYNSOL(STATE,D,NACT,NSTATE,ND,SIGN,DOPT,FOPT,
     +            TRANSF,NPRINT,DSOL,ANS)
      STOP
      END
      SUBROUTINE RETFUN(STATE,D,NACT,NST,ND,I,J,K,RTN)
      DIMENSION STATE(NST),D(ND)
      DIMENSION RFUN(4,3),XFUN(4)
      DATA XFUN/0.00,1.00,2.00,3.00/
      DATA RFUN/0.00,0.67,2.67,6.00,
     +          0.00,2.00,4.00,6.00,
     +          0.00,3.46,4.90,6.00/
      NPTS=4
C     BEGIN BINARY HALVING INTERPOLATION
      MIN=1
      MAX=NPTS
      X=D(K)
   10 CONTINUE
      NEXT=(MIN+MAX)/2
      IF(X.GT.XFUN(NEXT)) GOTO 20
      MAX=NEXT
      GOTO 30
   20 CONTINUE
      MIN=NEXT
   30 CONTINUE
      IF((MAX-MIN).GT.1) GOTO 10
      XFAC=(X-XFUN(MIN))/(XFUN(MIN+1)-XFUN(MIN))
      RFUN1=RFUN(MIN,I)
      RFUN2=RFUN(MIN+1,I)
      RTN=RFUN1 + XFAC*(RFUN2-RFUN1)
      RETURN
      END
```

Figure 6.8 Coding for the resource allocation problem and typical results

```
      SUBROUTINE TRANSF(STATE,D,NACT,NST,ND,I,J,K,JDS)
      DIMENSION STATE(NST),D(ND)
C DOWNSTREAM STATE IS SIMPLY INITIAL STATE LESS AMOUNT
C APPLIED TO THIS ACTIVITY.
      STDS=STATE(J) - D(K)
      CALL JSTATE(STATE,NST,STDS,JDS)
      RETURN
      END
      SUBROUTINE FEASBL(STATE,D,NACT,NST,ND,I,J,K,PEN)
      DIMENSION STATE(NST),D(ND)
C FEASIBLE IF ENOUGH CAPITAL AVAILABLE
      IF(D(K).LE.STATE(J)) PEN=0.0
      RETURN
      END

OPTIMUM COST OR RETURN=    .74600E+01
   XOPT( 1) =   0.
   XOPT( 2) =    .20000E+01
   XOPT( 3) =    .10000E+01

   AT STAGE   1
        STATE          DOPT          FOPT
      0.             0.            0.
       .50000E+00     .50000E+00    .33500E+00
       .10000E+01     .10000E+01    .67000E+00
       .15000E+01     .15000E+01    .16700E+01
       .20000E+01     .20000E+01    .26700E+01
       .25000E+01     .25000E+01    .43350E+01
       .30000E+01     .30000E+01    .60000E+01

   AT STAGE   2
        STATE          DOPT          FOPT
      0.             0.            0.
       .50000E+00     .50000E+00    .10000E+01
       .10000E+01     .10000E+01    .20000E+01
       .15000E+01     .15000E+01    .30000E+01
       .20000E+01     .20000E+01    .40000E+01
       .25000E+01     .25000E+01    .50000E+01
       .30000E+01     .30000E+01    .60000E+01

   AT STAGE   3
        STATE          DOPT          FOPT
      0.             0.            0.
       .50000E+00     .50000E+00    .17300E+01
       .10000E+01     .10000E+01    .34600E+01
       .15000E+01     .10000E+01    .44600E+01
       .20000E+01     .10000E+01    .54600E+01
       .25000E+01     .10000E+01    .64600E+01
       .30000E+01     .10000E+01    .74600E+01
```

Figure 6.8 — continued

```
C     MAIN PROGRAM TO SOLVE THE B.O.D. PROBLEM BY
C     DYNAMIC PROGRAMMING.  REFER TO LISTING OF "DYNAM" FOR
C     DEFINITION OF PARAMETERS.
      DIMENSION D(4),DOPT(3,5),FOPT(3,5)
      DIMENSION STATE(5),DSOL(3)
      EXTERNAL RETFUN,TRANSF,FEASBL
      NACT=3
      NSTATE=5
      ND=4
C     TO MINIMIZE COST...
      SIGN=-1.0
C     D() IS LEVEL OF TREATMENT
C     STATE() IS RESULTANT LEVEL OF B.O.D.
      DATA D     /0.0,1.0,2.0,3.0/
      DATA STATE/0.0,0.4,0.8,1.2,1.6/
      CALL DYNAM(STATE,D,NACT,NSTATE,ND,SIGN,DOPT,FOPT,
     +           FEASBL,TRANSF,RETFUN)
      NPRINT=1
      CALL DYNSOL(STATE,D,NACT,NSTATE,ND,SIGN,DOPT,FOPT,
     +            TRANSF,NPRINT,DSOL,ANS)
      STOP
      END
      SUBROUTINE RETFUN(STATE,D,NACT,NST,ND,I,J,K,RTN)
      DIMENSION STATE(NST),D(ND)
      DIMENSION COST(4,3)
C     RETURN IS COST INCURRED FOR ANY CHOSEN LEVEL OF TREATMENT
      DATA COST/  0.0,70.0,120.0,170.0,
     +            0.0,60.0,100.0,130.0,
     +            0.0,90.0,140.0,220.0/
      RTN=COST(K,I)
      RETURN
      END
      SUBROUTINE TRANSF(STATE,D,NACT,NST,ND,I,J,K,JDS)
      DIMENSION STATE(NST),D(ND)
      DIMENSION BOD(4,3)
C     TRANSFORMATION FUNCTION INVOLVES ADDITION OF CURRENT LEVEL
C     OF B.O.D. LOADING TO UPSTREAM STATE OF RECEIVING WATER
      DATA BOD/1.2, 0.8, 0.4, 0.0,
     +         0.8, 0.4, 0.0, 0.0,
     +         1.6, 1.2, 0.8, 0.0/
      STDS=STATE(J) + BOD(K,I)
      IF(STDS.LT.1.6001) GOTO 10
      JDS=0
      RETURN
   10 CONTINUE
      CALL JSTATE(STATE,NST,STDS,JDS)
      RETURN
      END
```

Figure 6.9 Coding for the waste-water treatment problem and typical results

```
      SUBROUTINE FEASBL(STATE,D,NACT,NST,ND,I,J,K,PEN)
      DIMENSION STATE(NST),D(ND)
      CALL TRANSF(STATE,D,NACT,NST,ND,I,J,K,JDS)
C  JDS.LT.0 IS IMPOSSIBLE...JDS.GT.5 IS NOT ALLOWABLE.
      IF(JDS.GT.0.AND.JDS.LE.5) PEN=0.0
      RETURN
      END

OPTIMUM COST OR RETURN=    .27000E+03
     XOPT( 1) =    .30000E+01
     XOPT( 2) =    .20000E+01
     XOPT( 3) =   0.

AT STAGE   1
          STATE              DOPT              FOPT
         0.                 0.                0.
          .40000E+00        0.                0.
          .80000E+00         .10000E+01        .70000E+02
          .12000E+01         .20000E+01        .12000E+03
          .16000E+01         .30000E+01        .17000E+03

AT STAGE   2
          STATE              DOPT              FOPT
         0.                  .10000E+01        .60000E+02
          .40000E+00         .20000E+01        .10000E+03
          .80000E+00        0.                 .17000E+03
          .12000E+01         .20000E+01        .22000E+03
          .16000E+01         .20000E+01        .27000E+03

AT STAGE   3
          STATE              DOPT              FOPT
         0.                 0.                 .27000E+03
          .40000E+00         .30000E+01        .32000E+03
          .80000E+00         .30000E+01        .39000E+03
          .12000E+01         .30000E+01        .44000E+03
          .16000E+01         .30000E+01        .49000E+03
```

Figure 6.9 — continued

6.7.5 The Pipeline Network Problem

Although this is not an allocation problem, a solution is possible by means of routines DYNAM and DYNSOL by appropriate choice of the decision and state variables. The coding and solution are shown in Figure 6.10. The directions are easily defined in terms of degrees azimuth; the states are simply rows 1 to 4 from furthest north to extreme south. Thus a decision to move at 45° will reduce the state level by 1, etc. Routine TRANSF is therefore simply expressed, with the constraint that the downstream state level must be in the range $1 \leqslant JDS \leqslant 4$. Routine FEASBL is slightly more complicated because 4 decisions are possible at stage 4 whereas d_k must take the values 45°, 90°, or 135° at the downstream stages. The return function (or cost) is defined as a series of cost matrices for each stage, the tabulated costs including the pump-station cost at the input state level and the pipeline cost resulting from the decision. For the special case of stage 1, the cost of the downstream pump-station is also included.

```
C     MAIN PROGRAM TO SOLVE THE PIPE NETWORK EXAMPLE BY
C     DYNAMIC PROGRAMMING.   REFER TO LISTING OF "DYNAM"
C     FOR DEFINITION OF PARAMETERS.
      DIMENSION D(4),DOPT(4,4),FOPT(4,4)
      DIMENSION STATE(4),DSOL(4)
      EXTERNAL RETFUN,TRANSF,FEASBL
      NACT=4
      NSTATE=4
      ND=4
C     TO MINIMIZE TOTAL COST...
      SIGN=-1.0
C     DECISION VARIABLE IS COMPASS ANGLE IN MULTIPLES OF
C     45 DEGREES
C     STATE() IS "LEVEL" ON THE NETWORK DIAGRAM
      DATA D     /45.,90.,135.,180./
      DATA STATE/1.0,2.0,3.0,4.0/
      CALL DYNAM(STATE,D,NACT,NSTATE,ND,SIGN,DOPT,FOPT,
     +            FEASBL,TRANSF,RETFUN)
      NPRINT=1
      CALL DYNSOL(STATE,D,NACT,NSTATE,ND,SIGN,DOPT,FOPT,
     +            TRANSF,NPRINT,DSOL,ANS)
      STOP
      END
      SUBROUTINE RETFUN(STATE,D,NACT,NST,ND,I,J,K,RTN)
      DIMENSION STATE(NST),D(ND)
      DIMENSION COST1(4,4),COST2(4,4),COST3(4,4),COST4(4)
      DATA COST1/ 0.,27.,25., 0.,
     +           23.,21.,25., 0.,
     +           22.,23.,25., 0.,
     +           30.,32., 0., 0./
      DATA COST2/ 0.,14.,15., 0.,
     +           22.,22.,20., 0.,
     +           25.,23.,22., 0.,
     +           20.,18., 0., 0./
      DATA COST3/ 0.,18.,17., 0.,
     +           21.,18.,17., 0.,
     +           19.,18.,17., 0.,
     +           21.,18., 0., 0./
      DATA COST4/12., 8., 7., 8./
      GOTO (10,20,30,40),I
   10 RTN=COST1(K,J)
      RETURN
   20 RTN=COST2(K,J)
      RETURN
   30 RTN=COST3(K,J)
      RETURN
   40 RTN=COST4(K)
      RETURN
      END
```

Figure 6.10 Coding for the pipeline network problem and typical results

```
      SUBROUTINE TRANSF(STATE,D,NACT,NST,ND,I,J,K,JDS)
      DIMENSION STATE(NST),D(ND)
      JDS=J + (K-2)
      IF(JDS.LE.0) JDS=0
      IF(JDS.GT.4) JDS=0
      RETURN
      END
      SUBROUTINE FEASBL(STATE,D,NACT,NST,ND,I,J,K,PEN)
      DIMENSION STATE(NST),D(ND)
      IF(I.LT.4) GOTO 10
      IF(J.EQ.2) PEN=0.0
      RETURN
   10 CONTINUE
      CALL TRANSF(STATE,D,NACT,NST,ND,I,J,K,JDS)
      IF(K.LE.3.AND.JDS.GT.0) PEN=0.0
      RETURN
      END
```

```
OPTIMUM COST OR RETURN=    .65000E+02
   XOPT( 1) =    .90000E+02
   XOPT( 2) =    .13500E+03
   XOPT( 3) =    .45000E+02
   XOPT( 4) =    .90000E+02

AT STAGE    1
          STATE            DOPT           FOPT
        .10000E+01       .13500E+03     .25000E+02
        .20000E+01       .90000E+02     .21000E+02
        .30000E+01       .45000E+02     .22000E+02
        .40000E+01       .45000E+02     .30000E+02

AT STAGE    2
          STATE            DOPT           FOPT
        .10000E+01       .13500E+03     .36000E+02
        .20000E+01       .13500E+03     .42000E+02
        .30000E+01       .90000E+02     .45000E+02
        .40000E+01       .45000E+02     .42000E+02

AT STAGE    3
          STATE            DOPT           FOPT
        .10000E+01       .90000E+02     .54000E+02
        .20000E+01       .45000E+02     .57000E+02
        .30000E+01       .13500E+03     .59000E+02
        .40000E+01       .90000E+02     .60000E+02

AT STAGE    4
          STATE            DOPT           FOPT
        .10000E+01       0.             .10000E+11
        .20000E+01       .90000E+02     .65000E+02
        .30000E+01       0.             .10000E+11
        .40000E+01       0.             .10000E+11
```

Figure 6.10 — continued

6.8 CONCLUDING REMARKS

6.8.1 Dynamic programming — A computational method

As Daellenbach and George[7] point out dynamic programming, unlike linear programming, is not a mathematical model which may be linked with a special algorithm and then programmed once and for all to solve every problem that satisfies the assumption of the model. Dynamic programming is a computational method which allows a complex problem to be broken up into a sequence of easier sub-problems by means of a recursive relation which may be evaluated by stages. For example, dynamic programming solves an n-variable problem sequentially in n stages where each stage involves optimization over one variable only.

6.8.2 Computational efficiency

The computational work involved in solving certain problems by dynamic programming is usually much less than that involved in solving similar problems by complete enumeration. The main reason for this is that all possible alternatives need not be examined. Computations increase roughly exponentially with the number of variables, but only linearly with the number of sub-problems or stages.

Consider a problem with n variables and assume that each variable can take 10 different values. The total number of combinations in a complete enumeration is 10^n. If there are only five variables (i.e. $n = 5$), the total number of alternatives is 10^5. If there are, however, twice as many activities ($n = 10$), then 10^{10} alternatives exist, which is 100 000 times more than in the case of five variables.

6.8.3 Problems with more than one state variable

All of the problems in this chapter have involved a single state variable.† However, in some situations more than one state variable is required to provide a complete description of the state of the process. Although there are no conceptual difficulties in dealing with more than one variable there are severe computational limitations on the viable number of state variables and it is hardly ever feasible to use more than three.

6.8.4 Second-best policies

All of the problems solved in this chapter have been directed towards finding optimal policies. However, in the real world of civil engineering there may be good reasons for not implementing an optimal policy. For example, the model adopted by the Systems Engineer may ignore certain characteristics which are difficult to quantify yet important and which may imply that the optimal policy is

† Do not confuse a state variable with a variable discussed in Section 6.8.2.

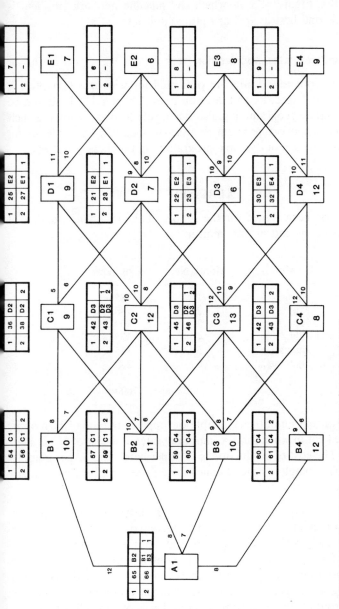

Figure 6.11 Best and second best policies for pipeline network problem

undesirable. Thus it may be useful to examine second-best policies which are not in any conflict with aspects ignored in the model.

Such an analysis is easily performed: at each stage evaluate the optimal and the next-best policy. See Figure 6.11 in which the pipeline network problem is reworked to find second-best as well as optimal policies.

6.8.5 Networks, critical paths, and decision trees

Later in this text various algorithms are presented for networks (Chapter 7), critical path analysis (Chapter 8), and decision trees (Chapter 11). It should be mentioned that dynamic programming also has a role to play in solving such problems. The pipeline network problem discussed in Section 6.2 provides useful insight into the possible use of dynamic programming for these problems.

6.9 EXERCISES

6.9.1 A contract has been signed for the supply of the following number of components at the end of each month:

Month	Month No.	No. of items
April	1	85
May	2	180
June	3	300
July	4	375
August	5	375
September	6	285
	Total	1600

Production during a month is available for supply at the end of the month, or it may be kept in stock for next month or later at a cost of $1 per item per month. The cost of production is $900 per batch and $2 per item. In what month is a batch to be made, and of what size, if the total costs are to be minimized?

6.9.2 A conduit has to be constructed to convey a given quantity of water from source A to demand area B. Whereas the starting point of the conduit can be located usually with a high degree of accuracy, the location of the other extremity of the conduit may often present a number of alternative choices. In addition to this, the terrain between source A and demand area B (its topography, geology, land use, etc.) may present a variety of alternative routes, each involving different levels of expenditure. This is shown graphically in Figure 6.12. In this figure, the water conduit (open channel or pipeline) must link A with either of the three terminal points B_1, B_2, or B_3. The possible points through which the conduit may pass are

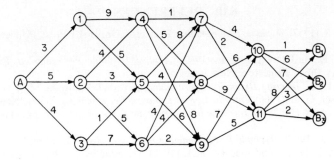

Figure 6.12 Water distribution network

numbered from (1) to (11), and the digit appearing on each link joining these various points represents the expenditure connected with the construction of this link (in monetary units). Find the path linking A with B_i ($i = 1, 2, 3$) requiring the minimum expenditure.

6.9.3 The pipe distribution system shown in Figure 6.13 is used to supply water to an irrigation region composed of 3 separate sub-regions. A total of 3

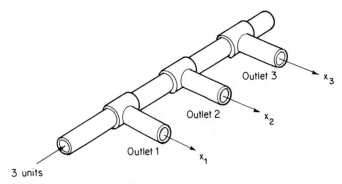

Figure 6.13 Pipe distribution system

units of water is available to the region. The table shows the benefits to each sub-region as a function of the water released to the sub-region:

Amount of water released at outlet x_i (units)	Benefits		
	Outlet 1 ($)	Outlet 2 ($)	Outlet 3 ($)
1	3	1	4
2	6	3	5
3	8	9	6

Allocate the water, using Dynamic Programming, so that benefits from the 3 units are maximized.

6.10 REFERENCES

1. Bellman, R. (1957). *Dynamic Programming*, Princeton University Press, Princeton, New Jersey.
2. Bellman, R., and Dreyfus, S. E. (1962). *Applied Dynamic Programming*, Princeton University Press, Princeton, New Jersey.
3. Aris, R. (1964). *Discrete Dynamic Programming*, Blaisdell, Waltham, Mass.
4. Nemhauser, G. L. (1966). *Introduction to Dynamic Programming*, Wiley, New York.
5. Jacobs, C. L. R. (1967). *An Introduction to Dynamic Programming*, Chapman and Hall, London.
6. White, D. J. (1967). *Dynamic Programming*, Oliver and Boyd, Edinburgh.
7. Daellenbach, H. G., and George, J. A. (1978). *Introduction to Operations Research Techniques*, Allyn and Bacon, Boston, Mass.

Chapter 7
Network Analysis

7.1 INTRODUCTION

The concept of a network is familiar in situations involving the delivery of any public service to a population distributed in heterogeneous fashion within a nation, region, or urban community. Networks of railroads or highways allow the movement of people and materials; networks of streets facilitate the development of cities with high-density population. In these same cities the delivery of electricity, gas, and water involves for each a network of conductors or conduits. Collection of liquid or solid waste to centralized processing centres involves a sewerage network or a network of vehicle routes. Communications for the purposes of telephony, radio, and television require networks of varying degrees of complexity and visibility.

In addition to these physical manifestations, networks may also be used to represent the relationship in time of a number of distinct but interdependent activities which occur in series and/or in parallel. These abstract networks provide an effective method of planning the activities and operations making up a complex project and show how delays or early completion at any particular stage will affect the remainder of the project. In short, the network is a classic illustration of a system in that it describes the interaction of many components to achieve a result which would be unobtainable by the sum of the individual contributions of the several components.

In the present chapter consideration is given to the modelling and analysis of networks which have a recognizable physical significance, such as traffic or pipe networks. In Chapter 8 some of the ideas developed here are applied to the special case of organizational networks for the purpose of construction management or project planning.

In the analysis of networks reference will be made to some of the concepts developed in earlier chapters. The calculation of the shortest path through a network may be done quite efficiently by methods closely related to dynamic programming (Section 6.3). Linear programming applications described in Chapter 4 (Section 4.2) are useful in solving transportation problems and the concept of the dual problem is of value in assessing network capacity.

7.2 ELEMENTARY GRAPH THEORY

A network — whether it is a physical system for water distribution or a means of describing the sequence of activities in a complex project — is usually represented by a linear graph. This does not mean a plot of function against argument but takes the form of a series of *nodes* (or vertices) connected by a set of *links* (or branches). Figure 7.1(a) shows a typical network in which every node is connected to at least one other node by a link; such a graph is termed *connected*. The graph is also

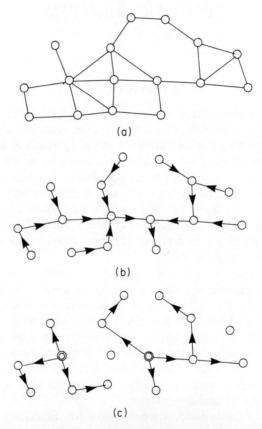

Figure 7.1 Cyclic networks and trees

planar because it can be represented in a plane with non-crossing links. If additional links were to be added so that there existed at least one link between every pair of nodes, the graph could then be said to be *complete* but clearly it would no longer be planar. None of the links in Figure 7.1(a) have a direction indicated by an arrowhead, implying that movement may occur in either direction; the graph is then said to be *undirected*.

The graph in Figure 7.1(b) has been derived from the previous one by deleting some of the links and by assigning a specific direction to each of the remaining links as indicated by the arrowhead. Notice that links have been removed so that there are no *loops* (or circuits). This graph is called a *tree* network because of its branching characteristics. For the selected directions of the links the system might represent a river with a number of tributaries or a sewerage system for the collection of waste water or storm water. A particular feature of Figure 7.1(b) is that every node is still connected; such a graph is called a *spanning tree*.

By contrast, the directed graph of Figure 7.1(c) shows two *sub-graphs* of the original network each of which is also a tree. The direction of some of the links has been changed and the arrangement might now represent a distribution system in which two centres (indicated by the double circles) supply the needs of the region. The supply centres are sometimes called *sources* or *origins* and the nodes to which the movement is directed are termed *sinks* or *destinations*. Notice that in the distribution system of Figure 7.1(c) every node can have no more than one link entering the node, whereas in the collection system of Figure 7.1(b) every node can have no more than one link leaving the node.

It is clear that a tree represents the minimal connection of the set (or sub-set) of nodes which it connects. The removal of any link from a tree would result in a further disconnection of the graph, while the addition of a link to the same set (or sub-set) of nodes will result in the creation of a loop. In general, therefore, the minimum-cost solution for a collection or distribution system will take the form of one or more trees emanating from the selected service centres.

Because many spanning trees may be identified for a particular set of nodes, it follows that although the minimum cost solution is a tree, a tree is not necessarily the minimum cost solution. Such a solution may not always be desirable in practice. For instance, in a water distribution network it is good practice to incorporate some additional links to form loops and thus increase the reliability of the supply to those nodes on the loop. It should be realized, however, that this increased reliability is obtained at the extra cost of adding the 'redundant' link(s).

7.3 NETWORK VARIABLES AND PROBLEM TYPES

7.3.1 Variables

Common to every problem in network analysis are variables of two distinct types:

(i) *Flow variables*. These describe a quantity which passes along or through the link in the appropriate direction.
(ii) *Potential variables*. These are necessary to define the change in state across the nodes at the start and finish of the link.

Although the potential variable may not be obvious in certain types of network problem, both types of variable must occur. Moreover, there exists in general a functional relationship between the flow and potential difference which must be known if the network problem involves the distribution of flow variables in the network.

Table 7.1 Flow and potential variables in different networks

Network type	Flow variable	Potential variable	Link equation
Electrical circuit	Current I	Voltage drop E	$E = IR$
Hydraulic network	Flow Q	Pressure drop h_f or energy drop ΔE	$h_f = KQ^n$
Solid waste collection	Tonnage T	Energy or cost C	$\Delta C = \phi(T)$
Traffic network	Flow f	Disutility cost c	$c = \phi(f)$

For various types of network, Table 7.1 shows the flow and potential variable together with a typical form of link equation. The application of Ohm's Law to a simple DC electrical circuit is well known. Similarly, the analysis of pipe networks by the Hardy–Cross or Newton–Raphson method is familiar to civil engineers. The movement of water through a pipe results in a drop in head or energy across the link, but the movement of solid waste in discrete containers does not have an obvious parallel. However, transportation of solid waste requires the expenditure of time, energy, and resources, all of which can be expressed as a cost, which typically is found to be a concave, non-linear function of the tonnage to be transported. Similarly, in a traffic network, the potential variable is usually the disutility as estimated by the traveller in terms of time, fuel, and wear and tear on both vehicle and driver. The example of Section 2.8 serves to illustrate the notion of route attractiveness as an inverse of travel time, the latter being a non-linear function of flowrate.

7.3.2 Kirchhoff's Laws

Two laws attributed to Kirchhoff are fundamental to the analysis of networks whether it concerns the flow of electric currents in a network, the distribution of water in a system of pipes, or the flow of traffic in a transportation network.

Kirchhoff's Node Law is a statement of the principle of conservation and states that the algebraic sum of all the flows entering or leaving a node must be zero. The law assumes steady flow and implies that the storage capacity of links and node is invariable with respect to time (e.g. zero).

Kirchhoff's Loop Law relates to the potential variable of the network and states that the algebraic sum of the potential differences around a closed loop must be zero. The loop law applies only to networks in which alternate routes exist between one or more origin–destination pairs and for which the link-law relating potential difference and flow variables must be satisfied.

Both laws involve the algebraic sum of quantities which in turn implies that a consistent sign convention must be adopted when defining link flows. The procedure is illustrated by the following simple example.

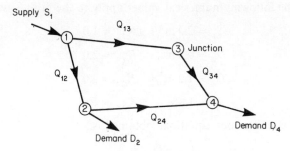

Figure 7.2 A simple water supply network

7.3.3 Example 7.1 — Water Supply Network

Figure 7.2 shows an elementary water distribution network in which a single source 1 supplies two sinks 2 and 4. Node 3 serves merely as a junction. It is assumed that the system is balanced, i.e.

$$S_1 = D_2 + D_4.$$

Flows into the system are assumed positive while flows abstracted from the system are negative. Similarly link flows into and out of a node are assumed to be positive and negative respectively. Kirchhoff's Node Law applied to each of the four nodes results in the following set of equations:

At node 1 $\quad S_1 - Q_{12} - Q_{13} = 0$

At node 2 $\quad Q_{12} - Q_{24} - D_2 = 0$

At node 3 $\quad Q_{13} - Q_{34} = 0$

At node 4 $\quad Q_{24} + Q_{34} - D_4 = 0.$

A brief inspection reveals that because the system is balanced the four equations are not independent and a further equation is required to uniquely define the link flows Q_{ij}.

Kirchhoff's Loop Law is applied assuming any arbitrary starting node and either clockwise or counter-clockwise direction around the loop. Thus, if the head loss from node i to node j is defined as ΔE_{ij} the loop equation is

$$\Delta E_{12} + \Delta E_{24} - \Delta E_{34} - \Delta E_{13} = 0.$$

Note that the negative sign is used when movement around the loop is against the direction of the flow variable. A link equation of the following general form may be defined in which n is an exponent, the value of which depends on the application. Thus

$$\Delta E_{ij} = K_{ij}(Q_{ij})^n. \tag{7.1}$$

The required equation is therefore

$$K_{12}(Q_{12})^n + K_{24}(Q_{24})^n - K_{34}(Q_{34})^n - K_{13}(Q_{13})^n = 0. \tag{7.2}$$

Assume that the following numerical values apply to the given system:

$$S_1 = 2.0 \text{ m}^3/\text{s} \quad K_{12} = 1.0$$
$$D_2 = 0.8 \text{ m}^3/\text{s} \quad K_{13} = 2.0$$
$$D_4 = 1.2 \text{ m}^3/\text{s} \quad K_{24} = 3.0$$
$$n = 1.85 \quad K_{34} = 1.5.$$

The system of simultaneous equations obtained is:

$$Q_{12} + Q_{13} = 2$$
$$Q_{13} = Q_{34}$$
$$Q_{12} - Q_{24} = 0.8 \quad (7.3)$$
$$Q_{24} + Q_{34} = 1.2$$
$$1(Q_{12})^{1.85} + 3(Q_{24})^{1.85} - 1.5(Q_{34})^{1.85} - 2(Q_{13})^{1.85} = 0.$$

In general, the solution to the set of simultaneous, non-linear equations may be obtained by an iterative technique such as the Newton–Raphson method (Section 2.8) or relaxation methods such as the Hardy–Cross method. In the present case substitution allows a function of a single variable (say Q_{13}) to be formed. Thus

$$(2 - Q_{13})^{1.85} + 3(1.2 - Q_{13})^{1.85} - 1.5(Q_{13})^{1.85} - 2(Q_{13})^{1.85} = 0. \quad (7.4)$$

This may be solved most easily by programming the function in a pocket calculator and finding Q_{13} by trial. This yields the following results:

$$Q_{13} = 0.76387 \text{ m}^3/\text{s}$$

therefore

$$Q_{12} = 1.23613 \text{ m}^3/\text{s}$$

and

$$Q_{24} = 0.43613 \text{ m}^3/\text{s}.$$

The potential difference variables can then be calculated by substituting the known values of K and the computed Q values in (7.1).

$$\Delta E_{1\text{-}3\text{-}4} = 1.2151 + 0.9113 = 2.1264 \text{ m}$$
$$\Delta E_{1\text{-}2\text{-}4} = 1.4802 + 0.6463 = 2.1265 \text{ m}.$$

Note that the absolute value of the potential variable (i.e. energy level) requires the specification of available or required energy level at one of the nodes.

7.3.4 Types of network problem

Of the many types of network problem only a few can be described within the scope of this text (see Smith and Tufgar[7], Smith[8]). It is useful, however, to classify network problems with respect to:

(a) the number of origins and destinations,
(b) the use or non-use of the link equation, and
(c) whether analysis or optimization is the goal.

Figure 7.3(a) illustrates a simple directed network with two sources (1, 2) and three sinks (3, 4, 5). For certain purposes it may be necessary to convert the

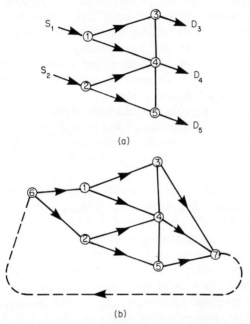

Figure 7.3 Forming a circularized network

problem into one with a single origin and destination. This is easily done by introducing a dummy origin (or 'super-source') and dummy destination (or 'super-sink') at nodes 6 and 7 respectively as shown in Figure 7.3(b). The links connecting the dummy nodes to the real network are pseudo-links in which the normal physical link equation does not apply. For example, if the potentials (e.g. energy levels) at nodes 1 and 2 are different then the link equations for links 6–1 and 6–2 will take the form

$$\begin{aligned}\Delta E_{6-1} &= E_6 - E_1 \\ \Delta E_{6-2} &= E_6 - E_2 \\ Q_{61} &= S_1 \\ Q_{62} &= S_2.\end{aligned} \quad (7.5)$$

In the same way, specified potential values at the sinks can be related to each other by means of the dummy destination 7 and to the source potential by means of a further dummy link (shown dotted) from 7 to 6. The circularized form of Figure 7.3(b) is useful in helping to define the total number of loops or cycles formed by the network and may be used to estimate the total capacity of a network.

The role of the link equation is often important in physical networks in which any change in flow variable is accompanied by a corresponding change in the potential difference across the link (e.g. electrical or hydraulic networks). In networks with discontinuous flows (such as traffic networks) the sensitivity of the potential variable may be significant only when the flow exceeds some threshold value. Travel time, for example, may be essentially independent of flowrate in sparsely travelled routes. In many transportation problems, therefore, the cost coefficients may be assumed to be constant and only Kirchhoff's Node Law need be used to ensure continuity at each node.

In other problems the aim is to identify the route of minimum cost between two nodes. It is assumed that this information will influence the decision of a single road network user, in an otherwise steady-state situation. The decision is unlikely to affect flow variables significantly and the disutility cost may reasonably be assumed to be constant.

Should the minimum cost route prove to be significantly cheaper than other alternatives it is possible that many users will be attracted by this, with consequent changes to the flow pattern and thus to the disutility costs. In this situation, the flows will stabilize when Kirchhoff's Loop Law (expressed in terms of the appropriate link equation) has been satisfied for all loops in the network.

As the network approaches saturation level it is likely that the transportation authority will enforce some form of link capacity constraints (e.g. by entrance ramp controls or turn prohibition). The user may then be forced to choose a route on the basis of feasibility rather than optimality and the loop laws will no longer be relevant. Thus, network capacity problems require a special treatment, to obtain flow values which simultaneously satisfy the node (continuity) law and the maximum flow constraints.

An important distinction must be made between problems of analysis of networks and those in which some form of optimization is attempted. In the illustrative example of Section 7.3.3 the system was defined in terms of the conveyance (or transportation) capacity by specifying the topology of the network and the parameters of the link equations K_{ij} and exponent n. The aim of the analysis is to determine the distribution of link flows for specified boundary conditions of inflow and outflow. For steady-state conditions which satisfy both node and loop laws, there is a single unique solution to this problem.

Consider now the problem of identifying the minimum-cost solution for the supply of water to nodes 2 and 4, given appropriate cost functions for the four links of the network. In general, the cost function for a link (i, j) will depend on the discharge Q_{ij}, the difference in elevation of the minimum pressure at nodes i and j, and the length L_{ij} and may involve sub-optimization of the pipe diameter/booster pump configuration. For simplicity the following cost functions will be assumed:

$$C_{12} = 100(Q_{12})^m$$
$$C_{13} = 200(Q_{13})^m$$
$$C_{24} = 300(Q_{24})^m$$
$$C_{34} = 150(Q_{34})^m.$$

The exponent m will normally be less than unity, thus exhibiting the economies of scale which one expects in the design of engineering components. For generality, however, and to demonstrate an important feature of non-linear objective functions, three cases will be considered in which $m = 1$, $m < 1$, and $m > 1$.

When $m = 1$ the problem is one of linear programming and may be stated as follows:

Minimize $z = 100Q_{12} + 200Q_{13} + 300Q_{24} + 150Q_{34}$

subject to

$$Q_{12} + Q_{13} = 2.0$$
$$Q_{12} - Q_{24} = 0.8 \qquad (7.6)$$
$$Q_{24} + Q_{34} = 1.2.$$

The solution must lie on a vertex of the feasible space and therefore one of the four flow variables must be zero. Since this problem is a particularly simple one it is easy to try different values for one particular variable (say Q_{13}) and to evaluate the other flows from the constraint equations and thus determine the total cost.

Table 7.2 shows for different values of exponent m ($m = 1.0$, 0.8, and 1.2) the total cost as a function of various trial values of the flow Q_{13}.

Table 7.2 Sensitivity of cost to exponent m and independent variable Q_{13}

	Cost		
Q_{13}	$m = 0.8$	$m = 1.0$	$m = 1.2$
0.0	521.22	560.00	603.11
0.2	556.62	550.00	553.19
0.4	564.76	540.00	521.85
0.6	562.84	530.00	501.87
0.8	552.62	520.00	492.14
1.0	532.78	510.00	493.49
1.2	488.61	500.00	512.11

For $m = 1.0$ the minimum cost is obtained when $Q_{13} = 1.2 \text{ m}^3/\text{s}$, i.e.

$$Q_{12} = 0.8 \text{ m}^3/\text{s} \quad Q_{13} = 1.2 \text{ m}^3/\text{s} \quad Q_{24} = 0.0 \quad Q_{34} = 1.2 \text{ m}^3/\text{s}$$

and the table confirms the linear nature of the relationship.

For the case of $m = 0.8$ the individual cost functions are concave and, moreover, the total cost is also concave with respect to Q_{13}. Thus, the solution must still lie at a vertex of the feasible space and is in fact identical with the linear case.

When $m = 1.2$ the cost functions and the total cost are now convex and it is apparent that the non-linear cost function has a minimum at a value of Q_{13} somewhere between the two extreme values of 0.0 and 1.2. In fact the minimum occurs with the solution

$$Q_{12} = 1.12 \text{ m}^3/\text{s} \quad Q_{13} = 0.88 \text{ m}^3/\text{s} \quad Q_{24} = 0.32 \text{ m}^3/\text{s} \quad Q_{34} = 0.88 \text{ m}^3/\text{s}.$$

Although a more elegant proof is both desirable and possible, the following rather significant conclusion can be drawn from this example.

When a network transportation problem involves the minimization of a separable, concave objective function, subject to the linear constraints of the node laws, the solution will always lie on a vertex of the feasible space. Since this means that there will be as many non-zero flow variables as there are nodes, the minimum cost solution must be a tree of the original graph.

Conversely, when the objective function of a network transportation problem is convex the minimum cost solution may be non-basic and may therefore involve loops in the final configuration. If the solution is then to be both optimal and feasible it is essential that the link equations relating potential difference to flow variable be satisfied in the design of the link components and subsequent evaluation of the link costs.

A further complication arises when dealing with non-linear objective functions. Although the concave minimization problem yields a solution at a vertex there is no guarantee that the solution is a global one. With the convex minimization problem, on the other hand, the solution may be non-basic but it can be proved that any solution found will be a global one.

7.4 MINIMUM-COST ROUTE

One of the most basic decisions in network analysis is the identification of the sequence of directed links between a single origin and destination which will minimize some objective function such as distance, time, or cost. The pipeline network problem discussed in Section 6.3 is somewhat similar with the difference that costs were incurred not only on the basis of route selected but for the 'state' position at which a pump station was to be built. Most 'cheapest route' problems ignore costs associated with the node (an exception is the inclusion of turn penalties) and, moreover, assume that the link costs are constant. In this section the problem will be formulated initially as a linear programming problem and then re-cast as the dual problem. This leads to a re-affirmation of the dynamic programming recursive principle of optimality and in conclusion a tree-building algorithm is described and illustrated with a simple example. The algorithm can be easily encoded and an appropriate subroutine TREE is included in Appendix A.

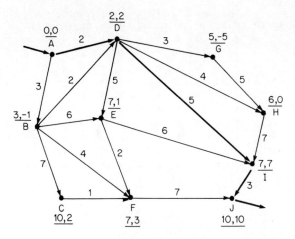

Figure 7.4 Shortest-path in a directed network

7.4.1 Linear Programming approach

The directed network in Figure 7.4 has the link cost c_{ij} marked as shown. The shortest (i.e. cheapest) route is to be found between origin A and destination J. Assume that unit flow enters at A and exits at J, and that this unit of flow cannot be sub-divided at branches. If the flow in a link (i, j) is x_{ij} which has values of 0 or 1 then Kirchhoff's node law can be applied at each of the ten nodes A, B, ..., J. The mathematical model is then

$$\text{Minimize } z = \sum_i \sum_j c_{ij} x_{ij} \tag{7.7}$$

subject to

$$\sum_j x_{ij} - \sum_i x_{ji} = 1 \quad \text{for } i = A$$

$$= 0 \quad \text{for } i = B, C, D, E, F, G, H, I$$

$$= -1 \quad \text{for } i = J$$

and

$$\text{all } x_{ij} \geq 0.$$

In matrix form the model statement is as follows:

$$\text{Minimize } z = \mathbf{c}^T . \mathbf{x} \tag{7.8}$$

where

$$\mathbf{c}^T = [3, 2, 7, 2, 6, 4, 1, 5, 3, 4, 5, 2, 6, 7, 5, 7, 3]$$

$$\mathbf{x} = [x_{AB}, x_{BC}, x_{AD}, x_{BD}, x_{BE}, x_{DE}, x_{BF}, x_{CF}, x_{EF},$$
$$x_{DG}, x_{DH}, x_{GH}, x_{DI}, x_{EI}, x_{HI}, x_{FJ}, x_{IJ}]^T$$

subject to

$$\mathbf{Ax} = \mathbf{b}$$

which is shown overleaf in expanded form.

$$\begin{array}{c}\text{link}\\ \begin{array}{c|ccccccccccccccccc|} & AB & BC & AD & BD & BE & DE & BF & CF & EF & DG & DH & GH & DI & EI & HI & FJ & IJ \\ \hline \text{node A} & 1 & 0 & 1 & 0 & 0 & 0 & 0 & 0 & 0 & 0 & 0 & 0 & 0 & 0 & 0 & 0 & 0 \\ B & -1 & 1 & 0 & 1 & 1 & 0 & 1 & 0 & 0 & 0 & 0 & 0 & 0 & 0 & 0 & 0 & 0 \\ C & 0 & -1 & 0 & 0 & 0 & 0 & 0 & 1 & 0 & 0 & 0 & 0 & 0 & 0 & 0 & 0 & 0 \\ D & 0 & 0 & -1 & -1 & 0 & 1 & 0 & 0 & 0 & 1 & 1 & 0 & 1 & 0 & 0 & 0 & 0 \\ E & 0 & 0 & 0 & 0 & -1 & -1 & 0 & 0 & 1 & 0 & 0 & 0 & 0 & 1 & 0 & 0 & 0 \\ F & 0 & 0 & 0 & 0 & 0 & 0 & -1 & -1 & -1 & 0 & 0 & 0 & 0 & 0 & 0 & 1 & 0 \\ G & 0 & 0 & 0 & 0 & 0 & 0 & 0 & 0 & 0 & -1 & 0 & 1 & 0 & 0 & 0 & 0 & 0 \\ H & 0 & 0 & 0 & 0 & 0 & 0 & 0 & 0 & 0 & 0 & -1 & -1 & 0 & 0 & 1 & 0 & 0 \\ I & 0 & 0 & 0 & 0 & 0 & 0 & 0 & 0 & 0 & 0 & 0 & 0 & -1 & -1 & -1 & 0 & 1 \\ J & 0 & 0 & 0 & 0 & 0 & 0 & 0 & 0 & 0 & 0 & 0 & 0 & 0 & 0 & 0 & -1 & -1 \\ \end{array}\end{array}$$

$$\begin{bmatrix} x_{AB} \\ x_{BC} \\ x_{AD} \\ x_{BD} \\ x_{BE} \\ x_{DE} \\ x_{BF} \\ x_{CF} \\ x_{EF} \\ x_{DG} \\ x_{DH} \\ x_{GH} \\ x_{DI} \\ x_{EI} \\ x_{HI} \\ x_{FJ} \\ x_{IJ} \end{bmatrix} = \begin{bmatrix} 1 \\ 0 \\ 0 \\ 0 \\ 0 \\ 0 \\ 0 \\ 0 \\ 0 \\ -1 \end{bmatrix}$$

The solution of such a linear programming problem could be lengthy as the number of variables and constraint equations may be large. In this particular case the solution is readily obtained by inspection, i.e.

$$x_{AD} = x_{DI} = x_{IJ} = 1$$

with all the other x terms being zero.

This problem can be formulated in terms of a dual problem and the reader should refer to (3.46) and (3.47) in Chapter 3 for a statement of the relationship between variables, coefficients, and stipulations. In this case the number of dual variables is equal to the number of equations for the nodes in the network. These variables are not restricted in sign as all constraints in the primal problem are equations and each of them represents the distance y_i from a common reference point to a node i. Therefore, for the given problem the dual may be defined as follows:

$$\text{Maximize } z' = y_A(1) + y_J(-1) \tag{7.9}$$

subject to

$$A^T \mathbf{y} \leqslant \mathbf{c}$$

i.e.

	node												
link	A	B	C	D	E	F	G	H	I	J			
AB	1	−1	0	0	0	0	0	0	0	0	y_A	≤	3
BC	0	1	−1	0	0	0	0	0	0	0	y_B		7
AD	1	0	0	−1	0	0	0	0	0	0	y_C		2
BD	0	1	0	−1	0	0	0	0	0	0	y_D		2
BE	0	1	0	0	−1	0	0	0	0	0	y_E		6
DE	0	0	0	1	−1	0	0	0	0	0	y_F		5
BF	0	1	0	0	0	−1	0	0	0	0	y_G		4
CF	0	0	1	0	0	−1	0	0	0	0	y_H		1
EF	0	0	0	0	1	−1	0	0	0	0	y_I		2
DG	0	0	0	1	0	0	−1	0	0	0	y_J		3
DH	0	0	0	1	0	0	0	−1	0	0			4
GH	0	0	0	0	0	0	1	−1	0	0			5
DI	0	0	0	1	0	0	0	0	−1	0			5
EI	0	0	0	0	1	0	0	0	−1	0			6
HI	0	0	0	0	0	0	0	1	−1	0			7
FJ	0	0	0	0	0	1	0	0	0	−1			7
IJ	0	0	0	0	0	0	0	0	1	−1			3

From the constraint equations it is clear that the dual variables have dimensions of cost or length travelled per unit flow. Also the dual problem may be changed to minimization by reversing the sign of the objective function and the constraints. Thus

$$\text{Minimize } z' = y_J - y_A = y_J \quad \text{for } y_A = 0 \quad (7.10)$$

subject to

$$A^T y \geqslant c$$

in which the origin (node A) is given an arbitrary distance of zero.

The solution, as shown in the following, is involved with a systematic search of the paths from source to sink assuming in each case that $y_A = 0$. As the problem concerns the determination of the minimum or shortest route, the links incident with this route and their length are required. If a link does not lie on the shortest route then it has a slack variable which will be defined as c.

Commencing at the source node A, a 'forward pass' is made through the network system as follows:

Node B: the distance from A to B is

$$y_B = y'_A + c_{AB} = 0 + 3 = 3$$

where y'_A is the minimum distance up to and including node A.

In this case the minimum distance from A to B is 3 as there is only one path available from the source to node B.

Node C: $y_C = y'_B + c_{BC} = 3 + 7 = 10$.

This is again the shortest route, being the only path available.

Node D: $y_D = y'_A + c_{AD} = 0 + 2 = 2$ (minimum $y_D = 2 = y'_D$).

$y_D = y'_B + c_{BD} = 3 + 2 = 5$.

Node E: $y_E = y'_B + c_{BE} = 3 + 6 = 9$

$y_E = y'_D + c_{DE} = 2 + 5 = 7$ (minimum $y_E = 7 = y'_E$).

Node F: $y_F = y'_B + c_{BF} = 3 + 4 = 7$

$y_F = y'_C + c_{CF} = 10 + 1 = 11$

$y_F = y'_E + c_{EF} = 7 + 2 = 9$ (minimum $y_F = 7 = y'_F$).

Node G: $y_G = y'_D + c_{DG} = 2 + 3 = 5$ (minimum $y_G = 5 = y'_G$).

Node H: $y_H = y'_D + c_{DH} = 2 + 4 = 6$

$y_H = y'_G + c_{GH} = 5 + 5 = 10$ (minimum $y_H = 6 = y'_H$).

Node I: $y_I = y'_D + c_{DI} = 2 + 5 = 7$

$y_I = y'_E + c_{EI} = 7 + 6 = 13$

$y_I = y'_H + c_{HI} = 6 + 7 = 13$ (minimum $y_I = 7 = y'_I$).

Node J: $y_J = y'_F + c_{FJ} = 7 + 7 = 14$

$y_J = y'_I + c_{IJ} = 7 + 3 = 10$ (minimum $y_J = 10 = y'_J$).

The minimum value for each node in the forward pass is placed in the left-hand box above that node as shown in Figure 7.4. The backward pass is then started by assuming that $y''_J = y'_J$, and the same procedure is repeated as before.

Node I: $y_I = y''_J - c_{IJ} = 10 - 3 = 7$ (maximum $y_I = 7 = y''_I$).

Node H: $y_H = y''_I - c_{HI} = 7 - 7 = 0$ (maximum $y_H = 0 = y''_H$).

Node G: $y_G = y''_H - c_{GH} = 0 - 5 = -5$ (maximum $y_G = -5 = y''_G$).

Node F: $y_F = y''_J - c_{FJ} = 10 - 7 = 3$ (maximum $y_F = 3 = y''_F$).

Node E: $y_E = y''_I - c_{EI} = 7 - 6 = 1$

$y_E = y''_F - c_{EF} = 3 - 2 = 1$ (maximum $y_E = 1 = y''_E$).

Node D: $y_D = y''_I - c_{DI} = 7 - 5 = 2$

$y_D = y''_H - c_{DH} = 0 - 4 = -4$

$y_D = y''_G - c_{DG} = -5 - 3 = -8$

$y_D = y''_E - c_{DE} = 1 - 5 = -4$ (maximum $y_D = 2 = y''_C$).

Node C: $y_C = y_F'' - c_{CF} = 3 - 1 = 2$ (maximum $y_C = 2 = y_C''$).
Node B: $y_B = y_F'' - c_{BF} = 3 - 4 = -1$
$y_B = y_C'' - c_{BC} = 2 - 7 = -5$ (maximum $y_B = -1 = y_B''$).
Node A: $y_A = y_D'' - c_{AD} = 2 - 2 = 0$
$y_A = y_B'' - c_{AB} = -1 - 3 = -4$ (maximum $y_A = 0 = y_A''$).

A branch lies on the shortest path if, on examining the forward and backward passes, it is seen to conform with the following conditions:

$$y_i = y_i' = y_i''$$
$$y_j = y_j' = y_j''$$
$$y_j' - y_i' = y_j'' - y_i'' = c_{ij}$$

where the subscripts i and j refer to the tail and arrow-head respectively of the branch.

In this particular problem it is seen that these conditions are met only along path A–D–I–J which is indicated by a heavier line in Figure 7.4.

7.4.2 A tree-building algorithm

In Section 7.4.1 it was shown that the solution of the dual problem involves a systematic search for the minimum cost route to any node j from any other node i such that the recursion equation is satisfied, i.e.

$$f^*(j) = \min\left[f^*(i) + c(i,j)\right] \tag{7.11}$$

where
$$f^*(n) = \text{minimum cost to node } n$$
and
$$c(i,j) = \text{cost of link } (i,j).$$

The similarity to the recursion equation of dynamic programming (Section 6.5.2) is apparent. Many 'cheapest-route' algorithms are based on successive applications of the recursion equation, starting from a specified origin or home-node. The end result will be a spanning tree of the original network which is constructed link by link. The process is best understood with reference to a specific example.

Figure 7.5(a) shows an undirected network of 11 nodes and 15 links, the costs marked against each link being appropriate for either direction. The origin is taken to be node 1 and is assigned an initial cost of zero.

Step 1 All the nodes directly connected to node 1 are examined to get the cost based on the cheapest route, and the corresponding upstream node

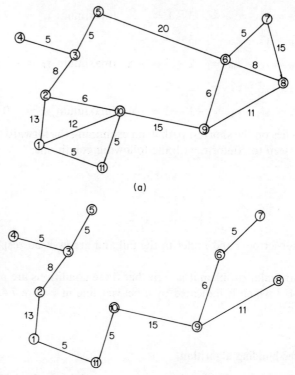

Figure 7.5 (a) Undirected network. (b) Spanning tree on node 1

i.e. for nodes j = 2   10   11

upstream nodes $u(j)$ = 1   1   1

minimum cost $f(j)$ = 13   12   5.

Step 2 Scan the list of $f(j)$ values and select the minimum $f(11) = 5$. This extends the tree from node 1 to 11.

Step 3 Delete node 11 from the table and add nodes which are directly connected to node 11

i.e. for nodes j = 2   10   10

upstream nodes $u(j)$ = 1   1   11

minimum cost $f(j)$ = 13   12   10.

Step 4 Repeat step 2 selecting node 10 as the next extension of the tree. Delete both references to node 10 and add nodes directly connected to 10

i.e. for nodes	j	= 2	2	9
upstream nodes	$u(j)$	= 1	10	10
minimum cost	$f(j)$	= 13	16	25.

The process continues until either all nodes in the network have been processed or until a particular origin–destination route has been added to the tree. The algorithm may be described in more general terms as follows:

Algorithm

1. Set $M = 10^{10}$.
2. Define origin k_0.
3. Set all $f^*(\cdot) = M$ to signify non-processed state.
4. Set $f^*(k_0) = 0$.
5. Identify all nodes j for which $c(i,j)$ is finite and for which $f^*(j) = M$.
6. For each j from step 5 tabulate the upstream node $u(j)$ and cost

$$f(j) = c(u(j), j) + f^*(u(j)).$$

7. Scan tables and determine node m such that $f(m) = f_{min}$.
8. If $f_{min} = M$, stop.
9. Set $f^*(m) = f_{min}$ and $u^*(m) = u(m)$.
10. Set $c(u(m), m) = c(m, u(m)) = 0$ to delete link from further consideration.
11. Set all $f(m) = M$ in table to delete node m from further consideration.
12. If m = specified destination node, stop.
13. Set $k_0 = m$.
14. If number of tabulated values is less than number of links go to step 5.
15. Go to step 7.

Continuing the analysis of the network in Figure 7.5 results in the following tables which illustrate the algorithm (Table 7.3). The upstream node corresponding to the minimum cost of any node may be used to trace the minimum cost route from the origin. For example, the cheapest route from node 1 to 6 is given by the chain

$$(6, 9), (9, 10), (10, 11), (11, 1).$$

A subroutine TREE is included in Appendix A which performs the algorithm described above. Apart from specifying the origin and destination (the latter is optional), the principal input is in the form of a square array which defines the costs of each link. The connectivity is also defined in a directed sense by setting coefficients equal to zero where no directed link exists. Thus if a directed link of cost 20 exists from node 5 to node 6, then

$$C(5, 6) = 20.0$$

but

$$C(6, 5) = 0.0.$$

Table 7.3   Results obtained from the tree-building algorithm

Table	Tentative nodal value													Opt.
j	2	10	11	10	2	9	3								3
$u(j)$	1	1	1	11	10	10	2								2
$f(j)$	M	M	M	M	M	25	21|							21
j	2	10	11	10	2	9	3	4	5						9
$u(j)$	1	1	1	11	10	10	2	3	3						10
$f(j)$	M	M	M	M	M	25	M	26	26						25
j	2	10	11	10	2	9	3	4	5	6	8				4
$u(j)$	1	1	1	11	10	10	2	3	3	9	9				3
$f(j)$	M	M	M	M	M	M	M	26	26	31	36				26
j	2	10	11	10	2	9	3	4	5	6	8				5
$u(j)$	1	1	1	11	10	10	2	3	3	9	9				3
$f(j)$	M	M	M	M	M	M	M	M	26	31	36				26
j	2	10	11	10	2	9	3	4	5	6	8	6			6
$u(j)$	1	1	1	11	10	10	2	3	3	9	9	5			9
$f(j)$	M	M	M	M	M	M	M	M	M	31	36	36			31
j	2	10	11	10	2	9	3	4	5	6	8	6	7	8	8
$u(j)$	1	1	1	11	10	10	2	3	3	9	9	5	6	6	9
$f(j)$	M	M	M	M	M	M	M	M	M	M	36	M	36	39	36
j	2	10	11	10	2	9	3	4	5	6	8	6	7	8 7| 7
$u(j)$	1	1	1	11	10	10	2	3	3	9	9	5	6	6 8| 6
$f(j)$	M	M	M	M	M	M	M	M	M	M	M	M	36	M 51	36

Clearly for networks of undirected links the array will be symmetrical about the leading diagonal.

7.4.3   A modified Travelling Salesman Problem

In the classic travelling-salesman problem a minimum-cost route is sought which passes through every node in a network once and only once, and which forms a closed loop or cycle by returning to the starting point. A much simpler problem results if the 'once and only once' constraint is replaced by the requirement that the route pass through each node *at least* once. The modified problem is of practical interest in planning solid-waste collection in which no significant penalty arises from re-visiting a node on a subsequent part of the route. The problem may be tackled by linear programming in the following way.

Figure 7.6 shows a connected network of 5 nodes. The 'cost' — in terms of distance or travel time — is on the links marked and is assumed to be independent of direction. (A problem with different costs in each direction could be posed in the same way.) The route is defined by flow variables in each direction of each link which take the value of 0 or 1.0. For convenience of notation the 14 flow variables are as indicated on the diagram.

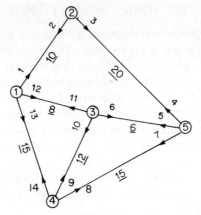

Figure 7.6 The travelling refuse-truck problem

The linear programming problem is set up with the constraints that the route must visit each node at least once and that at each node the algebraic sum of inflows and outflows must be zero. The mathematical model then takes the form

$$\text{Minimize } \sum c_i x_i \quad i = 1, 14 \tag{7.12}$$

subject to

$$\begin{aligned}
x_2 + x_{11} + x_{14} &\geq 1 \quad \text{at node 1} \\
x_1 + x_4 &\geq 1 \quad \text{at node 2} \\
x_5 + x_9 + x_{12} &\geq 1 \quad \text{at node 3} \\
x_7 + x_{10} + x_{13} &\geq 1 \quad \text{at node 4} \\
x_3 + x_6 + x_8 &\geq 1 \quad \text{at node 5}
\end{aligned}$$

and

$$\begin{aligned}
x_2 + x_{11} + x_{14} - x_1 - x_{12} - x_{13} &= 0 \quad \text{at node 1} \\
x_1 + x_4 - x_2 - x_3 &= 0 \quad \text{at node 2} \\
x_5 + x_9 + x_{12} - x_6 - x_{10} - x_{11} &= 0 \quad \text{at node 3} \\
x_7 + x_{10} + x_{13} - x_8 - x_9 - x_{14} &= 0 \quad \text{at node 4} \\
x_3 + x_6 + x_8 - x_4 - x_5 - x_7 &= 0 \quad \text{at node 5} \\
x_i &\geq 0 \quad \text{for all } i.
\end{aligned}$$

Even for this small network, the addition of necessary surplus and artificial variables increases the problem size to one of 29 variables and 10 constraints. For the standard travelling-salesman problem no analytical solution is yet available. Iterative solutions are possible and a good description of one method is given by Stark and Nicholls (1972).[1]

7.5 NETWORK CAPACITY PROBLEMS

An important class of network problems is concerned with the analysis of flow capacity when saturation is approached. As mentioned earlier, under such conditions the loop law may no longer be relevant since other capacity constraints over-ride the tendency for flows to be distributed until an equilibrium condition is reached. This problem is illustrated with reference to a typical system of one-way streets with maximum traffic flow constraints. The problem is examined first by means of the concept of network cuts. An alternative labelling technique is then introduced and finally the notion of a dual-graph is presented to facilitate the solution.

Figure 7.7(a) shows a portion of a street network within a city. The direction of allowable traffic movement and the maximum capacity of each street segment in

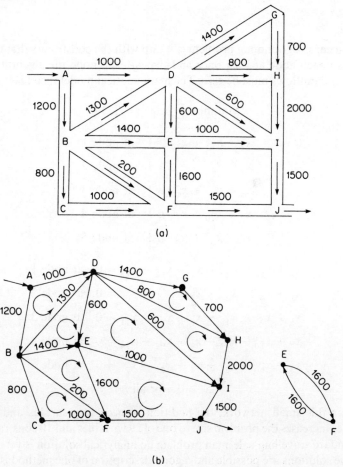

Figure 7.7 (a) A one-way street system. (b) The one-way street equivalent network

vehicles per hour is shown. The directed network corresponding to the physical problem can now be drawn as indicated in Figure 7.7(b).

The physical system may be altered by increasing or decreasing the capacity constraints of the streets, e.g. by widening the street and thus allowing more traffic to flow or allowing parking on one or both sides of the street and hence cutting down on the allowable flow. It is also possible that a 'one-way' street system be changed to accommodate 'two-way' traffic and hence an additional directed link be introduced as shown in the insert of Figure 7.7(b).

For the network of Figure 7.7(b) the *node incidence matrix* is given in Table 7.4.

Table 7.4 Node incidence matrix

Branch	AD	AB	BD	BE	DE	DG	DH	DI	GH	HI	EI	EF	IJ	BF	BC	CF	FJ
Node A	1	1	0	0	0	0	0	0	0	0	0	0	0	0	0	0	0
B	0	-1	1	1	0	0	0	0	0	0	0	0	0	1	1	0	0
C	0	0	0	0	0	0	0	0	0	0	0	0	0	0	-1	1	0
D	-1	-1	-1	0	-1	1	1	1	0	0	0	0	0	0	0	0	0
E	0	0	0	-1	-1	0	0	0	0	0	1	1	0	0	0	0	0
F	0	0	0	0	0	0	0	0	0	0	0	-1	0	-1	0	-1	1
G	0	0	0	0	0	-1	0	0	1	0	0	0	0	0	0	0	0
H	0	0	0	0	0	0	-1	0	-1	1	0	0	0	0	0	0	0
I	0	0	0	0	0	0	0	-1	0	-1	-1	0	1	0	0	0	0
J	0	0	0	0	0	0	0	0	0	0	0	0	-1	0	0	0	-1

The rows in this table represent Kirchhoff's Node Law and may be used in the solution of the problem. It should be noted that for each link represented by a column in the table, there can be only one set of coefficients (1, −1) corresponding to the initial and terminal nodes of the link. The remaining coefficients for that particular column will be zero. The node incidence matrix may be related to the flows **x** in each link and the net inflow or outflow (i.e. the stipulation **b**) by the constraint equation

$$A\mathbf{x} = \mathbf{b}. \tag{7.13}$$

In this case, the stipulations **b** are zero at all but nodes A and J.

A typical 'equation' for node E may be given as follows:

$$-Q_{BE} - Q_{DE} + Q_{EI} + Q_{EF} = 0 \tag{7.14}$$

where Q_{BE}, Q_{DE}, etc., are the flows (in this case vehicles per hour) in the links BE, DE, EI, and EF respectively.

It is obvious that if street segments BE and DE were used to capacity then there would still be an excess capacity of 600 vehicles per hour in the combined values of EI and EF.

The network of Figure 7.7(b) has ten nodes giving rise to ten rows representing the equations present in the node incidence matrix. In general, the node stipulations may not be zero and there may be two opposing directed links between some nodal points, in which case each equation becomes independent. If this were not the case then any one of these equations could be obtained by a linear combination of the remaining nine equations, leaving only nine independent equations.

To obtain the corresponding equations for the Loop Law we make use of the topology of the basic loops within the network. In this case ADB and BDE would be termed basic loops but ADEB would give extraneous information as it already encompasses the basic loops ADB and BDE. The number of basic loops in a network can be related to the number of nodes and links as follows. For a network of N nodes a spanning tree may be formed containing $(N - 1)$ links. If the network contains L links the $(L - N + 1)$ links not in the tree are called *chords*. For each chord added to the tree a loop or cycle is formed. Therefore the number of basic loops C is given by

$$C = (L - N + 1). \tag{7.15}$$

Here $C = 17 - 10 + 1 = 8$ which is easily confirmed by inspection. In a similar manner to the node incidence matrix a link coincident with the direction of the basic loop is denoted by the symbol 1 or -1 depending on its direction. The incident links for each loop are now entered into a row in the table as shown in Table 7.5. The coefficients of the loop equation based on Kirchhoff's Loop Law are now given by each row in the table. The application of the loop matrix will be seen later.

Table 7.5 Loop incidence matrix

Branch	AD	AB	BD	BE	DE	DG	DH	DI	GH	HI	EI	EF	IJ	BF	BC	CF	FJ
Node ABD	1	−1	−1	0	0	0	0	0	0	0	0	0	0	0	0	0	0
BDE	0	0	1	−1	1	0	0	0	0	0	0	0	0	0	0	0	0
DGH	0	0	0	0	0	1	−1	0	1	0	0	0	0	0	0	0	0
DHI	0	0	0	0	0	0	1	−1	0	1	0	0	0	0	0	0	0
DIE	0	0	0	0	−1	0	0	1	0	0	−1	0	0	0	0	0	0
BEF	0	0	0	1	0	0	0	0	0	0	0	1	0	−1	0	0	0
BFC	0	0	0	0	0	0	0	0	0	0	0	0	0	1	−1	−1	0
EIJF	0	0	0	0	0	0	0	0	0	1	−1	1	0	0	0	−1

7.5.1 Maximum flow–Minimum cut Theorem

In the traffic network problem in Figure 7.7(b) the traffic enters at source A and exits at sink J. The planners involved in the design of this project may be interested in determining the maximum throughput capacity of the street network, and if necessary improve on the vehicle throughput capacity of streets

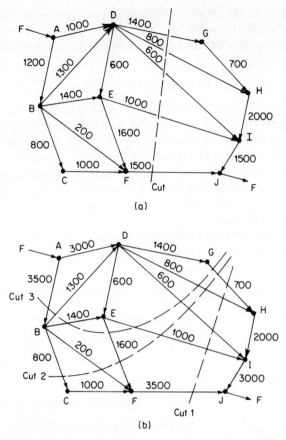

Figure 7.8 (a) Directed network with a single cut. (b) The modified directed network with three trial cuts

which could be potential bottlenecks. In Figure 7.8(a) the network is cut at some arbitrary section which severs the sink from the source along every path of positive capacity. The 'value' of the cut is found by the summation of the capacities of all the links at the cut, i.e.

$$Q_{DG} + Q_{DH} + Q_{DI} + Q_{EI} + Q_{FJ} = Q_{TOTAL} \tag{7.16}$$

where Q_{DG} is the maximum capacity of link DG, etc. In this case the value of the cut is found to be equal to

1400 + 800 + 600 + 1000 + 1500 = 5300 vehicles/hour.

This value of 5300 vehicles/hour represents the maximum capacity of the network at the cut intersections. However, the maximum throughput of this particular network may be controlled by the allowable traffic flow in links upstream or downstream of the particular cut shown in Figure 7.8(a) and therefore a systematic approach is necessary to determine the values of all possible cuts within the network.

It is obvious that a cut taken across links AD and AB will result in the minimum cut value, i.e.

$$1200 + 1000 = 2200.$$

Therefore, if the carrying capacity of the network is to be increased then the capacity of these links must be improved. Assume now that the local authority concerned have decided to increase the capacity of these two streets to 3000 and 3500 vehicles/hour for streets AD and AB respectively. The capacity at the source of the network is now much greater and if utilized fully would cause considerable traffic congestion at junctions I and F. Therefore, the next logical step is to increase the capacities of FJ and IJ to a value commensurate with those streets at the source of the network.

If this were done another investigation is required to find the value of various cuts. The modified capacities of the streets are now shown in Figure 7.8(b). Various cuts have been analysed, e.g.

Cut 1: $Q_{GH} + Q_{DH} + Q_{DI} + Q_{EI} + Q_{FJ} = 700 + 800 + 600 + 1000 + 3500$

$$= 6600 \text{ vehicles/hour.}$$

This value could be reduced by cutting the network along links EF, BF, and BC giving

Cut 2: $Q_{GH} + Q_{DH} + Q_{DI} + Q_{EI} + Q_{EF} + Q_{BF} + Q_{BC}$

$$= 700 + 800 + 600 + 1000 + 1600 + 200 + 800$$

$$= 5700 \text{ vehicles/hour.}$$

If this process is continued and all possible flow chains within the network are analysed, it is possible to determine the minimum cut. It should be noted that the directions of the flows must be taken into account when determining the sum of the minimum cut. In Figure 7.8(b) the cut 2 which crosses links GH, DH, DI, EI, EF, etc., has capacities with flows from the source to the sink and are designated as being positive values. However, in cut 3, which also crosses the first five branches of cut 2, there are two branches BE and BD which are effectively crossing the line of the cut in the opposite direction and must be designated zero values.† This results in the following:

Cut 3: $Q_{GH} + Q_{DH} + Q_{DI} + Q_{EI} + Q_{EF} - Q_{BE} - Q_{BD} + Q_{AB}$

$$= 700 + 800 + 600 + 1000 + 1600 - 0 - 0 + 3500$$

$$= 8200 \text{ vehicles/hour.}$$

† It should be noted that zero values are assigned to 'negative' links only because these values are upper limits and not actual flow values. If the results of an origin–destination survey were being analysed in this way the negative flows would have to be included with the appropriate value in an algebraic sum.

This is the maximum cut of the three investigated, and as all the possibilities have not yet been exhausted the procedure has to be continued until all the cut-sets have been evaluated. The minimum cut is also the maximum flow allowed in the network. In other words the maximum flow F' equals the minimum cut C' or

$$\text{Maximum } F = F' = C' = \text{Minimum } C.$$

In this example, a total of 40 legal cuts are shown in Figure 7.9. These are obtained by severing all possible flow paths from the source to the sink in the following systematic manner. Commencing from face AB, as shown in Figure 7.9(a), cut-set lines are drawn to each other face in the network e.g. AD, DG, GH, HI, and IJ so that the source and sink are no longer connected. These cuts are drawn around the only centre node E such that cut-sets 1 to 5 pass above the point and cut-sets 6 to 10 pass below the point. The same format is repeated for the other three faces resulting in the cut-sets in Figures 7.9(b), 7.9(c), and 7.9(d). It must be remembered that a cut-set is *legal* only if it cuts the network into *two* connected graphs, one of which contains the source and the other the sink node. The results of the forty cut-sets are shown in Table 7.6 and the minimum cut-set is number 3 in set (b) which has a value of 5100 vehicles/hour.

Each of the cuts may be interpreted in a slightly different sense that leads to a solution technique more easily implemented in a computer program. Cut 1 in set (a) isolates node A — the source. Cut 2 in set (a) isolates nodes A and D. Cut 3 in set (a) isolates nodes A, D, and G. Examination reveals that each cut isolates node A along with some combination of the nodes B to I (i.e. excluding the sink J). Now the cuts resulting from the isolation of any node are given by the element in the corresponding row of the incidence matrix A.

For example, inspection of the first row of the node incidence matrix reveals that isolation of node A alone corresponds to the cut set (AD, AB). By adding the rows corresponding to nodes A *and* B in Table 7.4 the cut set is found to be

$$(AD, AB, -AB, BD, BE, BF, BC)$$

or

$$(AD, BD, BE, BF, BC).$$

This may be confirmed by examining cut 1 of set (b).

The following algorithm may therefore be developed to obtain the cut-sets for all possible combinations of nodes B–I being isolated along with the source A. The minimum value may be obtained by scanning during the enumeration.

Cut-set algorithm

1. Set up the node incidence matrix A, where

$$A = A(i,j) \quad i = 1, N \quad N = \text{no. of nodes}$$
$$j = 1, L \quad L = \text{no. of links.}$$

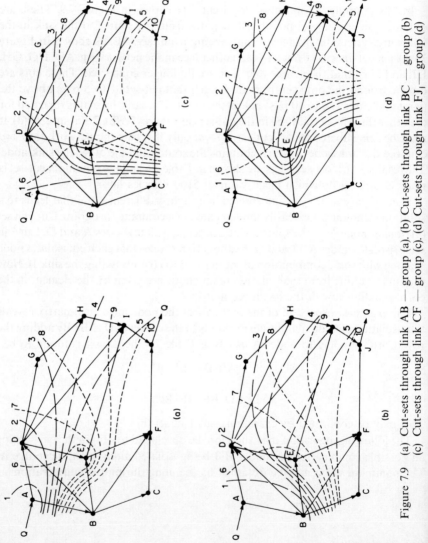

Figure 7.9 (a) Cut-sets through link AB — group (a). (b) Cut-sets through link BC — group (b) (c) Cut-sets through link CF — group (c). (d) Cut-sets through link FJ_1 — group (d)

Table 7.6 Road network cut-set results

		Cut-sets through AB — group (a)									
Street	Capacity	1	2	3	4	5	6	7	8	9	10
AB	3500	+	+	+	+	+	+	+	+	+	+
AD	3000	+					+				
BC	800										
BD	1300		−	−	−	−	−	−	−	−	−
BE	1400						−	−	−	−	−
BF	200										
CF	1000										
DE	600		+	+	+	+	−				
DG	1400		+					+			
DH	800		+	+				+	+		
DI	600		+	+	+			+	+	+	
EF	1600						+	+	+	+	+
EI	1000					−	+	+	+	+	
FJ	3500										
GH	700			+					+		
HI	2000				+					+	
IJ	3000					+					+
Cut-set capacity		6500	6900	6200	6700	7100	9100	8900	8200	8700	8100

		Cut-sets through BC — group (b)									
Street	Capacity	1	2	3	4	5	6	7	8	9	10
AB	3500										
AD	3000	+					+				
BC	800	+	+	+	+	+	+	+	+	+	+
BD	1300	−					+				
BE	1400	+	+	+	+	+					
BF	200	+	+	+	+	+	+	+	+	+	+
CF	1000										
DE	600		+	+	+	+	−				
DG	1400		+					+			
DH	800		+	+				+	+		
DI	600		+	+	+			+	+	+	
EF	1600						+	+	+	+	+
EI	1000				−		+	+	+	+	
FJ	3500										
GH	700			+					+		
HI	2000				+					+	
IJ	3000					+					+
Cut-set capacity		5400	5800	5100	5600	6000	7900	6400	5700	6200	5600

Table 7.6 Road network cut-set results (continued)

Street	Capacity	Cut-sets through CF — group (c)									
		1	2	3	4	5	6	7	8	9	10
AB	3500										
AD	3000	+					+				
BC	800										
BD	1300	+					+				
BE	1400	+	+	+	+	+					
BF	200	+	+	+	+	+	+	+	+	+	+
CF	1000	+	+	+	+	+	+	+	+	+	+
DE	600		+	+	+	+	−				
DG	1400		+					+			
DH	800		+	+				+	+		
DI	600		+	+	+			+	+	+	
EF	1600						+	+	+	+	+
EI	1000					−	+	+	+	+	
FJ	3500										
GH	700			+					+		
HI	2000				+					+	
IJ	3000					+					+
Cut-set capacity		6900	6000	5300	5800	6200	8100	6600	5900	6400	5800

Street	Capacity	Cut-sets through FJ — group (d)									
		1	2	3	4	5	6	7	8	9	10
AB	3500										
AD	3000	+					+				
BC	800										
BD	1300	+					+				
BE	1400	+	+	+	+	+					
BF	200										
CF	1000										
DE	600		+	+	+	+	−				
DG	1400		+					+			
DH	800		+	+				+	+		
DI	600		+	+	+			+	+	+	
EF	1600	−	−	−	−	−					
EI	1000						−	+	+	+	+
FJ	3500	+	+	+	+	+	+	+	+	+	+
GH	700			+					+		
HI	2000				+					+	
IJ	3000					+					+
Cut-set capacity		9200	8300	7600	8100	8500	8800	7300	6600	7100	6500

2. Set up the vector of maximum flow rates $\mathbf{C} = C(j), j = 1, L$.
3. Obtain the source node cut-set Q_s

$$Q_s = \sum_j A(s, j)C(j) \quad j = 1, L$$
$$s = \text{source node.}$$

4. Set $Q_{\min} = Q_s$.
5. Find any combination $k_1, k_2, k_3, \ldots$ of the set of $(N - 2)$ nodes between the source and sink.
6. Obtain the cut-set for this combination of intermediate nodes together with the source node as

$$Q_k = Q_s + \sum_j A(k_i, j)C(j) \quad j = 1, L$$
$$k_i = k_1, k_2, \ldots$$

7. If $Q_k < Q_{\min}$. Set $Q_{\min} = Q_k$.
8. Repeat steps 5 to 7 for all possible combinations of the $(N - 2)$ intermediate nodes.
9. $Q_{\min}$ is the minimum cut.

The automatic identification of combinations is not a trivial task and a subroutine COMBIN is included in Appendix A for use in this and other combinatorial search procedures. For this example $N = 10$ and thus $(N - 2) = 8$. The number of possible combinations (including all or none) of the 8 intermediate nodes is given by

$$1 + \sum \frac{8!}{m!\,(8 - m)!} \quad m = 1, 2, \ldots, 7$$

which is found to be 255 — considerably more than the 40 obtained by inspection. Typical of the cut-sets which were ignored would be the combined isolation of nodes A and C giving the cut-set (AD, AB, $-$BC, CF) = 7500.

Clearly, although such a procedure may be automated it is likely to be computationally expensive for large networks. On the other hand, manual identification of realistic cut-sets is both tedious and subject to error. Alternative strategies are developed in the following sections.

7.5.2 Labelling technique

A method of calculating the maximum flow in a directed network has been developed on the relation between maximum flow and minimum cut. The procedure involves a systematic investigation of the capacities of each link in the network and terminates when the existing flow reaches a maximum.

The nodes within the network are successively numbered by two numbers (h, k) where h is the number of the node preceding the present node and k denotes the

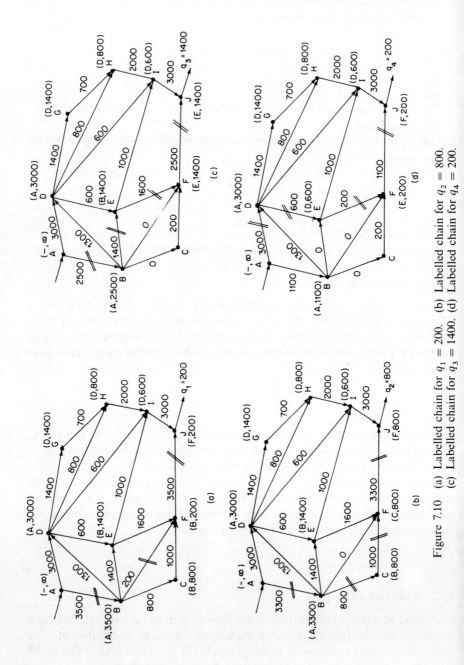

Figure 7.10 (a) Labelled chain for $q_1 = 200$. (b) Labelled chain for $q_2 = 800$. (c) Labelled chain for $q_3 = 1400$. (d) Labelled chain for $q_4 = 200$.

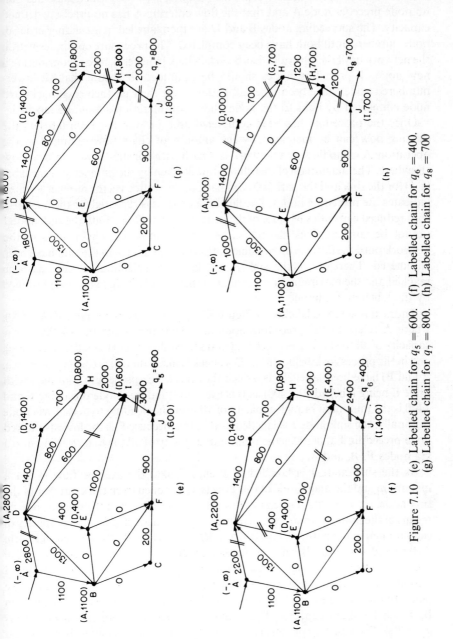

Figure 7.10 (e) Labelled chain for $q_5 = 600$. (f) Labelled chain for $q_6 = 400$. (g) Labelled chain for $q_7 = 800$. (h) Labelled chain for $q_8 = 700$

smallest positive capacity of all links preceding the present node. Starting with the source node A, in Figure 7.10(a), the labelling is $(-, \infty)$ which indicates that no node precedes node A and that the flow entering A has no precisely defined capacity. The succeeding nodes B and D are then labelled in ascending order of node numbers until both have been completed. The procedure continues within the network until the sink node has been labelled. In general, before considering a new node all preceding nodes should be examined to ensure that all lower-numbered nodes have been labelled. If this is not the case then each backward node connected by links to the node under consideration should be labelled.

Once the procedure has been completed and the sink has been labelled the existing flow can be increased by the amount of chain flow obtained in the operation. A chain flow in a network is a path with constant flow and zero flow elsewhere. The magnitude of this chain flow is given by the number k in the label (h, k) for the sink and the path is traced back to the source via the number h which indicates the preceding nodes. The labelling process is repeated for the network, with reduced capacities because of the deducted chain flows, until the sink node J cannot be labelled. This has the same effect of imposing chain flows in the network until the flows in all possible paths from the source to the sink have been eliminated. Therefore, if the sink is not labelled the minimum cut has been reached and the maximum flow is equal to the sum of all chain flows determined by the labelling technique.

The next node to be labelled in Figure 7.10(a) is B which is designated (A, 3500), where A refers to the preceding node and 3500 represents the smallest flow capacity of all the branches preceding node B. Node C is next labelled as (B, 800), and in the process of labelling node D a check is made that all preceding nodes (i.e. A and B) have been labelled, as indeed they have. The procedure continues with node E being labelled (B, 1400) as B is the lowest indexed node preceding E and 1400 is the minimum branch capacity of all branches leading to node E via node B. Finally, the sink node J is labelled as (F, 200), because F is the lowest-indexed node preceding J and 200 is the minimum capacity of all the links leading into J via nodes F, B, and A.

As the sink node J is labelled in Figure 7.10(a) the existing flow is not a maximum and the magnitude of this flow is determined from the k number of the sink node, i.e. $q_1 = 200$. The chain flow is then traced back to the source node by means of the number h and a chain flow $q_1 = 200$ is deducted from the capacity of each branch in the path, i.e. AB, BF, and FJ. The net result is indicated in Figure 7.10(b) and since link BF has no further capacity (in fact it has a net negative excess capacity but this is not relevant to this problem and need be considered only if circular chain flows exist in the network), then no further flow can take place. In this case the node F is labelled (C, 800) as no further flow can occur from B, which has zero capacity. The chain flow resulting from this operation is $q_2 = 800$. The labelling process is now repeated in its entirety after the first chain flow has been subtracted.

Following in this manner the chain flows $q_1, q_2, q_3, q_4, q_5, q_6, q_7$, and q_8 are calculated to give the maximum flow Q^* as

$$Q^* = q_1 + q_2 + q_3 + q_4 + q_5 + q_6 + q_7 + q_8 = 5100 \text{ vph}.$$

This value corresponds to cut-set number 3 in set (b) and on inspection of Figure 7.10(h) the cut-set can actually be deduced from the figure. Some interesting observations are possible from Figure 7.10(h). The link BD is effectively redundant and unless node B was created as a possible source there would be no point in having such a link. The links AB, DG, HI, EI, FJ, CF, and IJ all have excess capacities with the worst offenders in this case being AB and FJ. Therefore, if the objective was to optimize this road network the local authority concerned would have been better rewarded by increasing the capacities of links BF, DI, etc. by a commensurate amount and not improving the capacities of AB and FJ quite so much.

It is possible to execute the analysis in a tabulated form as shown in Table 7.7. The procedure is similar to that resulting in the series of networks shown in Figure 7.10 and also indicates the relationships between each succeeding and preceding node in the network. A dash indicates that there is no connection between the two nodes and a zero implies a connection but no capacity in that direction. The procedure commences by entering the first row A (i.e. $i = A$) which

Table 7.7 Solution of example by labelling technique

(a) $q_1 = 200$

i \ j	A	B	C	D	E	F	G	H	I	J	h	k
A	—	[3500]	—	3000	—	—	—	—	—	—	—	∞
B	0	—	800	1300	1400	[200]	—	—	—	—	A	3500
C	—	0	—	—	—	1000	—	—	—	—	B	800
D	0	0	—	—	600	—	1400	800	600	—	A	3000
E	—	0	—	0	—	1600	—	—	1000	—	B	1400
F	—	0	0	—	0	—	—	—	—	[3500]	B	200
G	—	—	—	0	—	—	—	700	—	—	D	1400
H	—	—	—	0	—	—	0	—	2000	—	D	800
I	—	—	—	0	0	—	—	0	—	3000	D	600
J	—	—	—	—	—	0	—	—	0	—	F	200

(b) $q_2 = 800$

i \ j	A	B	C	D	E	F	G	H	I	J	h	k
A	—	[3300]	—	3000	—	—	—	—	—	—	—	∞
B	0	—	[800]	1300	1400	0	—	—	—	—	A	3300
C	—	0	—	—	—	[1000]	—	—	—	—	B	800
D	0	0	—	—	600	—	1400	800	600	—	A	3000
E	—	0	—	0	—	1600	—	—	1000	—	B	1400
F	—	0	0	—	0	—	—	—	—	[3300]	C	800
G	—	—	—	0	—	—	—	700	—	—	D	1400
H	—	—	—	0	—	—	0	—	2000	—	D	800
I	—	—	—	0	0	—	—	0	—	3000	D	600
J	—	—	—	—	—	0	—	—	0	—	F	800

has links incident with nodes B and D and noting that $h = (—)$ and $k = \infty$. The next rows entered are B and D which are those indicated in row A. In row B ($i =$ B), $h =$ A, which is the precedence of node A, and $k = 3500$, which is the smaller of the coefficients in columns B and k of row A. Similarly, in row D ($i =$ D), $h =$ A, which is the precedence of node D, and $k = 3000$, which is the smaller of the coefficients in columns D and k. From an inspection of row B, the nodes incident with links from B are A, C, D, E, and F respectively, and since A and D are already labelled the analysis proceeds to nodes C, E, and F. Therefore rows C, E, and F are entered and inspected for corresponding nodes on incident links. Finally, the values of h and k in all rows, and $k = 200$ in the last row ($i =$ J), indicates that the chain flow is 200. The path of the chain flow can be traced for $h =$ F in the last row ($i =$ J) to $h =$ B in row F ($i =$ F) and finally to node A (i.e. $h =$ A in row $i =$ B). The capacity of each link within the path of the chain flow is marked by a square, thus enabling easy identification.

The chain flow, $q_1 = 200$, is then subtracted from the positive capacity of each branch in the path and added to the negative capacity (if pertinent) of that link. The result is shown in Table 7.7(b), e.g. capacity AB is now $3500 - 200 = 3300$. The same labelling technique is then applied to the modified table until a chain flow of $q_2 = 800$ is obtained in the last row. The process may be repeated until the sink node cannot be labelled and as in the previous analysis the chain flows are summed to give the maximum possible flow within the network.

7.5.3 Primal–dual graphs

In Section 7.5.1 the maximum flow was based on a comparison of cut-sets each of which bisect the original network such that source and sink are separated. The cut-sets are seen to be bunched together at the intersection of a link and then to diverge within the loop of which the link is a part. This suggests that alternative cuts may be viewed as alternative routes connecting the centroids of each basic loop. In order to formalize this treatment it is useful to introduce a return link connecting sink and source as illustrated in Figure 7.3(b). A new set of vertices is then constructed, one within each basic loop of the original network. These new vertices form the basis of a *dual-graph* which is essentially orthogonal to the original *primal graph* of Figure 7.9. The links of the dual graph each intersect a single link of the primal graph. The result is indicated in Figure 7.11. Each link in the dual graph is given a flow variable value equal to that of the upper flow limit in the corresponding link of the primal graph. Direction is defined by using a consistent transformation — e.g. positive directions in the dual are 90° out-of-phase (clockwise) from the direction in the primal.

The relationship between the primal and dual graphs is not simply geometric but has its roots in the primal–dual theory of linear algebra (see Section 3.11). Thus, whereas the primal involves maximization of flow variables subject to maximum capacity constraints, the dual involves minimization of potential differences or costs which are numerically equivalent to the capacity constraints.

In terms of network topology it is interesting to note that the loop matrix of the

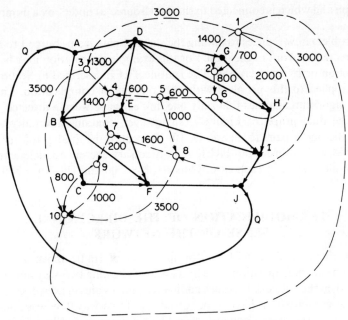

Figure 7.11 Primal and dual graphs of a network

Table 7.8 Loop incidence matrix

Link	AD	AB	BD	BE	DE	DG	DH	DI	GH	HI	EI	EF	IJ	BF	BC	CF	FJ	JA	Dual node
Loop																			
ADGHIJ	−1	0	0	0	0	−1	0	0	−1	−1	0	0	−1	0	0	0	0	1	1
DGH	0	0	0	0	0	1	−1	0	1	0	0	0	0	0	0	0	0	0	2
ABD	1	−1	−1	0	0	0	0	0	0	0	0	0	0	0	0	0	0	0	3
BDE	0	0	1	−1	1	0	0	0	0	0	0	0	0	0	0	0	0	0	4
DEI	0	0	0	0	−1	0	0	1	0	0	−1	0	0	0	0	0	0	0	5
DHI	0	0	0	0	0	0	1	−1	0	1	0	0	0	0	0	0	0	0	6
BEF	0	0	0	1	0	0	0	0	0	0	0	1	0	−1	0	0	0	0	7
EFJI	0	0	0	0	0	0	0	0	0	0	1	−1	1	0	0	0	−1	0	8
BCF	0	0	0	0	0	0	0	0	0	0	0	0	0	1	−1	−1	0	0	9
	0	1	0	0	0	0	0	0	0	0	0	0	0	0	1	1	1	−1	10

primal becomes the node incidence matrix of the dual. Table 7.8 is a modified version of the loop matrix of Table 7.5, whereby the order of the rows has been modified so that the loops correspond to the vertices as numbered in Figure 7.11. The eight loops of the original primal have been augmented by adding a ninth loop due to the introduction of the recirculating link (JA). Finally, a tenth row is added such that the elements of each column sum to zero. This tenth row corresponds to the node incidence equation of a tenth node which is outside the dual graph and which is connected to the dual 'source' at node 1 by a dummy link (10, 1).

The problem now reduces to finding the shortest path from source 1 to sink 10 in the directed network of the dual. For this example the solution may be found by inspection or by enumeration. The routine TREE described in Section 7.4.2 may be applied to this problem with only one minor complication. The tree-building algorithm assumes that only one link exists between a directed pair of nodes. The dual graph of Figure 7.11, however, contains alternative routes between the nodal pairs (1, 2) and (9, 10). To reduce the dual graph to a form compatible with subroutine TREE it is necessary merely to delete the less attractive alternative in each case. Thus $c(1, 2) = 700$ and $c(9, 10) = 800$.

7.6 MODIFICATION OF THE DIRECTIONAL SENSE OF THE NETWORK

If the planners involved with the optimization of traffic movement in the particular network in question have the choice of allowing two-way flow in some branches then the network becomes undirected and a series of networks showing the various combinations have to be solved, e.g. if two-way flow was allowed in segment EF of the primal network then two solutions would be required, one with directed flow from E to F and the other with flow from F to E.

The dual graph is ideal for this purpose and is extremely useful in the analysis of networks where the direction cannot be assigned a priori for the flow in some of the branches. If the branch DE is to be a one-way street but the optimum direction is not known, then the influence of this on the final result can be analysed as follows. In the dual graph the chain 4–5 would have an arrow indicating traffic flow from node 4 to node 5 and this would result in the maximum capacity of the network being increased by 600 and hence the logical direction for traffic to flow in branch DE is as it stands.

7.7 EXERCISES

7.7.1 A conduit has to be constructed to convey a given quantity of water from source A to demand area B. Whereas the starting point of the conduit can be located usually with a high degree of accuracy, the location of the other extremity of the conduit may often present a number of alternative choices. In addition to this, the terrain between source A and demand area B (its topography, geology, land use, etc.) may present a variety of alternative routes, each involving different levels of expenditure. This is shown graphically in Figure 7.12. In this figure, the water conduit (open channel or pipeline) must link A with either of the three terminal points B_1, B_2, or B_3. The possible points through which the conduit may

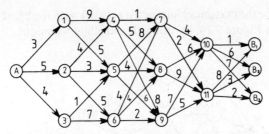

Figure 7.12 A conduit routing problem

pass are numbered from (1) to (11), and the digit appearing on each link joining these various points represents the expenditure connected with the construction of this link (in monetary units). Find the path linking A with B_i ($i = 1, 2, 3$) requiring the minimum expenditure.

7.7.2 (a) Discuss and contrast the various techniques available for determining the maximum flow in a directed network.
(b) A street network is depicted in the directed network of Figure 7.13 where S is the source node, n is the sink node, and the numbers along the arcs denote the capacities of the traffic flow.

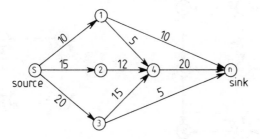

Figure 7.13 A directed street network

Illustrate the following notions with an example from the network shown:

(i) A path connecting the source and the sink.
(ii) A cut separating the source and the sink.
(iii) The capacity of a cut.

Find the maximum flow from the source to the sink using the labelling method. Find the maximum cut, and verify the maximum flow–minimum cut theorem.

7.7.3 Determine the maximum flow in the network with the branch capacities as shown in Figure 7.14, by using the labelling technique.

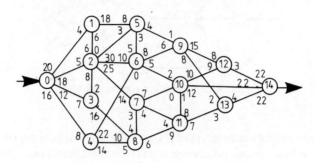

Figure 7.14 A capacitated network

7.7.4 Solve the network shown in Figure 7.14 by
 (a) the minimum cut–maximum flow method and
 (b) the primal-dual graph method.

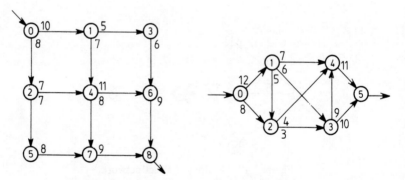

Figure 7.15 Capacitated network of Example 7.7.5

7.7.5 Determine the maximum flow in the network of Figure 7.15 with branch capacities shown in the figures for each problem by constructing flow chains from the source to the sink.

7.8 REFERENCES

1. Stark, R. M., and Nichols, R. L. (1972). *Mathematical Foundations for Design: Civil Engineering Systems*, McGraw-Hill, New York.
2. Au, T., and Stelson, E. (1969). *Introduction to Systems Engineering, Deterministic Models*, Addison-Wesley, Reading, Mass.

3. Seshu, S., and Read, M. B. (1961). *Linear Graphs and Electrical Networks*, Addison-Wesley, Reading, Mass.
4. Berge, C., and Ghouila-Houri, A. (1965). *Programming, Games and Transportation Networks*, Wiley, New York.
5. Biswas, A. K. (1976). *Systems Approach to Water Management*, McGraw-Hill Kogakusha, New York.
6. Thomann, R. V. (1972). *Systems Analysis and Water Quality Management*, McGraw-Hill, New York.
7. Smith, A. A., and Tufgar, R. H. (1977). *Planning regional network systems for minimum cost*, Can. J. Civil Eng., vol. 4, pp. 202–213.
8. Smith, A. A. (1982). *Compact formulation of pipe network problems*, Can. J. Civil Eng., vol. 4, pp. 611–623.

Chapter 8
Critical Path Analysis

8.1 INTRODUCTION

In the previous chapter it was mentioned that problems of planning and scheduling engineering projects can be represented by networks which indicate the various activities in the proper order of their execution. An effective method of planning must provide a clear picture of the relationship between the activities and show how delays at any particular stage will affect the remainder of the project.

More traditional forms of planning such as the 'bar chart' or 'Gantt chart' are good methods for illustrating progress, but they provide no direct indication of the relationship between the various operations, nor do they show the extent to which delays will affect the remainder of the project.

The critical path method (CPM) utilizes the techniques of network analysis to represent a project plan by a schematic diagram which represents the sequence and inter-relationship of the activities or operations within the project. Delays in the completion of an activity can be analysed to determine the effect on the overall completion time of the whole project. With this information it is possible to re-allocate the resources available (e.g. money, equipment, or manpower) and compensate at least in part for the unforeseen delay.

8.2 CONSTRUCTION OF THE NETWORK

The initial synthesis of the network requires the division of the project into clearly defined activities or tasks. For example, in conducting a traffic survey for the planning of an urban mass transit system, the entire project might be envisaged as consisting of the following major activities:

A. Define the scope of the survey.
B. Establish the procedure of the survey.
C. Design questionnaires for the survey.
D. Hire and organize staff.
E. Train the staff to be familiar with the questionnaires.
F. Select sampling stations along the proposed routes.

G. Assign the staff to sampling stations.
H. Conduct the field survey.

The project is then illustrated graphically as a network of arrows, with each arrow representing an activity. These activities, if following each other during the execution of the project, are drawn as arrows in sequence with the direction of the arrow indicating the natural progression of the project. If activities are allowed to occur concurrently then they are drawn as arrows in parallel. A complex project will be made up of many such sequences and the inter-relationship of these sequences is illustrated by the direction of the arrows that connect the sequences. Once the network has been completed, it is then possible to assign details of duration, cost, and resources to each activity. Also, as described later, it will be possible to examine, schedule, and control rate of progress, costs, and the allocation of resources.

8.2.1 Activities and events

As already mentioned, a project must be broken down into well defined units before a plan or schedule can be prepared. These units are referred to as *activities*. The nodes that mark the completion and start of each activity are called *events*.

An *activity* represents a job or task that has to be carried out and forms an integral part of the complete project. During the construction of a house one such activity might be to 'pour the foundations', and later during the advertising campaign for selling the houses an important activity would be 'preparation of press release and photographs'. An activity may also represent the passing of time with no actual work being carried out, e.g. 'await delivery of materials' or 'setting and hardening of concrete'.

In general, the division of activities for a network depends also to some extent on the level of management which makes use of it. For top management, a simple network representing major activities of the project may be sufficient, but a project engineer who is directly responsible for its performance must know to a greater depth the details of the activities. However, it is wise to start the analysis with a relatively simple network and if this proves insufficient in the proper scheduling of events then recourse can be made to a more detailed network. If the work is all of one category and is to be planned and supervised by one person, then obviously that person, with no additional assistance, should collect the information and prepare the network. Where the project consists of a multiplicity of categories of work with separate groups responsible for sections of the project, it may be necessary for each group to be represented when the network is prepared. The need for proper liaison between the representatives of each group cannot be overstressed.

Generally, there will be many amendments made to the network between the preliminary planning stage and the preparation of the final plan. Even then the network is not necessarily complete as alterations may be required or desirable as the project gets under way.

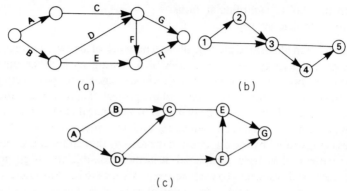

Figure 8.1   Activity- and event-oriented networks: (a) activity-on-branch; (b) event-on-node; (c) activity-on-node

The terminology used in CPM networks may best be described with reference to Figure 8.1. In Figure 8.1(a) the activities A to H are placed on the branches connecting the nodes and the network is therefore referred to as being *activity-oriented*. In Figure 8.1(b), the activities are again 'on-branch' but in this case the identifiers are *event-oriented*. In Figure 8.1(c) the activities are placed on the nodes with the branches simply denoting the precedence relationships of the activities. The *event* referred to in the previous sentence is an occurrence at a point in time which marks the start or completion of one or more activities. Examples of such *events* are 'start digging for foundations', 'start roof tiling', 'complete concreting slab', and 'completion of project'. In the remainder of this chapter the activity-oriented or activity-on-branch method will be used.

Once a project has been suitably broken down into activities, it may be illustrated graphically by means of a network of arrows representing the activities. The events are symbolized by circles drawn at the tail and arrowhead respectively representing the commencement and completion of the activity. In the case of an activity immediately following another the arrows are drawn sequentially as in Figure 8.2(a). This indicates that the foundation concrete cannot be poured until the foundations have been dug. The events have been described in this instance but, as is often the case, single events frequently mark the start or completion of several activities and are therefore usually unnamed in order to avoid confusion. For instance, in Figure 8.2(b) the network indicates that neither the unloading of passengers or luggage can begin until the car has come to a halt.

To design a network that will satisfy all the constraints requires a great deal of skill. In some cases, managerial decisions are extremely difficult to formulate in a diagram and errors may consequently occur. However, it is reasonably simple to test a given network and so it is easier to commence the analysis with a relatively coarse network and refine it later. These concepts may best be presented through

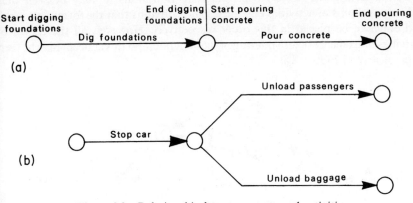

Figure 8.2 Relationship between events and activities

the description of a typical problem which has the following preliminary list of activities:

A. Site clearing.
B. Dig foundations.
C. Place formwork.
D. Pour concrete.
E. Obtain steel reinforcement.
F. Cut and bend steel reinforcement.
G. Lay out foundations.
H. Obtain concrete.
I. Place steel reinforcement.

These activities are listed in the order in which they spring to mind and are obviously not in the proper sequence necessary for a detailed network analysis. An examination of the list shows that an accurate grouping is obvious and, if considering physical constraints only, the precedence relationships of activities are obtained as shown in Table 8.1.

Table 8.1 Activities for reinforced concrete foundation

Activity	Description	Prerequisite	Duration
A	Site clearing	—	6
B	Lay out foundations	A	4
C	Obtain steel reinforcement	A	8
D	Obtain concrete	A	6
E	Dig foundations	B	6
F	Cut and bend steel	C	2
G	Place formwork	E	3
H	Place steel reinforcement	F, G	6
I	Place concrete	D, H	2

On inspecting the project from different viewpoints, it may be seen that individual chains of activities emerge, e.g. it is obvious that the formwork, steel, and concrete be placed in this order respectively. However, all the chains of activities must merge before pouring of the concrete, and this concept may be seen quite clearly in Figure 8.3.

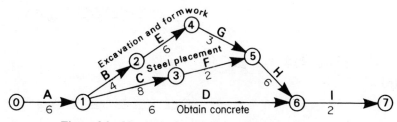

Figure 8.3 Networks for reinforced concrete foundation

The predecessor–successor relationships of various activities on branches can be given by the matrix in Table 8.2 which is sometimes referred to as the *implicit precedence matrix*. Note that the activities on the branches (i, j) are arranged in ascending order of *i* first. The predecessors of an activity in a column heading are

Table 8.2 Implicit precedence matrix

Predecessor	A (0, 1)	B (1, 2)	C (1, 3)	D (1, 6)	E (2, 4)	F (3, 5)	G (4, 5)	H (5, 6)	I (6, 7)
A(0, 1)		×	×	×					
B(1, 2)					×				
C(1, 3)						×			
D(1, 6)									×
E(2, 4)							×		
F(3, 5)								×	
G(4, 5)								×	
H(5, 6)									×
I(6, 7)									

located by the rows in which the *x*'s are marked in that column. Conversely, the successors of an activity in a row heading are given by the columns in which the *x*'s are found in that row. Thus, all the predecessors and successors of any activity (i, j) can be obtained by following through immediate predecessors and successors step by step.

8.2.2 Dummy activities

In certain circumstances it is necessary to show that one event cannot take place until a previous event has occurred although there may be no physical interaction between the two. Therefore, in order to demonstrate the sequential ordering of the two events a *dummy arrow* is drawn between them. This dummy activity requires neither time nor resources although it is dealt with in the same way as a standard activity.

Figure 8.4 shows a section of a network giving an example of the use of dummy arrows. This diagram suggests that the installation of the window frames has to

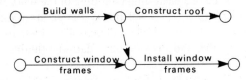

Figure 8.4   Example of a dummy arrow (activity)

await the building of the walls of the house and the fabrication of the frames for the windows. However, the construction of the roof need not necessarily await the completion of the fabrication or installation of the window frames but can proceed as soon as the walls have been completed.

If the dummy arrow is omitted then one could surmise, wrongly in this case, that the window frames could be installed before the walls are built, which is obviously incorrect.

Dummy arrows are also used to prevent ambiguity, so that two activities sharing the same start and end events can be identified separately. The need for this becomes evident when events are numbered for scheduling and analysis. An example to demonstrate this principle is given in Figure 8.5.

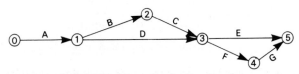

Figure 8.5   Example of a bottleneck

Figure 8.5 indicates that neither activity E or F can commence until both activities C and D have been completed. This may be an unreasonable restriction of the network and it might be more desirable if F could start as soon as D alone is completed and need not necessarily await the completion of C. However, E may still not commence until both C and D are completed as previously described in Figure 8.5. Therefore, the network now requires modification and this may be done by the suitable insertion of a dummy variable as shown in Figure 8.6. The unique designations of activities E and F are now seen to be preserved.

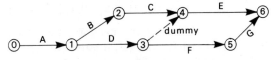

Figure 8.6   Modified 'bottleneck'

8.3 FORMULATING THE SCHEDULING PROBLEM

The scheduling problem for a given network plan is to determine an optimal schedule which is consistent with the given logical sequences and durations of various activities within the network, and one which requires a minimum time for the completion of the project. Even if the planner goes no further than the preceding steps, he or she will, by the very act of breaking down the project and preparing the network, have a clear understanding of the work involved and the problems that are likely to arise. Nevertheless, the greatest value of the CPM comes from the application of simple calculations which introduce the all-important element of time, and provide information valuable during both the planning and implementation stages of the project.

The introduction of the time element enables the planner to determine the overall time of the project, the sequence of activities that control the length of the project, and the timing of intermediate deadlines and objectives.

Consider a network with $n + 1$ nodes, the initial event being 0 and the last event being n. The event times at these nodes are $x_0, x_1, x_2, \ldots, x_n$, respectively, which increase in magnitude as they are units of time. These event times are constrained by the duration of the activities between the events.

The duration of any activity (i, j) between events i and j is denoted by $D_{ij} (\geqslant 0)$. If the events in the directed network are arranged such that the event x_i at the tail end is always smaller than the event x_j at the arrowhead of every activity (i, j), we have $i < j$. Then the difference between event times x_j and x_i must be greater than or equal to D_{ij}. Hence we can now formulate the scheduling problem in order to minimize the completion time z as follows:

$$\text{Minimize } z = x_n - x_0 \tag{8.1}$$

subject to

$$x_j - x_i - D_{ij} \geqslant 0 \quad \text{for all } (i, j) \tag{8.2}$$

$$x_j > x_i \quad \text{for all } j > i, \quad i, j = 0, 1, 2, \ldots, n. \tag{8.3}$$

The events are unrestricted in sign as the reference point of the time scale has not yet been chosen. Usually, the initial event x_0 starts at time zero and therefore all other values of x must be positive. If we now introduce a slack variable x_{ij} for each activity (i, j) then the inequality constraints may be replaced by equality constraints. Therefore, the problem may now be re-stated as

$$\text{Minimize } z = x_n \tag{8.4}$$

subject to

$$x_j - x_i - D_{ij} - x_{ij} = 0 \quad \text{for all } (i, j) \tag{8.5}$$

$$x_{ij} \geqslant 0 \quad \text{for all } (i, j) \tag{8.6}$$

$$x_j > x_i \geqslant 0 \quad i, j = 0, 1, 2, 3, \ldots, n$$

Having finished the construction of the network, the planner should now estimate as closely as possible the most likely duration of each activity within the project. There should be at least a fifty-fifty chance of the activity being completed within the duration or expected time. In general, any over- or under-estimations should even themselves out over the duration of the project as a whole. At this stage, the desirable completion time for the project need not be taken into account. In fact, the presence of 'deadlines' should not influence the estimates of durations of various activities, ending in unrealistic quotations being made with everyone being unhappy!

8.3.1 Earliest event times

Once the activities have been assigned realistic time durations D_{ij} it is possible to calculate the *earliest event times*. As the name implies, this is the earliest time by which that event can take place, and is based on the length of time required to execute the chain of activities that culminate in that event.

In Figure 8.7, the earliest time that event 2 can take place is day 3, and the earliest that event 3 can take place is day 4 (i.e. 3 + 1).

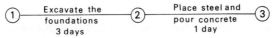

Figure 8.7 Event times for two activities

Where two or more chains of activities lead up to an event, the earliest time at which that event can take place is the time of the chain taking the longest time to complete.

In Figure 8.8, the earliest time at which event 2 can take place is after 1 day has elapsed; event 4 after 1 + 3 = 4 days have elapsed; and event 3 after 6 days.

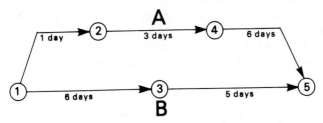

Figure 8.8 Event times in a two-chain network

However, what happens to event 5? This has two chains of events leading up to it, chain A via events 1–2–4–5, and chain B via events 1–3–5. The first chain, i.e. A, is completed after 1 + 3 + 6 = 10 days, but event 5 cannot take place until chain B, which is of duration 6 + 5 = 11 days, is also complete. Hence, the earliest time that event 5 can take place is after 11 days and this termed as being the *earliest event time* (EET) for event 5. These times are then entered on the network framed by a square as shown in Figure 8.9.

If it is now stipulated that event 4 is unable to commence until event 3 has been completed then a dummy activity must be introduced into the network. This is

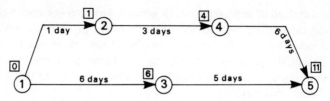

Figure 8.9 □ = Earliest even time: the earliest time at which the event can take place

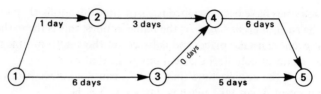

Figure 8.10 Insertion of a dummy activity

indicated in Figure 8.10. The dummy activity 3–4 takes zero time but its presence will influence the earliest event times of the network as follows. Event 2 is not influenced by the presence of the dummy activity and hence its earliest event time remains unaltered. However, event 4 can now be attained via two chains of activities:

1–2–4, which takes $1 + 3 = 4$ days

1–3–4, which takes $6 + 0 = 6$ days.

Therefore, the earliest time at which event 4 can now take place has been changed from 4 to 6 days. Activity 4–5 commences at this time and will not be completed until after $6 + 6 = 12$ days.

The chain of events 1–3–5 remains unaltered at $6 + 5 = 11$ days, and therefore the EET for event 5 is now 12 days as shown in Figure 8.11. This is also the earliest

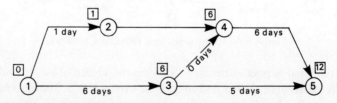

Figure 8.11 Effect of dummy activity on earliest event times

time by which the project as a whole can be completed. It is obvious that it is unnecessary to trace the chains of activities from the start of the project each time an earliest event time is to be determined. This may be done by adding the durations D_{ij} of the activities to the previously determined EET at the start of the activity, e.g. the EET for event 5 may be found by adding to the earliest event times of events 4 and 3 the durations of activities 4–5 and 3–5 respectively: $6 + 6 = 12$ and $6 + 5 = 11$, the longer time, 12 days, being the EET of event 5.

8.3.2 Latest event times

Once the earliest event times have been found the next logical step is to calculate the *latest event time* (LET) for each event. This is defined as being the latest time at which an event can take place if the earliest time for project completion as a whole is not to be exceeded. In this case the calculations commence with the project end event and a *backward pass* is made through the network, deducting the duration D_{ij} of each activity from the latest event time of its end event.

Referring to Figure 8.11, it is seen that if the earliest project time of 12 days is not to be exceeded, then the LET for event number 5 must also be the twelfth day. The LET is then placed within a triangle immediately to the right of the square representing the earliest event time (Figure 8.12).

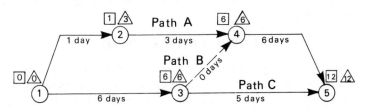

Figure 8.12 Network showing EET and LET values

Following the events in descending order back through the network we next arrive at event number 4. The LET for event 4 is obtained by deducting from the LET of event 5 the duration of activity 4–5, i.e. 6 days. This results in a LET of $12 - 6 = 6$th day for event 4.

Now, event number 3 has two chains of events leading back to it. Therefore, if the project time is not to be exceeded, event 3 must take place in sufficient time to allow the longest sequence of activities to take place, in this case sequence 3–4–5 which takes $0 + 6 = 6$ days as against sequence 3–5 which takes only 5 days. Hence, event 3 must take place 6 days before the twelfth day, i.e. $12 - 6 = 6$th day.

As mentioned previously, for the calculation of the earliest event times it is unnecessary to work through the whole network each time the LET needs to be found. It is sufficient to consider only the activities leading from the event and the latest even times of the events at the end of the activities. Hence, the LET for event 2 is (Latest event time 4) – (Duration of activity 2–4) $= 6 - 3 = 3$ days.

Finally, event 1 has two activities emanating from it. The activity in the longest chain determines the latest time by which the event can take place if the project completion time is not to be delayed. The two choices are as follows:

(Latest event time of event 2) − (Duration of activity 1–2) = 2 days
or
(Latest event time of event 3) − (Duration of activity 1–3) = 0.

The longest time path is therefore via activity 1–3 and the latest time by which event 1 can take place is day zero.

The LET for the start event must always be zero if this method is followed. The end result of adding up the durations of the activities in the longest chain to obtain the earliest event time of the last event, and then taking away the durations of the same activities to arrive at the LET of the start event, will always be zero.

Alternatively, if the project time of 12 days is not to be exceeded and the longest chain of activities is of 12 days' duration, then the latest time the project can start is day zero.

This provides a check on the calculations made for determining the latest event times. If the latest event time for the start event is greater than zero then an error has been made in the calculations. The final result for this particular problem is given in Figure 8.12. Once this stage has been reached it is then feasible to determine the *critical path* and carry out a complete time analysis of the programme of events.

8.4 THE CRITICAL PATH

After the *earliest event times* (EET) and the *latest event times* (LET) for each event have been determined it is possible to calculate the critical path or paths for the network. An event lies in the critical path if its EET (x'_i) and its LET (x''_i) are identical, as any delay in satisfying this event will automatically violate its latest event time (which was based on the EET for project completion), and the project is delayed.

8.4.1 Total float

If the EET for any event is less than its LET, a certain delay in completing that event is allowable which will not delay the completion of the overall project. For example, referring to Figure 8.12, activities 1–2 and 2–4 have two spare days. Either of them could take two days longer to complete without the path becoming critical. This extra time could be divided between them, e.g. one day to activity 1–2 and one day to activity 2–4, or any other combination provided that the total spare time does not exceed two days. On the other hand, event 4 is a critical event lying on the critical path, because x'_4 and x''_4 are equal, and no tolerance is permissible in satisfying this event if the project completion time of 12 days is to be satisfied. An analysis of Figure 8.12 indicates that events 1, 3, 4, and 5 are critical and lie on the critical path. Event 2 is a non-critical event and therefore does not lie on the critical path. It is a central event in the activities 1–2 and 2–4 which are known to have a certain flexibility in their start and finish times. This spare time is known as the *total float* of the two activities and can be used in either one or both of the activities but must not exceed a total time of two days.

Therefore, the total float is the amount of spare time attributed to an activity and if absorbed within the activity, resulting in a lengthening of the duration D_{ij}, then the floats in both previous and succeeding activities may be reduced. If the duration of activity 1–2 is increased by one day from 1 to 2 days, i.e. one day of the float is used, then the total time for path A becomes 11 days. The spare time has been reduced to one day and activities 1–2 and 2–4 now have total floats of only one day. However, a further increase of one day in the duration of activity 1–2 would make all paths in the network critical, as the EET and LET of event 2 would now be equal.

It is unnecessary to calculate the total duration of each path to determine the total float of any particular activity on the path concerned. The total float may be found very easily by means of the leeway between the start and finish events of the activity. Consider activity 2–4 as shown in Figure 8.13.

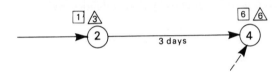

Figure 8.13   Determination of total float

The information already calculated and entered on the diagram is as follows:

Duration D_{24}	= 3 days
Earliest event time, event 2	= day 1
Latest event time, event 2	= day 3
Earliest event time, event 4	= day 6
Latest event time, event 4	= day 6.

This information shows that activity 2–4 may start on day 1 (earliest event time for event 4), but does not have to be finished until day 6 (latest event time for event 4). Therefore, there are five days in which to carry out this activity, but as it requires only three days there is a *total float* of two days. The total float may therefore be calculated directly from the information shown on the diagram by deducting the EET of the start event and the duration from the LET of the end event.

8.4.2   Other types of float

In addition to total float, other types of float can also be defined. The practical value of these often does not warrant the additional calculation, but there may be instances where this information would be of value. The definitions of the other types of float are as follows.

Let t_{ij} be defined as the start of an activity (i,j), and T_{ij} the finish time of the same activity. The earliest time that an activity (i,j) may start without interfering with completion time of its predecessors is given by t'_{ij} and the latest time that this activity (i,j) can start without interfering with the completion time of its successors is denoted by t''_{ij}. Similarly, the earliest finish time T'_{ij} and the latest finish time T''_{ij} are also defined. These activity start and finish times can be related to the earliest event times x'_i or x'_j, and to the latest event times, x''_i or x''_j as follows:

$$t'_{ij} = x'_i, \qquad (8.7a)$$

$$T'_{ij} = x'_i + D_{ij}, \qquad (8.7b)$$

$$T''_{ij} = x''_j, \qquad (8.7c)$$

and

$$t''_{ij} = x''_j - D_{ij}. \qquad (8.7d)$$

Therefore, the *total float* in an activity (i,j) is defined by

$$x^t_{ij} = T''_{ij} - T'_{ij} = t''_{ij} - t'_{ij} \qquad (8.8)$$

and for a critical activity

$$x^t_{ij} = 0.$$

Unless reasons are given specifically to the contrary, it is always advisable to follow a schedule of the earliest event times, i.e. to start and finish all activities as early as possible. Such a schedule is called the early time schedule. However, once a float in an activity is used it may interfere with the floats of its predecessors. The duration of an activity with *free float* can be increased by the amount of float without affecting the progress of succeeding activities. This will, however, reduce the amount of total float available for preceding activities. A *free float* occurs only in the last of a series of activities leading to a critical path and may be calculated by deducting the EET of the start event and the duration from the EET of the end activity, i.e.

$$x^f_{ij} = x'_j - x'_i - D_{ij}. \qquad (8.9)$$

Therefore, for activity 2–4 in Figure 8.13,

Free float $(2, 4) = 6 - 1 - 3 = 2$ days.

If a portion of the *total float* in the activity (i,j) is entirely independent of successors or predecessors, that portion is called the *independent float* and may be defined as follows:

$$x^i_{ij} = x'_j - x''_i - D_{ij}, \quad x^i_{ij} \geqslant 0. \qquad (8.10)$$

In other words, the independent float is calculated by deducting the latest event time of the start event and the duration of the activity from the earliest event time of the end event, e.g. activity 3–5 in Figure 8.14.

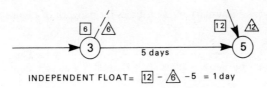

INDEPENDENT FLOAT = ⌑12⌑ - ⟋6⟍ -5 = 1 day

Figure 8.14 Determination of independent float

8.5 ACTIVITY TIMES

Also of importance to the user of the CPM are the times which activities, as opposed to events, can take place. In fact, this information is often of greater use than the event times, since it is more usual to designate the start and finish dates of activities than the occurrence of events when scheduling work operations.

The activity times may be calculated as follows:

(a) *Earliest start time* (t'_{ij})

The earliest time at which an activity can start is immediately the start event has occurred. The earliest start time is the earliest event time of the start event; thus, the earliest start time for activity 2–4 = day 1.

(b) *Earliest finish time* (T'_{ij})

To enable an activity to finish as early as possible then it must start at the earliest start time. In other words, the earliest finish time is equal to the earliest start time plus the duration of the activity; thus, the earliest finish time for 2–4 is = 1 + 3 = day 4.

(c) *Latest start time* (t''_{ij})

This is the latest time at which the activity can start without delaying the finish of the project, i.e. the latest start time is the latest event time for the end event minus the duration; thus, the latest start time for 2–4 = 6 − 3 = day 3.

(d) *Latest finish time* (T''_{ij})

The latest time at which an activity can finish is immediately before the latest time the end event takes place, i.e. the latest finish time is equal to the latest event time of the end event and the latest event time for 2–4 = day 6.

As stated in the previous section the differences between the two start times and between the two finish times are the same and equal to the total float of the activity, e.g. for 2–4, (3 − 1) = (6 − 4) = (2 days of total float). When this difference is zero, there is zero total float and the activity is therefore on the critical path. For the purpose of scheduling and control these details regarding

Table 8.3 Summary of event times for network of Figure 8.12

Activity	Duration	Start		Finish		Total float	Critical activities (*)
		Earliest	Latest	Earliest	Latest		
1–2	1	0	2	1	3	2	
1–3	6	0	0	6	6	0	*
2–4	3	1	3	4	6	2	
3–4	0	6	6	6	6	0	*
3–5	5	6	7	11	12	1	
4–5	6	6	6	12	12	0	*

activity times may be entered into a table to provide an invaluable aid for planning operations. Table 8.3 shows such a table for the network of Figure 8.12.

8.6 THE CRITICAL TIME PATH

The minimum time required for the completion of all activities in a project is governed by the longest time path linking the initial event and the terminal event of the project. Hence, the solution of the critical scheduling problem is equivalent to the determination of the longest time path in a network having a unit time flow from the source to the sink. If a branch (i, j) is an integral part of a path constituting the longest time path between the source and sink, it must satisfy the condition $x_{ij} = 0$. In other words, it must be completed on time if a minimum duration is given to complete the project. Therefore,

$$x_j - x_i = D_{ij}.$$

If we now start to construct a path from node 0 (the initial event) to every other node (in increasing order of node), then the value of x_i corresponding to the longest of all possible paths from node 0 to node i is denoted by x'_i. Starting from node i, the time path to each of the nodes j on the other ends of all bursting branches (i.e. all branches emanating from i) incident to node i is given by

$$x_j = x'_i + D_{ij}.$$

This value of x_j for any other branch (i, j) incident to i, along with those for other branches incident to j, will be investigated later in order to determine the maximum value of x_j (i.e. x'_j) for node j. The procedure can now be repeated for all nodes within the system until the resulting $x'_n = $ maximum x_n.

As the procedure moves in an ascending manner from node to node it is referred to as a *forward pass*. The set of x_i thus obtained is denoted by x'_i ($i = 0, 1, \ldots, n$) and each x_i represents the earliest time that event i may take place and is the earliest event time for event i.

If the same procedure is used starting with node n and then moving in a backward direction to every other node i, then the value of x_n for node n is taken to be $x_n'' = $ maximum x_n. At any node x_j where one or more bursting branches meet, the value of x_j corresponding to the longest of all possible time paths leading back from the sink n to node j (i.e. the minimum value of x_j) is denoted by x_j''. Therefore, moving backward through the network from node j, the time path to each node i on the other ends of all merging branches incident to node j is given by

$$x_i = x_j'' - D_{ij}.$$

This value of x_i for any branch (i, j) incident on j will be examined later along with those for other branches incident to i in determining minimum $x_i = x_i''$ for node i. In a similar manner to the *forward pass*, the procedure can be repeated in descending order of node magnitude until $i = 0$, with $x_0'' = 0$. The procedure is referred to as a *backward pass*, and the set of x_i obtained is denoted by x_i'' ($i = 0, 1, 2, \ldots, n$), commencing with $x_n'' = $ maximum x_n. In this case, the values of x_i'' represent the latest time that event i may take place, i.e. the *latest event time* for event i.

Once the *backward* and *forward passes* have been calculated it is possible to determine whether a branch (i, j) lies within the longest time path maximum x_n on validating the following conditions:

$$x_i = x_i' = x_i'' \tag{8.11a}$$

$$x_j = x_j' = x_j'' \tag{8.11b}$$

and

$$x_j' - x_i' = x_j'' - x_i'' = D_{ij}. \tag{8.11c}$$

The computation involved in this procedure is explained in detail in the following example.

8.7 A SWIMMING POOL PROJECT

A residential owner has decided to have a swimming pool constructed in his rear garden in order that his family may enjoy more fully the dubious pleasures of bathing during the British summer. What was first thought to be a relatively straightforward project turns out, on closer analysis, to be a rather complicated procedure. This is due mainly to the relative inaccessibility of the garden from the main road and the lack of space for storing the raw materials on site, e.g. it is impossible to machine-dig the excavation and so two labourers are required. Similarly, the soil cannot be conveniently removed and so this will have to be distributed around the site and in adjacent gardens (assuming good relationships between the owner and his neighbours).

Each labourer works a maximum 35 hours per week (or 7 hours per day for 5 days) and the dimensions of the swimming pool are given in Figure 8.15. The construction of the retaining wall is illustrated in Figure 8.16 with a reinforced concrete ring beam around the top perimeter of the pool to prevent damage by freezing. Solar panels are used to heat the pool water.

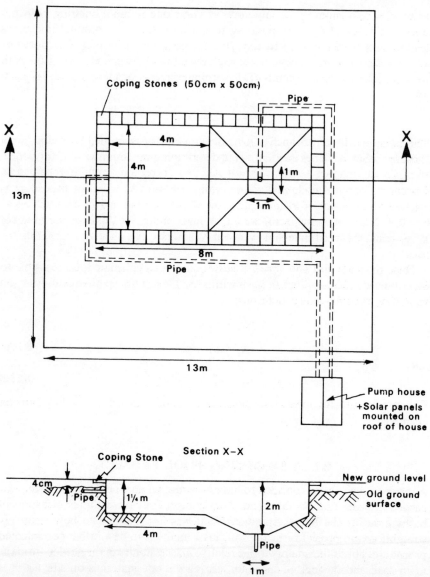

Figure 8.15 Plan and sectional view of a swimming pool

It is now possible to break down the schedule of events in a logical manner and hence determine the minimum time necessary to complete the pool. This is achieved by assigning realistic times to the different tasks and constructing a network of all the activities.

A description of the activities involved is given in Table 8.4, and as may be seen these are not in the correct sequence of events. The network shown in Figure 8.17

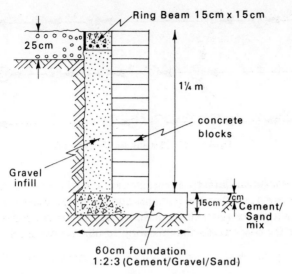

Figure 8.16 Detail of retaining wall

Table 8.4 Activities in the swimming pool project

Activity (i,j)	Description	Duration (days)
1, 2	General site clearance and levelling of swimming pool area	$1\frac{1}{2}$
2, 4	Setting out the dimensions of the pool for excavating	$1\frac{1}{2}$
4, 6	Removal of soil and levelling around garden	7
6, 8	Excavation of deep end	2
8, 9	Construction of footings for retaining walls	2
7, 10	Placement of pipes for drainage and re-circulating system	$1\frac{1}{2}$
1, 8	Construction of pump house	3
1, 3	Placement of solar panels for heating water	1
3, 5	Connecting pipework for solar panels	$1\frac{1}{2}$
9, 11	Construction of retaining wall	4
11, 14	Plastering of retaining wall	1
14, 16	Laying and floating of pool floor	$\frac{1}{2}$
16, 19	Fitting of pool liner	1
10, 12	Fitting of underwater light and electrical connections	$1\frac{1}{2}$
18, 19	Laying of coping and paving stones around pool edge	3
12, 18	Testing of electrical system	1
6, 13	Constructing of retaining wall for increased garden level	1
5, 7	Placement of pipes and connecting heat pump	2
7, 8	Connecting pipe network for filtering and re-circulating	2
13, 15	Backfilling of gravel infill	$1\frac{1}{2}$
15, 17	Construction of reinforced concrete ring beam	1
17, 19	Returfing of garden area	1

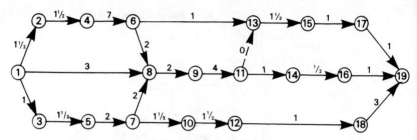

Figure 8.17 Setting up the network

is one form of solution but is by no means unique. In the network the durations of the activities are marked on the branches and knowing these it is now possible to proceed with a *forward pass* through the system. In the forward pass, $x'_1 = 0$ for node 1. For the other nodes the following results are obtained:

$i = 2,\quad x_2 = x'_1 + D_{1,2} = 0 + 1\tfrac{1}{2} = 1\tfrac{1}{2}\quad (\max x_2 = 1\tfrac{1}{2} = x'_2)$

$i = 3,\quad x_3 = x'_1 + D_{1,3} = 0 + 1 = 1\quad (\max x_3 = 1 = x'_3)$

$i = 4,\quad x_4 = x'_2 + D_{2,4} = 1\tfrac{1}{2} + 1\tfrac{1}{2} = 3\quad (\max x_4 = 3 = x'_4)$

$i = 5,\quad x_5 = x'_3 + D_{3,5} = 1 + 1\tfrac{1}{2} = 2\tfrac{1}{2}\quad (\max x_5 = 2\tfrac{1}{2} = x'_5)$

$i = 6,\quad x_6 = x'_4 + D_{4,6} = 3 + 7 = 10\quad (\max x_6 = 10 = x'_6)$

$i = 7,\quad x_7 = x'_5 + D_{5,7} = 2\tfrac{1}{2} + 2 = 4\tfrac{1}{2}\quad (\max x_7 = 4\tfrac{1}{2} = x'_7)$

$i = 8,\quad x_8 = x'_6 + D_{6,8} = 10 + 2 = 12$
$ x_8 = x'_1 + D_{1,8} = 0 + 3 = 3 \quad\bigg\}\ (\max x_8 = 12 = x'_8)$
$ x_8 = x'_7 + D_{7,8} = 4\tfrac{1}{2} + 2 = 6\tfrac{1}{2}$

$i = 9,\quad x_9 = x'_8 + D_{8,9} = 12 + 2 = 14\quad (\max x_9 = 14 = x'_9)$

$i = 10,\quad x_{10} = x'_7 + D_{7,10} = 4\tfrac{1}{2} + 1\tfrac{1}{2} = 6\quad (\max x_{10} = 6 = x'_{10})$

$i = 11,\quad x_{11} = x'_9 + D_{9,11} = 14 + 4 = 18\quad (\max x_{11} = 18 = x'_{11})$

$i = 12,\quad x_{12} = x'_{10} + D_{10,12} = 6 + 1\tfrac{1}{2} = 7\tfrac{1}{2}\quad (\max x_{12} = 7\tfrac{1}{2} = x'_{12})$

$i = 13,\quad x_{13} = x'_6 + D_{6,13} = 10 + 1 = 11$
$ x_{13} = x'_{11} + D_{11,13} = 18 + 0 = 18\ \bigg\}\ (\max x_{13} = 18 = x'_{13})$

$i = 14,\quad x_{14} = x'_{11} + D_{11,14} = 18 + 1 = 19\quad (\max x_{14} = 19 = x'_{14})$

$i = 15,\quad x_{15} = x'_{13} + D_{13,15} = 18 + 1\tfrac{1}{2} = 19\tfrac{1}{2}\quad (\max x_{15} = 19\tfrac{1}{2} = x'_{15})$

$i = 16,\quad x_{16} = x'_{14} + D_{14,16} = 19 + \tfrac{1}{2} = 19\tfrac{1}{2}\quad (\max x_{16} = 19\tfrac{1}{2} = x'_{16})$

$i = 17,\quad x_{17} = x'_{15} + D_{15,17} = 19\tfrac{1}{2} + 1 = 20\tfrac{1}{2}\quad (\max x_{17} = 19\tfrac{1}{2} = x'_{17})$

$i = 18,\quad x_{18} = x'_{12} + D_{12,18} = 7\tfrac{1}{2} + 1 = 8\tfrac{1}{2}\quad (\max x_{18} = 8\tfrac{1}{2} = x'_{18})$

$$i = 19, \quad \begin{aligned} x_{19} &= x'_{16} + D_{16,19} = 19\tfrac{1}{2} + 1 = 20\tfrac{1}{2} \\ x_{19} &= x'_{17} + D_{17,19} = 20\tfrac{1}{2} + 1 = 21\tfrac{1}{2} \\ x_{19} &= x'_{18} + D_{18,19} = 8\tfrac{1}{2} + 3 = 11\tfrac{1}{2}. \end{aligned} \right\} (\max x_{19} = 21\tfrac{1}{2} = x'_{19})$$

In the backward pass, the procedure commences by designating $x''_{19} = \max x_{19} = 21\tfrac{1}{2}$. The values at the remaining nodes are obtained as follows:

$i = 18$, $x_{18} = x''_{19} - D_{18,19} = 21\tfrac{1}{2} - 3 = 18\tfrac{1}{2}$ (min $x_{18} = 18\tfrac{1}{2} = x''_{18}$)

$i = 17$, $x_{17} = x''_{19} - D_{17,19} = 21\tfrac{1}{2} - 1 = 20\tfrac{1}{2}$ (min $x_{17} = 20\tfrac{1}{2} = x''_{17}$)

$i = 16$, $x_{16} = x''_{19} - D_{16,19} = 21\tfrac{1}{2} - 1 = 20\tfrac{1}{2}$ (min $x_{16} = 20\tfrac{1}{2} = x''_{16}$)

$i = 15$, $x_{15} = x''_{17} - D_{15,17} = 20\tfrac{1}{2} - 1 = 19\tfrac{1}{2}$ (min $x_{15} = 19\tfrac{1}{2} = x''_{15}$)

$i = 14$, $x_{14} = x''_{16} - D_{14,16} = 20\tfrac{1}{2} - \tfrac{1}{2} = 20$ (min $x_{14} = 20 = x''_{14}$)

$i = 13$, $x_{13} = x''_{15} - D_{13,15} = 19\tfrac{1}{2} - 1\tfrac{1}{2} = 18$ (min $x_{13} = 18 = x''_{13}$)

$i = 12$, $x_{12} = x''_{18} - D_{12,18} = 18\tfrac{1}{2} - 1 = 17\tfrac{1}{2}$ (min $x_{12} = 17\tfrac{1}{2} = x''_{12}$)

$i = 11, \quad \begin{aligned} x_{11} &= x''_{14} - D_{11,14} = 20 - 1 = 19 \\ x_{11} &= x''_{13} - D_{11,13} = 18 - 0 = 18 \end{aligned} \right\} (\min x_{11} = 18 = x''_{11})$

$i = 10$, $x_{10} = x''_{12} - D_{10,12} = 17\tfrac{1}{2} - 1\tfrac{1}{2} = 16$ (min $x_{10} = 16 = x''_{10}$)

$i = 9$, $x_9 = x''_{11} - D_{9,11} = 18 - 4 = 14$ (min $x_9 = 14 = x''_9$)

$i = 8$, $x_8 = x''_9 - D_{8,9} = 14 - 2 = 12$ (min $x_8 = 12 = x''_8$)

$i = 7, \quad \begin{aligned} x_7 &= x''_{10} - D_{7,10} = 16 - 1\tfrac{1}{2} = 14\tfrac{1}{2} \\ x_7 &= x''_8 - D_{7,8} = 12 - 2 = 10 \end{aligned} \right\} (\min x_7 = 10 = x''_7)$

$i = 6, \quad \begin{aligned} x_6 &= x''_{13} - D_{6,13} = 19 - 1 = 18 \\ x_6 &= x''_8 - D_{6,8} = 12 - 2 = 10 \end{aligned} \right\} (\min x_6 = 10 = x''_6)$

$i = 5$, $x_5 = x''_7 - D_{5,7} = 10 - 2 = 8$ (min $x_5 = 8 = x''_5$)

$i = 4$, $x_4 = x''_6 - D_{4,6} = 10 - 7 = 3$ (min $x_4 = 3 = x''_4$)

$i = 3$, $x_3 = x''_5 - D_{3,5} = 8 - 1\tfrac{1}{2} = 6\tfrac{1}{2}$ (min $x_3 = 6\tfrac{1}{2} = x''_3$)

$i = 2$, $x_2 = x''_4 - D_{2,4} = 3 - 1\tfrac{1}{2} = 1\tfrac{1}{2}$ (min $x_2 = 1\tfrac{1}{2} = x''_2$)

$i = 1, \quad \begin{aligned} x_1 &= x''_3 - D_{1,3} = 6\tfrac{1}{2} - 1 = 5\tfrac{1}{2} \\ x_1 &= x''_8 - D_{1,8} = 12 - 3 = 9 \\ x_1 &= x''_2 - D_{1,2} = 1\tfrac{1}{2} - 1\tfrac{1}{2} = 0 \end{aligned} \right\} (\min x_1 = 0 = x''_1).$

In order to determine the critical activities, the relationship defined in (8.11) is used. Thus, each branch (i, j) is checked in which $x'_i = x''_i$ and $x'_j = x''_j$. For this particular problem the solid lines in Figure 8.18 show all the branches the earliest

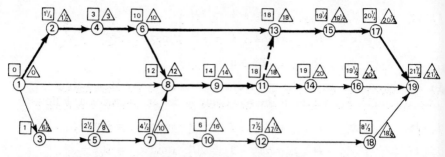

Figure 8.18 Completed network for the swimming pool project

and latest event times of which satisfy equations (8.11a) and (8.11b). Equation (8.11c) is now used to determine whether path 13 to 6, or path 13–11–9–8–6 lies on the critical path.

For path 13–6

$$x'_j - x'_i = x''_j - x''_i = D_{ij}$$
$$x'_j - x'_i = 18 - 10 = 8$$
$$x''_j - x''_i = 18 - 10 = 8$$
$$D_{ij} = 1.$$

Hence branch 6–13 does not lie on the critical path. Inspection of the network shown in Figure 8.18 indicates that path 13–11–9–8–6 does satisfy all of equation (8.11) and therefore does lie on the critical path.

8.8 COMPUTER SOLUTION

From the previous example it is clear that for projects of any significant size the repetitious nature of the work is tedious and presents great scope for human error. The method is ideally suited to computer solution and many sophisticated and efficient programs have been written for the application of Critical Path analysis. Appendix A contains a subroutine CPM which illustrates how the logic described in the previous sections can be encoded in a computer program. The routine is intended for educational value and is therefore not necessarily suited for practical application. This section shows how the routine may be used to solve the swimming pool project described in Section 8.7.

Reference should be made to the listing of routine CPM in Appendix A for a description of the parameters employed. The input data comprise the description of the activities in terms of origin and destination nodes (events) and the definition of activity durations. In addition, to allow the arrays to be dimensioned dynamically, the number of nodes and activities (including dummy activities) must be specified.

It is clear therefore that the user must go through the process of setting up the network as described in the early sections of this chapter. This process cannot be automated, although it may be assisted by computer use. The data may be assigned to the appropriate arrays either by reading from a file, by assignment statements, or DATA statements. For generality, a data file will be used here, as illustrated in Table 8.5.

Table 8.5 Data file for the network of Figure 8.17 (the swimming pool project)

19	Nodes	
23	Activities	
1	2	1.5
2	4	1.5
4	6	7.0
6	8	2.0
8	9	2.0
7	10	1.5
1	8	3.0
1	3	1.0
3	5	1.5
9	11	4.0
11	14	1.0
14	16	0.5
16	19	1.0
10	12	1.5
18	19	3.0
12	18	1.0
6	13	1.0
5	7	2.0
7	8	2.0
13	15	1.5
15	17	1.0
17	19	1.0
11	13	0.0

A simple driving program to read the data and call routine CPM is shown in Figure 8.19. It is assumed that the data file is assigned to unit 1 and automatic output of a table of results is obtained by setting NW = 6. The arrays of the calling program are dimensioned for arbitrary values of NACT (100) and NNODE (80) to allow a range of problems to be solved. Figure 8.19 also shows the results obtained and it will be left to the reader to confirm agreement with the results of Section 8.7.

8.9 RESOURCE ALLOCATION

Resource allocation is carried out to ensure that the most economic use is made of the resources available. These may be of a very diverse nature with widely differing characteristics.

```
      DIMENSION NUS(100),NDS(100),DUR(100),STMIN(100),STMAX(100)
      DIMENSION FTMIN(100),FTMAX(100)
      DIMENSION EVTMIN(80),EVTMAX(80),FF(80),TF(80)
C  REWIND DATA FILE ON UNIT 1.
      REWIND 1
C  READ SIZE OF NETWORK
      READ (1,10) NACT
      READ (1,10) NNODE
   10 FORMAT(I5)
C  READ ACTIVITIES
      DO 20 IACT = 1,NACT
      READ (1,30)NUS(IACT),NDS(IACT),DUR(IACT)
   30 FORMAT(2I5,F10.3)
   20 CONTINUE
C  OUTPUT RESULTS TO UNIT 6
      NW=6
C
      CALL CPM(NUS,NDS,DUR,NACT,NNODE,STMIN,STMAX,
     +         FTMIN,FTMAX,EVTMIN,EVTMAX,FF,TF,NW)
      STOP
      END
```

	NUS	NDS	DURATN	STMIN	FTMIN	STMAX	FTMAX	FF()	TF()
1	1	2	1.5	0.0	1.5	0.0	1.5	0.0	0.0
2	2	4	1.5	1.5	3.0	1.5	3.0	0.0	0.0
3	4	6	7.0	3.0	10.0	3.0	10.0	0.0	0.0
4	6	8	2.0	10.0	12.0	10.0	12.0	0.0	0.0
5	8	9	2.0	12.0	14.0	12.0	14.0	0.0	0.0
6	7	10	1.5	4.5	6.0	14.5	16.0	0.0	10.0
7	1	8	3.0	0.0	3.0	9.0	12.0	9.0	9.0
8	1	3	1.0	0.0	1.0	5.5	6.5	0.0	5.5
9	3	5	1.5	1.0	2.5	6.5	8.0	0.0	5.5
10	9	11	4.0	14.0	18.0	14.0	18.0	0.0	0.0
11	11	14	1.0	18.0	19.0	19.0	20.0	0.0	1.0
12	14	16	0.5	19.0	19.5	20.0	20.5	0.0	1.0
13	16	19	1.0	19.5	20.5	20.5	21.5	1.0	1.0
14	10	12	1.5	6.0	7.5	16.0	17.5	0.0	10.0
15	18	19	3.0	8.5	11.5	18.5	1.5	10.0	10.0
16	12	18	1.0	7.5	8.5	17.5	18.5	0.0	10.0
17	6	13	1.0	10.0	11.0	17.5	18.0	7.0	7.0
18	5	7	2.0	2.5	4.5	8.0	10.0	0.0	5.5
19	7	8	2.0	4.5	6.5	10.0	12.0	5.5	5.5
20	13	15	1.5	18.0	19.5	18.0	19.5	0.0	0.0
21	15	17	1.0	19.5	20.5	19.5	20.5	0.0	0.0
22	17	19	1.0	20.5	21.5	20.5	21.5	0.0	0.0
23	11	13	0.0	18.0	18.0	18.0	18.0	0.0	0.0

Figure 8.19 Main program and results for the swimming pool project

The main steps in resource allocation are:

(i) Estimating and scheduling of resources required.
(ii) Determination of available resources.
(iii) Allocation of resources.
(iv) Re-planning.

Before proceeding with a resource allocation analysis, it is essential that the main categories of resources and their different characteristics are recognized. These may take the form of *capital* which could take the form of a lump sum at the commencement of the project or at required intervals during the project. *Manpower*, unlike capital, cannot be carried forward from one day to the next. If an error of judgement is made in hiring more men than is necessary on a particular day, then that loss is irredeemable. However, manpower does have the flexibility of being movable from one location to another provided the work is in the same category. Other forms of resource are *plant and equipment* which may either be hired or owned by the firm carrying out the project. As with *manpower* unused *plant and equipment* represents a direct loss either in terms of hire charges, depreciation, or loss of earning power.

Materials are a form of resource which, if purchased too early during the life of the project, result in a loss in terms of interest on the capital expended, but, unlike *plant and equipment*, can be carried forward from one period to another. Finally, *space* is an important resource, the correct allocation of which is often neglected. This may take the form of accommodation for manpower or equipment, storage space for materials, or working space which may not be required once the project has been completed.

Therefore, irrespective of which resource is being allocated, the problem is basically that of scheduling the various activities of the project such that the requirements of key resources are *levelled* to relatively constant rates. If there are no restrictions on resources then the process of levelling involves only the adjustment of the schedule of the non-critical activities within the project completion time as determined by the critical path method. Alternatively, if, as is more likely, the availability of a resource is limited then the levelling process may require the adjustment of activity durations and/or project completion.

Each of the previously mentioned resources must be analysed sequentially in their order of importance. There is no established procedure for finding an optimal solution for the problem, but once a trial solution has been obtained then its relative merit can be assessed. The resource required per day is denoted by q_{ij} for the particular activity (i, j) and x_{ij}^k is the occurrence of that activity (i, j) on the kth day of the project such that

$$x_{ij}^k = \begin{cases} 1, & \text{if } (i,j) \text{ occurs on the } k\text{th day} \\ 0, & \text{if it does not.} \end{cases} \quad (8.12)$$

Let Q_k be the total requirement of the resource on the kth day for all the activities in the project, and n the number of time units necessary to complete the project. Hence,

$$Q_k = \sum_{ij} q_{ij} x_{ij}^k, \quad k = 1 \text{ to } n. \quad (8.13)$$

Thus, the average daily requirement of the resource is

$$Q_{\text{ave}} = \frac{1}{n} \sum_{k=1}^{n} Q_k. \qquad (8.14)$$

An indication of the uniformity of Q_k over the duration of the project is given by summing the squares of the differences $(Q_{\text{ave}} - Q_k)$. Hence, a solution to the problem would be to determine the least squares form of the sum S, i.e.

$$\text{Minimize } S = \sum_{k=1}^{n} (Q_{\text{ave}} - Q_k)^2. \qquad (8.15)$$

Therefore, when min $S = 0$, the daily requirement is a constant. From equation (8.14), in which Q_{ave} is a constant, we have

$$\text{Minimize } S = \sum_{k=1}^{n} Q_{\text{ave}}^2 - 2 Q_{\text{ave}} \sum_{k=1}^{n} Q_k + \sum_{k=1}^{n} Q_k^2$$

$$= n Q_{\text{ave}}^2 - 2 Q_{\text{ave}} (n Q_{\text{ave}}) + \sum_{k=1}^{n} Q_k^2$$

$$= \sum_{k=1}^{n} Q_k^2 - n Q_{\text{ave}}^2.$$

As nQ_{ave}^2 is a constant, the minimization procedure requires only

$$\text{Minimize } S' = \sum_{k=1}^{n} Q_k^2. \qquad (8.16)$$

Therefore, once a number of trial solutions have been obtained, their relative merit may be judged by means of (8.16).

One technique of obtaining trial solutions of resource levelling problems is to list the project activities along with their daily resource requirements. Then, a comparison should be made of all possible S' values by scheduling non-critical activities for various times as allowed by the floats. The earliest time schedule is often used as the starting point for such analyses as it immediately removes any concern regarding the completion time of the activities involved in the project. It should be borne in mind that although the total floats do not influence their predecessors or successors, the dependent floats may do and can only be used if this is ensured not to be the case.

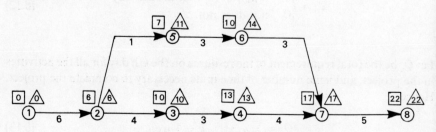

Figure 8.20 Network for resource allocation

An example problem is given to demonstrate the applicability of the technique. A network as shown in Figure 8.20 has been obtained for a project. The resources estimated for each activity are as follows:

Activity	Manpower	Man/days	Comments
1–2	8	48	Critical activity
2–3	4	16	Critical activity
2–5	2	2	Non-critical
3–4	6	18	Critical
4–7	4	16	Critical
5–6	3	9	Non-critical
6–7	4	12	Non-critical
7–8	4	20	Critical

Using the earliest starting time schedule, it may be seen that activity (6–7) will terminate at day 13 of the project leaving a float of 4 days available. Figure 8.21 shows a bar chart of the labour force required for each day of the project.

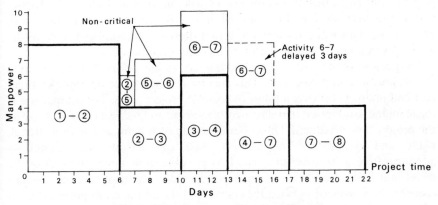

Figure 8.21 Resource allocation diagram

The level of resource required varies from 4 men for days 18–22 up to a maximum of 10 men for the period day 11 to day 13. The labour force required for the critical activities may not be tampered with unless the project time is to be increased. Therefore, the value of S' for this first trial is

$$8^2 + 8^2 + 8^2 + 8^2 + 8^2 + 8^2 + 6^2 + 7^2 + 7^2 + 7^2 + 10^2 + 10^2$$
$$+ 10^2 + 4^2 + 4^2 + 4^2 + 4^2 + 4^2 + 4^2 + 4^2 + 4^2 + 4^2$$

i.e.
$$S' = 1011.$$

However, the labour force employed for activity 6–7 could be moved from days 11–13 to days 14–16, as shown by the dotted lines in Figure 8.21. This would result in the following value for S':

$$8^2 + 8^2 + 8^2 + 8^2 + 8^2 + 8^2 + 6^2 + 7^2 + 7^2 + 7^2 + 6^2 + 6^2 + 6^2$$
$$+ 8^2 + 8^2 + 8^2 + 4^2 + 4^2 + 4^2 + 4^2 + 4^2 + 4^2$$

i.e.
$$S' = 963.$$

As this is a relatively simple problem the optimum solution has been obtained in two trial solutions. In fact the simplicity of the problem is constricting in the sense that very little room for manoeuvre is allowed. With a more realistic problem the permutations increase tremendously and much thought would have to be exercised in the analysis.

8.9.1 Time–cost trade-offs

In the preceding sections the *normal* durations of activities in a project have been assumed in the application of the critical path method. It is often possible to reduce the durations of many, if not all, of the activities in the network by incurring the penalty of increased cost. If an activity is reduced to its lowest possible time scale, then it is known as a *crash* duration, and when all activities within the network are thus reduced it is referred to as an *all-out crash* programme. This will obviously result in a decrease in completion time of the whole project, but at a commensurate increase in project cost.

The purpose of the *least-cost* method is to determine the optimum project time for the lowest total project cost. This may be achieved by comparing the direct cost and indirect cost over a range of possible project times. The comparison is made on the assumption that *direct* costs increase and *indirect* costs decrease as the project time is reduced. Direct costs include such items as labour, material, plant, and equipment. The indirect costs can include such things as administrative costs, overheads, etc. It is possible to minimize the sum of the direct and indirect costs at a certain project time. The least-cost method can therefore be visualized as being the converse of the all-out crash programme, in which the minimization of the project time is the objective.

In general, the relation between the duration and cost of an activity follows the general shape of the curve shown in Figure 8.22. There is usually a practical limit

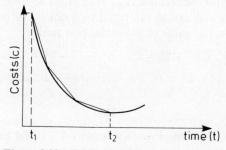

Figure 8.22 Typical duration–cost function

t_1, below which the activity cannot be completed regardless of the cost, and, similarly, there is a practical duration t_2, above which the cost curve may start to increase.

In the significant range of the time–cost trade-off curve, the shape is usually convex and may be approximated by piecewise continuous linear segments as shown in the diagram. If one of these segments is examined in detail, as shown in Figure 8.23, then it is possible to arrive at a mathematical relationship between the costs and durations for various project completion times.

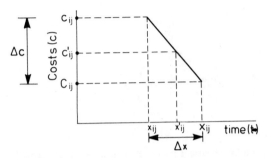

Figure 8.23 Segment of a duration–cost curve

Let X_{ij} and x_{ij} be the normal and crash durations respectively for activity (i, j), and let C_{ij} and c_{ij} be the associated costs. If the rate of change of cost with respect to time, i.e. the slope of the linear segment, is denoted by M_{ij}, then

$$M_{ij} = \left| \frac{\Delta c}{\Delta x} \right|.$$

Therefore, for any duration x'_{ij}, the linear time–cost relationship is given by

$$c'_{ij} = C_{ij} + M_{ij}(X_{ij} - x'_{ij}).$$

In a time–cost trade-off problem, an attempt is made to vary the activity durations so that the project completion time and its associated cost can be determined. In this way numerous combinations can be examined and eventually the minimum-cost solution for the specified time can be determined.

8.9.2 Example of time–cost trade-off

A contractor has given estimates for erecting a temporary hut, as listed in Table 8.6. The activities can run concurrently except that activity A must precede all others and activity E must follow all others. It is required to construct a graph showing the relationship between the completion of the project and its total cost. The precedence–successor relationships for this problem are given in Figure 8.24.

The critical path for the problem is given by 1–2–5–6 and is of 14 hours

Table 8.6 Duration–cost data for temporary hut erection

Activity	Description	Normal duration (hr)	Cost (£)	Crash duration (hr)	Cost (£)	$\Delta C/\Delta T$
A (1, 2)	Materials to site	5	30	4	40	10
B (2, 5)	Erect hut	6	12	2	20	2
C (2, 3)	Install electricity	4	10	3	18	8
D (2, 4)	Install plumbing	5	12	3	20	4
E (5, 6)	Connect services	3	16	—	—	—

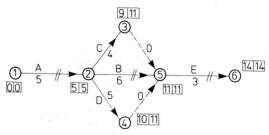

Figure 8.24 Critical path network for data of Table 8.6

duration. There is a total float of 2 hours along path 2–3–5. and a total float of 1 hour along path 2–4–5.

The total cost for this particular problem, in which all durations are normal, is £80. If we now decide to decrease the duration of the project then it is necessary to select the activity which has the lowest cost rate, i.e. activity B (2, 5). This should give us the minimum-cost solution for the particular project completion time. We now have to decrease the duration of activity B either by its full range (i.e. normal duration − crash duration), or the lowest float value along a sub-critical path — whichever is the smaller. In this case the maximum range is four hours in activity B, i.e. the erection of the hut could be reduced from six to two hours. However, if this activity is crashed to its minimum duration, then the lowest value of float along a sub-critical path will have been exceeded and the critical path network needs to be completely re-analysed. Therefore, we choose to reduce the duration of activity B by *one* hour only, i.e. a duration corresponding to the lowest value of float along a sub-critical path.

The new cost for the project will now be equal to

$$£80 + (1 \text{ hour} \times £2/\text{hour}) = £82.$$

This reduction in the duration of activity B now renders path 2–4–5 critical, as the LET for nodes 3, 4, and 5 is now 10 hours.

The remaining float of 1 hour in activity C must now be brought into the reckoning if the project duration is to be minimized further. At the moment, the possibility of crashing activity A is still uneconomical at the rate of £10/hour. Again, the float value of 1 hour is less than the maximum range of 3 hours in activity B and so the project duration is decreased by 1 hour.

Thus, we can reduce the LET of node 3 by a further hour, which makes path 2–3–5 critical. In fact, all the activities are now critical. However, it is also necessary to reduce the durations of activities D and B by one hour if the overall project completion time is to be reduced from 13 to 12 hours.

The total cost for the project is now:

$$£82 + (1 \times £(4 + 2)) = £88.$$

All three branches, i.e. activities B, C, and D, must now be reduced simultaneously if any further reduction in project time is required. This would be achieved at a total cost of £14/hour. However, crashing activity A is less expensive at £10 per hour, and so the total project cost commensurate with 11-hour completion time is

$$£88 + (1 \cdot \times £10) = £98.$$

Finally, the project time may be reduced by one further hour if activities, B, C, and D are reduced by one hour. Thus, the total cost for a 10-hour completion time is

$$£98 + (1 \times £(2 + 8 + 4)) = £112.$$

The network for the rapid durations is now given in Figure 8.25.

The project completion cost curve for various completion times may now be drawn and is shown in Figure 8.26. This solution considers only *direct* costs which must increase the total cost of the project. See Exercise 8.10.4 for consideration of indirect costs.

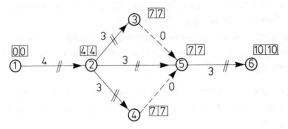

Figure 8.25 Revised network

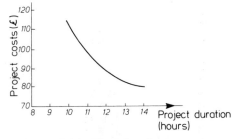

Figure 8.26 Final duration–cost curve

8.10 EXERCISES

8.10.1 (a) Construct a network for a project (such as an annual maintenance) which consists of the following activities:

Activity		Activity time (days)	Resources (no. of men)
Initial node	Terminal node		
0	1	1	1
1	2	2	1
1	4	3	1
2	3	4	1
3	4	16	2
3	5	14	1
3	6	14	2
4	7	12	2
4	8	14	3
4	9	10	1
5	10	0	0
6	10	0	0
7	10	5	1
8	10	4	2
9	10	6	1
10	11	3	2

(b) Calculate the critical path(s) and the total duration of the project.
(c) How many paths (of all kinds) are there?
(d) How many men are actually required on each day of the project, given that there are 6 available?
(e) Is the critical path altered if there are only 5 men available? If so, what are the new critical events, what is the new project duration, and what are the new floats?

8.10.2 Construct an activity-on-branch network from the precedence relationships of activities in the following table, and define the critical path and its length:

Activity	Predecessors	Duration
A	—	2
B	A	5
C	None	4
D	A	7
E	A	12
F	B	6
G	C, D	3
H	B	9
I	E, F, G	4
J	E, F, G	1
K	C, D	8
L	None	11
M	J, K, L	6

8.10.3 In the excavation of a building foundation, the total volume is subdivided into eight smaller zones as shown in the figure. Two excavators are used, with the first one assigned to zones 1, 2, 7, and 8 consecutively, and the second one assigned to zones 3, 4, 5, and 6 consecutively. The excavation of each zone is considered as a separate activity. Hence, the sequence of events is governed by two factors: (1) the locations of the zones, and (2) the order of excavation specified for the excavators.

Given that the excavations for zones 1, 2, ..., 8 are designated as activities 1, 2, ..., 8 respectively, and that the durations and precedence relationships of these activities are shown in the table below, determine the minimum completion time for the project.

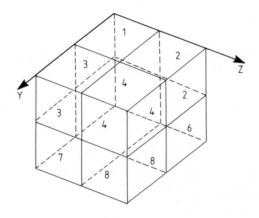

Activity	Description (Excavate zone)	Predecessors	Duration (hr)
1	1	—	23
2	2	1	68
3	3	—	17
4	4	3	50
5	5	1, 4	25
6	6	2, 5	75
7	7	3, 2	34
8	7	4, 7	102

8.10.4 The example shown in Figure 8.2.4 only considered direct costs. If *indirect* costs are significant it is possible that savings due to a reduction in overall project time will more than offset the increased direct costs. Show that if indirect costs are £7 per hour then the optimum completion time of the project is 12 hours.

8.10.5 Given the following network data, develop a schedule for minimum equipment requirements:

Activity		Duration	Mode	Manpower	Equipment		
I	J				A	B	C
1	2	8	Continuous	4	1	1	0
1	3	7	Continuous	10	0	0	1
1	5	12	Intermittent	5	0	1	0
2	3	4	Continuous	6	0	0	0
2	4	10	Intermittent	7	1	0	0
3	4	3	Continuous	5	0	1	0
3	5	5	Intermittent	8	0	0	0
3	6	10	Intermittent	7	0	1	0
4	6	7	Continuous	11	0	1	1
5	6	4	Continuous	3	1	0	0

8.10.6 From the following network data, determine the critical path, starting and finishing times, total and free floats:

Activity	Description	Duration
1–2	Excavate stage 1	4
1–8	Order and deliver steelwork	17
2–3	Formwork stage 1	4
2–4	Excavate stage 2	5
3–4	Dummy	0
3–5	Concrete stage 1	8
4–6	Formwork stage 2	2
5–6	Dummy	0
5–9	Backfill stage 1	3
6–7	Concrete stage 2	8
7–8	Dummy	0
7–9	Dummy	0
8–10	Erect steel work	10
9–10	Backfill stage 2	5

8.11 REFERENCES

1. Antill, J. M., and Woodhead, R. W. (1970). *Critical Path Methods in Construction Practice*, 2nd ed., Wiley Interscience, New York.
2. Armstrong-Wright, A. T. (1969). *Critical Path Method*, Longman, London.
3. Lockyer, K. G. (1964). *An Introduction to Critical Path Analysis*, Pitman, London.
4. O'Brien, J. J. (1965). *C.P.M. in Construction Management*, McGraw-Hill, New York.
5. Battersby, A. (1964). *Network Analysis for Planning and Scheduling*, Macmillan, London.

Chapter 9
Economic Aspects of Systems Engineering

9.1 TIME AND MONEY

9.1.1 Cash flow

Cash flow is the summation over a period of time of the income and expenditures associated with a project. Both income and expenditure are commonly expressed in monetary terms (e.g. £ or $) and represent the value of the physical consequences of some course of action. Although the assignment of a monetary value to physical consequences is not always straightforward it will be assumed for the moment that this is so. A common way of illustrating cash flow processes graphically is by means of the *cash flow diagram*. Figure 9.1 shows a typical example in which time is plotted along the horizontal axis. The usual convention is to plot income as an upward pointing arrow above the axis and expenditure as a downward pointing arrow below the axis. Each arrow represents the income or expenditure during the year (or other period of time) immediately prior. Figure

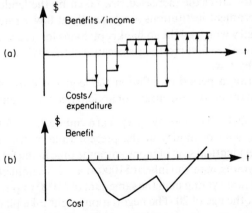

Figure 9.1 (a) A typical cash flow diagram. (b) The corresponding cash position diagram

309

9.1(a) illustrates a typical commercial venture in which capital expenditure is phased over two stages. The income shows a short growth period immediately after completion of the main investment and then a period of steady income until stage two is complete.

An alternative view of the same information is shown in the *cash position diagram* of Figure 9.1(b). As the name implies, this shows the financial situation at any point in time and is simply the time integral of the cash flow diagram. The structural engineer will recognize the analogy with load and shear force diagrams on a horizontal beam.

The information contained in these diagrams could be conveyed in words or in a table of figures but the graphic nature of the diagrams is sometimes useful for visualization of the economic consequences of a course of action. Another convention in the use of cash flow diagrams is the use of arrows pointing towards the time axis. These are taken to mean either income or expenditure and are useful in describing general relations over time. This device will be used in the sections which follow to illustrate the notion of interest.

9.1.2 Interest and equivalence of time

Interest, in the monetary sense, is the reward for making capital available over a period of time. The rate of interest is commonly expressed as a percentage of the capital sum per time period, e.g. 8% per year, $1\frac{1}{2}$% per month. Notice that 12% per year computed annually does differ from 1% per month computed monthly. Interest rates are affected by many factors but in general reflect inversely the willingness of the owner of the capital to make it available on loan. Three typical influences are (other things being equal):

(i) The state of the economy. In an expanding economy there is a good market for lenders of money since there is a high demand for capital and thus a high interest rate.
(ii) The risk involved. Loans which are less certain to be paid back carry a higher interest rate to reflect the increased risk taken by the lender. For example loans to government institutions are generally at lower rates than loans to provide industry which might go bankrupt: loans for a second mortgage are more expensive than first mortgages because of the higher priority accorded to the latter by law.
(iii) Inflation. During a period of inflation the lender will usually increase the interest rate in order to compensate for anticipated loss of purchasing power.

Accepting the fact that money may earn interest, it follows that the attractiveness of a sum of money at the present time is not the same as the attractiveness of an identical sum of money at some time in the future. The certain knowledge that a young man will inherit $1000 at age 21 is unlikely to persuade a banker to lend the same young man the same sum of $1000 to purchase a second-hand motorbike at the age of 20. The negotiation might take place in one of two ways.

a) The banker agrees to lend $1000 now but requires a lump sum of $1100 repaid in a year's time.
b) The banker agrees to accept the $1000 when it becomes available in one year's time but is prepared to lend only $909 for the present.

Under option (a) the notion of 10% interest per annum is clearly seen to apply. With option (b) the same interest rate results in the present value of the inheritance being only $909. The practice of discounting future amounts in order to transform them into equivalent present-day values is fundamental to economic analysis. To make realistic investment decisions, it is vital that each monetary value be identified or described by both amount and time. Amounts at different times cannot be directly combined or compared, since they are not in common units.

In addition to lump sums committed or recovered at different points in time, most projects involve to a greater or lesser degree the recurring income or expenditure of annual amounts. These may be constant over a period of many years or may vary with time in a regular or complex fashion and might represent such things as operation and maintenance costs, income from sales or services, repayment of portions of a loan, or accumulative savings for an anticipated future expenditure. In the discussions which follow the calculations will frequently involve some or all of the following terms:

$P = $ A capital sum at the 'present' time.
$F = $ A capital sum at some 'future' time.
$A = $ An annual sum assumed constant over a period of time unless otherwise specified.

The terms 'present' and 'future' are merely relative and need not apply to the real time of the economic analysis.

In addition to the quantities P, F, and A the calculations will generally involve the following two factors.

$i = $ The interest rate per time period (years unless otherwise specified) expressed as a decimal quantity (e.g. 0.08 and *not* 8%).
$N = $ The number of time periods (years unless specified).

Thus the calculation of equivalent economic values will usually involve any four of the five quantities P, F, A, i, and N. In the following sections the particular functional forms of these relations are developed.

9.2 COMPOUND INTEREST FACTORS

In this section six different functional relationships are developed involving the variables P, F, A, i, and N. Wherever annual amounts are involved it is assumed that these are constant over the N time-periods. In all of the functions the dependent variable is either P, F, or A; a convenient notation is therefore to express the factors as a ratio of any two of the amounts P, F, or A with the numerator as the dependent variable and with i and N as parameters, e.g.

$$\frac{F}{P}(i, N)$$

represents the factor used to define a future sum F in terms of a present sum P and the parameters i and N. For convenience the time intervals are assumed to be years, but any value may be used. Compound interest is assumed in all cases, i.e. interest earned in any year is added and used to compute interest in the following year.

9.2.1 Compound interest F/P (i, N)

This factor is used to compute a future sum F which will result from the investment of a present sum P at interest rate i for a period of N years.

After 1 year, $\quad F_1 = P + iP = (1 + i)P$

After 2 years, $\quad F_2 = P + iP + i(P + Pi) = (1 + i)^2 P$

$\vdots$

After N years, $\quad F_N = (1 + i)^N P$

or

$$F/P\,(i, N) = (1 + i)^N. \tag{9.1}$$

For example $F/P\,(8\%, 20) = 4.6610$ so that a sum of $100 invested at 8% for 20 years will be worth $100 \times 4.661 = \$466.10$. The function is sometimes referred to as the Single-Payment Compound Interest Factor. Figure 9.2 illustrates the relationship between F and P on a cash flow diagram.

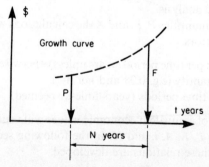

Figure 9.2 Single-payment amounts

9.2.2 Present value P/F (i, N)

This factor is used to calculate the equivalent value at some earlier time of a specified future sum, assuming a constant interest rate. The factor is obviously the reciprocal of the single-payment compound interest factor and is usually described as the Discount Factor. By the same reasoning as before

$$\frac{P}{F}(i, N) = (1 + i)^{-N}. \qquad (9.2)$$

Using the same example as previously a sum of \$466.10 to be paid in 20 years has a present value of \$100 using a constant discount rate of 8%.

9.2.3 Uniform series compound interest $F/A\ (i, N)$

This factor is used to calculate the terminal value F of a series of uniform payments A over a period of N years. Using the compound interest factor the future value of each annual payment may be calculated as follows:

1st year's payment $\quad A(1 + i)^{N-1}$

2nd year's payment $\quad A(1 + i)^{N-2}$

$\vdots$

Last year's payment $\quad A$

$$F = A + A(1 + i) + \ldots + A(1 + i)^{N-2} + A(1 + i)^{N-1}. \qquad (9.3)$$

Multiplying (9.3) by $(1 + i)$

$$F(1 + i) = A(1 + i) + A(1 + i)^2 + \ldots + A(1 + i)^{N-1} + A(1 + i)^N. \qquad (9.4)$$

Subtracting the two series yields

$$Fi = -A + A(1 + i)^N$$

therefore

$$F = A[(1 + i)^N - 1]/i$$

or

$$\frac{F}{A}(i, N) = [(1 + i)^N - 1]/i. \qquad (9.5)$$

As an example consider a fund into which \$1000 is contributed each year for 15 years. With an interest rate of 6% the terminal value will amount to

$$\$1000 \times (1.06^{15} - 1)/0.06 = \$1000 \times 23.2760$$

$$= \$23\ 276.$$

9.2.4 Sinking fund $A/F\ (i, N)$

This is the inverse calculation to the uniform series compound interest factor and involves the determination of the annual payment which will be necessary in order to accumulate a terminal (i.e. future) sum of a specified amount. The sinking fund factor is thus given by

$$\frac{A}{F}(i, N) = i/[(1 + i)^N - 1]. \qquad (9.6)$$

Consider, for example, a fund set up to replace a piece of equipment which is expected to wear out in 15 years' time and cost $20 000 at that time. With an interest rate of 6% the annual payment required is given by

$$A = \$20\,000 \times \frac{A}{F}(6\%, 15)$$

$$= \$20\,000 \times 0.04296$$

$$= \$859.26.$$

Figure 9.3 illustrates in cash flow form the relationship involved in uniform series calculations involving a future sum.

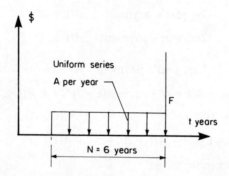

Figure 9.3 Uniform series — future-sum cash flow

9.2.5 Uniform series present value P/A (i, N)

Uniform series calculations may have to be carried out in relation to a capital sum valued at the commencement of the time period. This could be done by means of the functions developed earlier, but it is convenient to define and evaluate specific functions for this task. With reference to the cash flow diagram of Figure 9.4 the uniform series of payments of value A may be seen to accumulate to

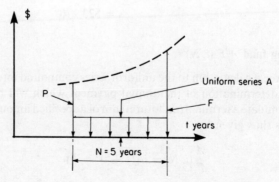

Figure 9.4 Uniform series — present-value cash flow

a terminal sum F in N years. The present value of F is, however, represented by P where

$$\frac{F}{A}(i, N) = \frac{[(1 + i)^N - 1]}{i}$$

and

$$\frac{P}{F}(i, N) = (1 + i)^{-N}.$$

Thus the Uniform Series Present Value factor is given by

$$\frac{P}{A}(i, N) = \frac{P}{F}(i, N) \times \frac{F}{A}(i, N)$$

$$= \frac{[(1 + i)^N - 1]}{i(1 + i)^N}. \tag{9.7}$$

It is worth noting that as N tends to infinity the factor $P/A\,(i, N)$ tends to $1/i$.

For example, calculate the present value of capital investment which a prospective housebuyer can afford if he or she intends to commit \$10 000 per year for the next 20 years at an interest rate of 10%.

$$\frac{P}{A}(10\%, 20) = 8.514$$

$$P = \$85\,140.$$

9.2.6 Capital recovery factor $A/P\,(i, N)$

If, as is more common, a prospective purchaser wishes to calculate the effect of his indebtedness on the annual income for the next N years, the calculation is simply the inverse of the preceding factor. Thus

$$\frac{A}{P}(i, N) = \frac{i(1 + i)^N}{[(1 + i)^N - 1]}. \tag{9.8}$$

Once again it should be noted that as $N \to \infty$

$$\frac{A}{P}(i, \infty) = i.$$

This factor is of chief importance in paying back loans for capital investment. A new sewage works costing \$4.3 million is to be paid off at 4% over 30 years. The annual payment is given by

$$A = 4.3 \times 10^6 \times \frac{A}{P}(4\%, 30)$$

$$= 4.3 \times 10^6 \times 0.05783$$

$$= \$248\,669.$$

9.3 NON-UNIFORM SERIES CASH FLOWS

The compound interest factors developed in the previous section assume that the series of annual payments — where one exists — is uniform. In some cases similar types of cash flow calculation must be carried out for series which are non-uniform. Typically the series of annual amounts may vary with a constant gradient or may follow a monotonically but not uniformly increasing time stream of costs or benefits. The series may vary geometrically or be accelerated or decelerated to promote or defer growth during the investment period. Appropriate formulae for several cases have been developed. An excellent review of these factors together with tables of values is given by James and Lee.[1]

9.3.1 Gradient series

The series illustrated in Figure 9.5 may result from a uniformly increasing series of payments from which the base A_1 has been removed — A_1 being the amount paid in year 1.

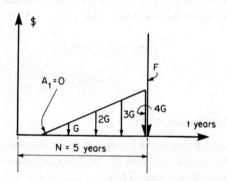

Figure 9.5 A uniform-gradient series

For the series shown with $N = 5$,

$$F = G(1 + i)^3 + 2G(1 + i)^2 + 3G(1 + i) + 4G$$

$$F(1 + i) = G(1 + i)^4 + 2G(1 + i)^3 + 3G(1 + i)^2 + 4G(1 + i).$$

Subtracting gives

$$Fi = G[(1 + i)^4 + (1 + i)^3 + (1 + i)^2 + (1 + i) + 1] - 5G$$

or

$$Fi + 5G = G[(1 + i)^4 + (1 + i)^3 + (1 + i) + 1].$$

Multiplying by $(1 + i)$ yields

$$(Fi + 5G)(1 + i) = G[(1 + i)^5 + (1 + i)^4 + (1 + i)^3 + (1 + i)^2 + (1 + i)]$$

and again subtracting results in

$$(Fi + 5G)i = G[(1 + i)^5 - 1]$$

or

$$Fi = G\frac{(1 + i)^5 - 1}{i} - 5G.$$

Thus

$$\frac{F}{G} = \frac{1}{i}\left[\frac{(1 + i)^5 - 1}{i} - 5\right].$$

In general

$$\frac{F}{G}(i, N) = \frac{1}{i}\left[\frac{(1 + i)^N - 1}{i} - N\right]. \tag{9.9}$$

Such a series may be converted into a uniform series by the following manipulation. Consider a uniform series of annual payments A over a period of N years such that the future equivalent sum F is the same as given by the uniform gradient series, as shown in Figure 9.6.

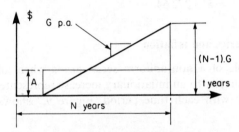

Figure 9.6 Equivalent uniform series

Now

$$\frac{F}{A}(i, N) = \frac{(1 + i)^N - 1}{i}$$

so that

$$\frac{F}{G}(i, N) = \frac{1}{i}\frac{F}{A}(i, N) - \frac{N}{i}$$

or

$$\frac{A}{F}(i, N)\frac{F}{G}(i, N) = \frac{1}{i} - \frac{N}{i}\frac{A}{F}(i, N)$$

or

$$\frac{A}{G}(i, N) = \frac{1}{i} - \frac{N}{i}\frac{A}{F}(i, N).$$

However

$$\frac{A}{F}(i, N) = \frac{i}{(1 + i)^N - 1}$$

or

$$\frac{A}{G}(i, N) = \frac{1}{i} - \frac{N}{(1+i)^N - 1}. \tag{9.10}$$

Example 9.1

For a series of annual payments of $1000 increasing by $50 over 20 years find the equivalent annual uniform series and present value. Discount at 8%.

$$A_{equ} = 50 \left[\frac{1}{0.08} - \frac{20}{(1.08)^{20} - 1} \right]$$

$$= 50 \times 7.037 = 351.85$$

therefore

$$A = 1000 + 351.85 = 1351.85$$

therefore

$$P = A \frac{P}{A}(8\%, 20) = 1351.85 \times \frac{(1.08)^{20} - 1}{0.08(1.08)^{20}}$$

$$= \$13\,272.64.$$

9.3.2 Geometric series and inflation

A common example of a non-uniform series arises when annual payments (usually costs) are subject to an inflationary increase represented by a constant percentage increase with each time period. Figure 9.7 shows the cash flow

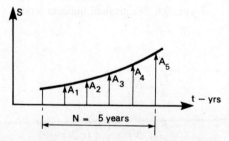

Figure 9.7 Geometric series of annual payments

diagram for a series of payments over a five-year period, subject to a rate of inflation of $r\%$. That is,

$$A_2 = A_1(1 + r)$$
$$A_3 = A_2(1 + r) = A_1(1 + r)^2$$
$$\vdots$$
$$A_5 = A_1(1 + r)^4.$$

The future value F may then be obtained in the same manner as in Section 9.2.3.

$$F = A_5 + A_4(1+i) + A_3(1+i)^2 + A_2(1+i)^3 + A_1(1+i)^4$$
$$= A_1[(1+r)^4 + (1+i)(1+r)^3 + (1+i)^2(1+r)^2$$
$$+ (1+i)^3(1+r) + (1+i)^4]$$
$$= A_1(1+r)^4\left[1 + \left(\frac{1+i}{1+r}\right) + \left(\frac{1+i}{1+r}\right)^2 + \left(\frac{1+i}{1+r}\right)^3 + \left(\frac{1+i}{1+r}\right)^4\right].$$

Multiplying by $(1+i)/(1+r)$ and subtracting yields

$$F\left[\left(\frac{1+i}{1+r}\right) - 1\right] = A_1(1+r)^4\left[\left(\frac{1+i}{1+r}\right)^5 - 1\right].$$

Now the present value P is related to the future value F by the relation

$$P = F(1+i)^{-5}$$

$$\therefore P = A_1 \frac{(1+r)^4}{(1+i)^5} \frac{\left[\left(\frac{1+i}{1+r}\right)^5 - 1\right]}{\left[\left(\frac{1+i}{1+r}\right) - 1\right]}$$

$$\therefore \frac{P}{A_1} = \frac{\left[\left(\frac{1+i}{1+r}\right)^5 - 1\right]}{(i-r)\left(\frac{1+i}{1+r}\right)^5}.$$

Thus in general the present value of a geometric series with initial payment A_1 with inflation rate r and discounting rate i is given by

$$\frac{P_0}{A_1}(i, r, N) = \frac{\left[\left(\frac{1+i}{1+r}\right)^N - 1\right]}{(i-r)\left(\frac{1+i}{1+r}\right)^N}. \tag{9.11}$$

Example 9.2

Annual payments of $100 per year are assumed to be subject to a rate of inflation of 9% per annum. Determine the present value of payments over a 20-year period using a discounting rate of 13%.

$$\frac{P}{A_1}(13\%, 9\%, 20) = \left[\left(\frac{1.13}{1.09}\right)^{20} - 1\right] \bigg/ \left[(0.13 - 0.09)\left(\frac{1.13}{1.09}\right)^{20}\right]$$

$$= 12.8409$$

$$\therefore P = \$1284.09.$$

Note that the effective discounting rate is given by

$$i_{\text{eff}} = \frac{i-r}{1+r} = \frac{0.19 - 0.09}{1.09}$$

$$= 0.0367 \quad \text{or} \quad 3.67\%.$$

However, if this rate is used in computing a present value in the previous example the result is

$$\frac{P}{A}(3.67\%, 20) = 13.9966.$$

The corresponding figure of $1399.66 represents the inflated value at the end of the first year, i.e. $1284.09 × 1.09 = $1399.66.

9.4 DEPRECIATION AND SALVAGE VALUE

Frequently capital expenditures result in the creation of some asset which has a finite economic life. At the end of that economic life, the asset may possess some finite residual value which may be re-converted into a benefit at some future time. If for some reason the asset is disposed of prior to the end of its economic life the benefit which results will have some intermediate value between the initial 'new' value and the final salvage value. The calculation of this value depends on the method employed for calculating the depreciation. Two basic methods are considered here.

9.4.1 Straight-line depreciation

The method is very simple and virtually self explanatory. Figure 9.8(a) shows the situation of an asset purchased for capital sum P and which has an expected life N at the end of which the salvage value is defined as $S = xP$ ($0 \leqslant x \leqslant 1.0$). After a period $M < N$ the depreciated value is given by

$$F = S + (P - S)(N - M)/N. \tag{9.12}$$

9.4.2 Declining-balance depreciation

An alternative method of calculating the depreciation of an asset is to compute annual depreciation as a constant fraction or percentage of the depreciated book-value at the end of the preceding year. Thus with reference to Figure 9.8(b) the value in successive years will be given by

$$P$$
$$P(1-d)$$
$$P(1-d)^2$$
$$\vdots$$

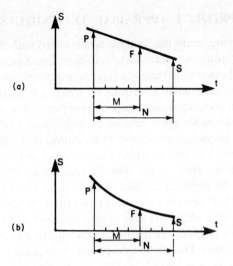

Figure 9.8 (a) Straight-line depreciation. (b) Declining-balance depreciation

so that after N years the salvage value is given by

$$S = P(1 - d)^N \tag{9.13}$$

where d is the depreciation rate. The rate may be defined explicitly or alternatively may be defined implicitly by specifying a finite salvage value S after N years, i.e.

$$d = 1 - (S/P)^{1/N}. \tag{9.14}$$

Example 9.3

A piece of machinery is replaced for a cost of $10 000 which has an expected life of 12 years. The economic life of the project, however, will terminate in 8 years. Calculate the depreciated value at the end of project life if the salvage value of the machinery is $1000.

Using straight line depreciation the value will be

$$F = 1000 + (10\,000 - 1000)(12 - 8)/12$$

$$= \$4000.$$

Using constant rate of book-value the depreciation rate is defined implicitly by the relation

$$1000 = 10\,000(1 - d)^{12}$$

$$\therefore \quad d = 0.1746 \quad \text{or} \quad 17.46\%.$$

Thus after 8 years the depreciated value is given as

$$F = 10\,000(1 - 0.1746)^8 = \$2154.43.$$

9.5 PROJECT APPRAISAL TECHNIQUES

As applied to engineering, economic analysis is concerned with the estimation of the real cost of using resources —natural, technological, and financial —in order to establish priorities between alternative or competing proposals. It is important to distinguish between alternative proposals by which a desired goal might be achieved by more than one means and competing proposals in which two or more distinct but equally desirable goals require in total more resources than are available. Alternative proposals are almost by definition mutually exclusive and discounting or interest rates are dependent on external influences; competing proposals may however result in the interest rate being determined by the opportunity afforded by the 'best' project to make a financial profit.

In making a choice between possible alternatives the engineer must be able to use an appraisal technique which will lead to an evaluation of their real costs. As discussed earlier in this chapter, the timing of expenditure and income is a vital factor in such comparisons. This section described various methods in common use.

9.5.1 Present-value analysis

As the name suggests this technique involves the discounting of all income and expenditure at any time in the future to an equivalent value at a single point in time. It is usually convenient to use the present time or the time of commencement of the project which is under review. The choice of project is based on the largest present value of the discounted algebraic sum of benefits less costs over the economic life of the project. The following guidelines apply to present-value calculations:

(a) All present values should be discounted to the same time base even if alternatives are to be initiated at different times.
(b) The same discount rate should be used throughout even if different sources of funds are to be used.
(c) Calculations should be based on a common period of analysis, even if projects may have a different economic life. This may involve calculating the cost of extending a short-lived project or the residual (salvage?) value of a project prior to the end of its maximum life.
(d) Projects with a negative present value should be rejected. This is an indication that either the project is economically unsound or that some intangible benefit has been omitted from the calculation.

Example 9.4

Two projects are to be compared over an economic life of 25 years using a discounting rate of 15%. The expenditures and incomes are:

Project	Capital cost	Annual income
A	90 000	15 000
B	70 000	13 000

The income may be represented as a present-value capital sum using the factor

$$P/A\ (15\%, 25) = (1.15^{25} - 1)/(0.15 \times 1.15^{25}) = 6.4642.$$

Thus

$$PV_A = -90\,000 + (6.4642 \times 15\,000) = +6963$$
$$PV_B = -70\,000 + (6.4642 \times 13\,000) = +14\,034.$$

Clearly project B is preferable, other things being equal.

If the discount rate is only 8% find the effect on the analysis. Now

$$P/A\ (8\%, 25) = 10.6748$$

and

$$PV_A = -90\,000 + (10.6748 \times 15\,000) = +70\,122$$
$$PV_B = -70\,000 + (10.6748 \times 13\,000) = +68\,772.$$

The lower discounting rate greatly increases the present value of future income and project A is now marginally better.

Example 9.5

Mechanical plant in a production unit may be of two types. Type A costing $10 000 has an expected life of 10 years; type B costs $14 000 but has a life of 15 years. Net income from the process is $2000 p.a. If the interest rate is 10% find the best choice of plant.

A period of 30 years is chosen for the analysis since this conveniently covers multiples of both plant life values. The cash flow diagram for the two alternatives is shown in Figure 9.9.

The income is the same for both proposals and is calculated as

$$PV_1 = \$2000 \times \frac{P}{A}(10\%, 30) = 2000 \times 9.4269 = \$18\,853.$$

The cost of type A plant is computed as follows:

$$10\,000 + 10\,000 \times \frac{P}{F}(10\%, 10) + 10\,000 \times \frac{P}{F}(10\%, 20)$$

$$= 10\,000(1.0 + 0.385543 + 0.148644)$$

$$= 15\,342.$$

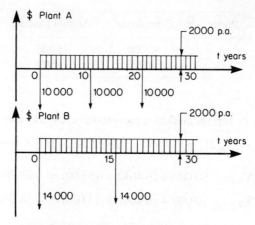

Figure 9.9 Cash flows for Example 9.5

Similarly, the cost of type B plant is given by

$$14\,000 + 14\,000 \times \frac{P}{F}(10\%, 15) = 14\,000(1.0 + 0.239392)$$

$$= 17\,351.$$

Therefore

$$PV_A = 18\,853 - 15\,342 = 3512. \qquad \leftarrow \text{Adopt}$$

$$PV_B = 18\,853 - 17\,351 = 1503.$$

Example 9.6

Storage tanks and pumps are to be provided for a growing demand. Two alternative proposals are to be considered.

A. Construct Tank 1 with Pump in Year 1 at capital cost of $400 000 and annual operating charges of $30 000. Then in Year 12 construct Tank 2 with Pump at additional capital cost of $500 000 and operating costs increased from $30 000 to $55 000 p.a.
B. Construct Tank 3 with Pump in Year 1 at capital cost of $650 000 and running charges of $30 000 p.a. Then in Year 12 add extra pump capacity at cost of $50 000 and increased running costs of $55 000 p.a.

Use a discount rate of 8% p.a.

The cash flow diagrams for the two proposals are shown in Figure 9.10.

Conversion of the capital sums to a present value at year 1 is done in a similar manner as before. Likewise the conversion of the operating annual cost series from year 1 to year 12 to a present value at year 1 is quite straightforward. Running costs after year 12 are treated as follows.

Assume that the life of the system is unlimited and convert the annual series

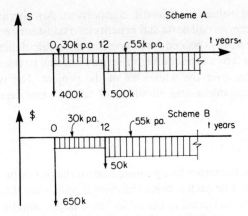

Figure 9.10 Cash flows for Example 9.6

from year 12 to year ∞ to a capital sum at year 12. Then convert this capital sum to a present value at year 1.

Therefore the present value of running cost from year 12 to year ∞ is

$$PV = \$55\,000 \times \frac{P}{A}(8\%, \infty) \times \frac{P}{F}(8\%, 12).$$

Recall that $P/A\,(i, \infty) = 1/i$. Thus

$$-PV_A = 400\,000 + 30\,000 \times \frac{P}{A}(8\%, 12) + 500\,000 \times \frac{P}{F}(8\%, 12)$$

$$+ 55\,000 \times \frac{P}{A}(8\%, \infty) \times \frac{P}{F}(8\%, 12)$$

$$= 400\,000 + (30\,000 \times 7.5361) + (500\,000 \times 0.397\,114)$$

$$+ (55\,000 \times 12.5 \times 0.397\,114)$$

$$= 400\,000 + 226\,083 + 198\,557 + 273\,016$$

$$= 1\,097\,656$$

Similarly

$$-PV_B = 650\,000 + 226\,083 + (50\,000 \times 0.397\,114) + 273\,016$$

$$= 1\,168\,955.$$

Note that both schemes have a negative present value since no income or benefits have been included in the analysis. Assuming that the benefits have an equivalent value greater than the cost of either scheme, Proposal A is the preferred one.

9.5.2 Annual-cost comparison

In the present-value method of economic analysis the various incomes and expenditures occurring at different times in the life of a project were converted to equivalent present values for realistic comparison. An alternative method of comparison is to convert all costs and benefits into constant annual equivalents. This is simply the inverse procedure to present-value calculations and results in each item, whether it be a capital sum or a series of annual payments, being spread as a uniform series over the whole life of the project. The two methods are mathematically equivalent and should lead to the same result in any given situation.

Example 9.7

Three alternative schemes are being considered for the provision of machinery for a pumping station. For each scheme the capital cost, annual running cost, and salvage value are as indicated in the table. Most important, the expected life of each scheme is different. Determine the most economical proposal, assuming a constant interest rate of 8%, to provide a pumping facility for an indefinite number of years.

Scheme	A	B	C
Capital cost	$25 000	$50 000	$35 000
Annual cost	$ 3000	$ 2000	$ 2500
Salvage value	$ 5000	$ 7000	$ 6000
Life (years)	10	29	16

In contrast with Example 9.5 the use of present-value comparison is impractical since the lowest common multiple of the three life-times is 2320. Using annual-cost comparison the two capital sums are reduced to equivalent annual costs, e.g. for scheme A the annual cost is given by

$$AC_A = -25\,000 \times \frac{A}{P}(8\%, 10) \quad \text{(capital)}$$

$$-3000 \quad \text{(annual)}$$

$$+5000 \times \frac{A}{F}(8\%, 10). \quad \text{(salvage)}$$

Now

$$\frac{A}{P}(8\%, 10) = \frac{0.08(1.08)^{10}}{(1.08)^{10} - 1} = 0.1490$$

$$AC_A = -3725.74 - 3000.00 + 345.15$$

$$= -6380.59.$$

In a similar fashion

$$AC_B = -4480.93 - 2000.00 + 67.33$$

$$= -6413.60$$

$$AC_C = -3954.19 - 2500.00 + 197.86$$
$$= -6256.33.$$

Thus scheme C is preferable.

Example 9.8

Three alternative designs for a new bridge have costs as shown and also an estimated probability of failure based on the return period of the flood for which they have been designed.

	Scheme		
	A	B	C
Height above normal water level	5'	8'	12'
Return period (yr) for flood of that height	10	40	160
Capital cost	320 000	400 000	490 000
Annual cost	9500	11 700	13 600
Rebuilding cost	285 000	320 000	400 000
Life = 40 years Interest 4% Zero salvage			

The equivalent annual cost of each design is made up of three components:

(a) The capital cost expressed as an equivalent annual cost over the anticipated life of 40 years.
(b) The annual maintenance cost.
(c) The expected annual damage cost, evaluated from the cost of reconstruction divided by the average number of years between recurrences of the design flood.

Thus for scheme A

$$\text{Annual cost} = 320\ 000\ \frac{0.04(1.04)^{40}}{(1.04)^{40} - 1}$$
$$+ 9500$$
$$+ \tfrac{1}{10}(285\ 000)$$
$$= 16\ 168 + 9500 + 28\ 500 = 54\ 200.$$

Similarly for scheme B

$$\text{Annual cost} = 20\ 209 + 11\ 200 + 8000 = 39\ 909$$

and for scheme C

$$\text{Annual cost} = 24\ 757 + 13\ 600 + 2500 = 40\ 857$$

On the basis of these figures, scheme B is best, although only marginally preferable to scheme C.

9.6 OTHER METHODS

Several other techniques of project appraisal are in general use but it is beyond the scope of this chapter to deal with these because of certain subtleties of interpretation which require fairly detailed consideration. The benefit–cost ratio method for instance employs the ratio of all benefits to all costs, both quantities being calculated either as present values or as annual equivalents. Some doubt may arise as to whether certain items should be considered as benefits or as negative costs (or vice versa) and this decision may significantly alter the result. Moreover, the method fails as a means of appraisal when benefits cannot be evaluated with certainty.

The rate-of-return method is a technique in which the discounting rate is evaluated which will result in zero present value. The method is computationally more complex and is open to misinterpretation especially when net benefits change sign in successive years. The method has advantages in allowing easy comparison with other investment opportunities and is of particular use in ranking projects which compete for a limited amount of capital. A full treatment of these and other methods can be found in the texts by James and Lee[1] and by Woods.[5]

9.7 EXERCISES

9.7.1 A contractor requires an additional 1000 sq. m of warehouse space. A reinforced concrete building will cost $220 000 whereas the same space can be constructed with a galvanized building for $150 000. The life of the concrete building is estimated at 50 years with annual maintenance of $1200; the life of the metal building is 30 years and annual maintenance is $1800. Average annual property taxes are currently 1.2% of the new value for the concrete building and 0.5% of the new value for the metal building. Assume that the metal building has a salvage value based on a depreciation rate of 7% per year.

Another building with approximately the same space is located 2 km from the central site and can be leased for $10 000 per year, the lease being re-negotiated after each five-year period. In addition, the management estimates that the extra cost in material handling will amount to $480 per month.

Recommend the best course of action. Assume an interest rate of 14% per annum. Assume also that all continuing costs, such as maintenance, lease renewal, taxes, and material handling costs, are subject to an inflationary increase of 10% per annum.

9.7.2 An irrigation project is being planned which requires a capital investment of $100 000 in year 1 followed by annual operating and maintenance costs which commence in year 2 at $5000 per year and increase uniformly thereafter at a rate of $500 per year. The yield from the project is not expected to commence until year 4 when an income of $15 000 is anticipated. Due to inflation this profit is expected to increase by 9% per year. If the project life is 20 years from initial investment and the discounting (interest) rate is 12.5% determine if the scheme is financially justifiable. What should the profit in year 4 be if the scheme is to break even over the 20-year period?

9.7.3 Treatment plant is to be provided for a period of 25 years, and the merits of staging the construction are to be examined. In addition, the operating and maintenance costs may be carried directly by the authority or subcontracted.

If plant for 100% capacity is provided, the present capital cost is $250 000. Plant for 50% capacity would cost $150 000 but installation of the second stage in year 10 would be subject to an inflation rate of 3% per year.

The plant has an expected life of 25 years and the salvage value of the second stage may be estimated on the basis of linear depreciation.

Operation and maintenance if done 'in-house' will cost $8000 in year 1 and increase at a uniform rate of $1200 per year. On the other hand, if subcontracted, the cost for the first 10 years would be $16 000 per year and for the last 15 years the cost would be $32 000 per year. Operating costs are independent of whether or not the construction is done in stages.

If the interest rate for the 25-year period is estimated to be 8% per annum, make recommendations to the authority on the most economical course of action.

9.8 REFERENCES

1. James, L. D., and Lee, R. R. (1971). *Economics of Water Resources Planning*, McGraw-Hill, New York.
2. Sanders, M., and Dublin, S. W. (1973). *Civil Engineering and Economics and Ethics for Professional Engineering Examinations*, Hayden, Rochelle Park, NJ.
3. Riggs, J. L. (1977). *Engineering Economics*, McGraw-Hill, New York.
4. A.S.C.E. Programmed Learning Courses. *Engineering Economy*, A.S.C.E., 345 East 47th St., New York, N.Y. 10017.
5. Woods, D. R. (1975). *Financial Decision Making in the Process Industry*, Prentice-Hall, Englewood Cliffs, NJ.

Chapter 10
Modelling and Simulation

10.1 INTRODUCTION

Models are imitations of reality and may be employed to represent a real-life situation in such a way as to allow study of some particular aspect or the emphasis of some special characteristic under conditions of control which are not possible in real life. Usually models are simpler than the situations they represent. Models may be classified as iconic, analogue, or symbolic.

Iconic models are scaled-down versions of the prototype. Typical examples are photo-elastic models for the study of stress distribution, spline models for the structural behaviour of plane frames, small-scale hydraulic models, and aerodynamic models to study the interaction of wind and tall structures in the atmospheric boundary layer.

Analogue models use one set of properties to represent another set which obeys the same natural laws. For example, electric current and potential difference may be used to represent ideal fluid flow, ground-water movement, heat conduction, or torsion, all of which may be represented by Poisson's equation.

Symbolic models employ an abstract representation of variables and their relationships or interaction. Although frequently of relatively simple form, the implementation of such models may involve many repetitive calculations and use is generally made of digital computer facilities.

This chapter is concerned with symbolic modelling and will consider two groups:

(i) *Deterministic models* in which relatively complex systems are modelled by computational means to facilitate experimental (i.e. trial and error) optimization.
(ii) *Probabilistic models* in which certain components of a system are subject to some degree of uncertainty which makes analytical study difficult or impossible.

10.2 DETERMINISTIC MODELS

Models of this type are frequently *representational* — i.e. the engineer attempts to analyse and describe the outcome of a certain set of circumstances by using state-of-the-art relationships. The model may comprise simply an automated sequence of analytical steps to predict the value of one or more dependent parameters as a function of a set of specified input or independent parameters.

For instance, a computer program may allow the user to specify the properties of a plane frame and obtain member forces as a function of external loads. On a much simpler level, the resistance moment RM of a singly reinforced rectangular concrete section may be calculated as a function of the dimensions b and d, the steel area A_{st}, and the allowable stresses f_c and f_y, which are respectively the ultimate compressive stress in the concrete and the yield stress of the steel. That is,

$$RM = \phi(b, d, A_{st}, f_c, f_y). \qquad (10.1)$$

Although this is a simple everyday design calculation, it will take on some of the qualities of a 'model' if implemented in an elementary computer program and executed in a time-sharing environment which allows the user to manipulate the values of one or more input parameters and observe the sensitivity of the result (RM) to these changes.

10.2.1 Example 10.1: Simulating the carrying capacity of a beam

A rectangular singly reinforced concrete beam has the dimensions and allowable stresses shown in Figure 10.1. Determine to what extent the allowable uniformly distributed live load (UDLL) is sensitive to the beam width.

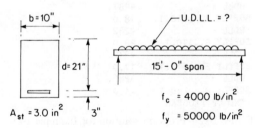

Figure 10.1 Simply-supported singly reinforced concrete beam

Algorithm

1. Specify d, A_{st}, f_c, f_y.
2. Set trial value for b.
3. Calculate resistance moment (RM).
4. Calculate dead load of section (DL).
5. Get allowable live load BM using appropriate load factors.
6. Calculate and display allowable live load (LL).

The calculation of resistance moment in step (3) may be achieved by a standard routine SRMULT (*S*ingly *R*einforced *M*oment by *ULT*imate load theory) listed

```
C     PROGRAM TO FIND ALLOWABLE LIVE
C     LOAD ON S.R. RECT. BEAM (EX 10.1)
      REAL LLBM
      SPAN=15.0
      D=21.0
      B=12.0
      AST=3.0
      FC=4000.0
      FY=50000.0
   10 CONTINUE
      WRITE(6,20)
   20 FORMAT(" SUPPLY B...F10.1     ")
      READ(5,30)B
   30 FORMAT(F10.1)
      IF(B.LT.0.0) STOP
      CALL SRMULT(B,D,AST,FC,FY,RM)
      RM=RM/12.0
      UDDL=150.0*24.0*B/144.0
      DLBM=UDDL*SPAN*SPAN/8.0
      ADLBM=1.4*DLBM
      ALLBM=0.9*RM-ADLBM
      LLBM=ALLEBM/1.7
      UDLL=8.0*LLBM/(SPAN*SPAN)
      WRITE(6,40)UDLL
   40 FORMAT("   UDLL =  ",F16.0)
      GOTO 10
      END
 SUPPLY B...F10.1      12.0
 UDLL =                4262.
 SUPPLY B...F10.1      14.0
 UDLL =                4282
 SUPPLY B...F10.1      16.0
 UDLL =                4287.
 SUPPLY B...F10.1      18.0
 UDLL =                4282
 SUPPLY B...10.1       20.0
 UDLL =                4270.
 SUPPLY B...F10.1      17.0
 UDLL =                4286.
 SUPPLY B...F10.1      -1.0
 END OF PROGRAM
```

Figure 10.2 Program solution for Example 10.1

in Appendix A. Figure 10.2 shows a FORTRAN program to solve this problem and a set of typical results.

More complex deterministic models may be encountered when economic considerations enter into the decision-making process. Instead of proportioning a section to make maximum use of the allowable stresses, the problem assumes the form of a mathematical model in which the objective function is expressed in monetary terms and the technological requirements take the form of constraints. When the number of significant design variables is small (i.e. ≤ 3), the solution may be obtained by experimental optimization.

10.2.2 Example 10.2: Minimum-cost design of a reinforced concrete slab

Minimum-cost design of a singly reinforced concrete slab may be based on the use of the function SRMULT described in the previous example. For specified forces, however, the resistance moment must be calculated for trial values of effective depth and steel area until the resistance moment, RM, exceeds or equals the applied bending moment, BM. Additional constraints may be introduced with regard to

(i) minimum slab thickness,
(ii) minimum steel ratio,
(iii) maximum deflection,
(iv) maximum steel ratio.

The following internal memorandum from the costing department of Messrs Crummley Concrete Products illustrates such a problem.

CRUMMLEY CONCRETE PRODUCTS

Memorandum:
To: Ben Tover, Drawing Office
From: Les X. Pensive, Costing Dept.
Re: *New material Costs in Highrise Floors.*

I have received information that costs of premixed structural concrete will experience a significant upward revision later this year following anticipated relaxation of the Anti-Inflation Board guidelines. Projected prices delivered on site are noted below, to which a further $45.00 per cu. yd must be added to cover plant and labour for placing.

$$\begin{array}{lll} \text{Strength } f_c' & 3000 \text{ psi} & \$28.75/\text{cu. yd} \\ & 4000 \text{ psi} & \$31.35/\text{cu. yd} \\ & 5000 \text{ psi} & \$34.75/\text{cu. yd.} \end{array}$$

Steel prices will also affect the cost of reinforcing (fixed) although the cost spread is smaller; e.g.

$$\begin{array}{lll} \text{Yield strength } f_y & 40\,000 \text{ psi} & \$0.30/\text{lb} \\ & 50\,000 \text{ psi} & \$0.32/\text{lb} \\ & 60\,000 \text{ psi} & \$0.34/\text{lb.} \end{array}$$

In view of these increases it is essential that the design of the one-way spanning slab units for the Cloud Nine building be reviewed. For your information, the details of the design specification are attached. It is some time since this design was prepared and you should check that all details of the specification and design are in compliance with the Building Code.

Please let me have your report by the end of the month, together with a technical appendix in case the client requests this.

Les X. Pensive.

CRUMMLEY CONCRETE PRODUCTS

Memorandum:
To: Les X. Pensive, Costing Dept.
From: Ben Tover, Drawing Office
Re: *Technical specification: Cloud Nine Building.*

Simply supported slabs for floors:

Simply supported span (one-way)		18 ft
Sustained load for partitions		15 lb/sq. ft equivalent
Live load		100 lb/sq. ft
Load factors	dead load	1.4
	live load	1.7
Steel ratio	minimum	$200/f_y$
	maximum	0.75 of balanced design ratio
Deflection		span/360 for live and sustained loads only

Although a fully automatic design procedure might be developed, a relatively simple trial and error procedure is used which allows the slab thickness and steel area to be optimized for any selected allowable stresses and costs for steel and concrete. The algorithm might take the form shown in Figure 10.3.

1. Select material stresses; note costs.
2. Select trial slab thickness d_0
3. Check $d_0 \geqslant$ minimum slab thickness. If not go to 2.
4. Calculate Dead Load (DL).
5. Calculate BM using appropriate load factors.
6. Select trial steel area A_{st}.
7. Check $A_{st}/bd \geqslant$ minimum steel ratio. If not, go to 6.
8. Calculate balanced design steel ratio ρ_b.
9. Check $A_{st}/bd \leqslant 0.75 \rho_b$. If not, go to 2.
10. Calculate resistance moment RM.
11. Check $RM \geqslant BM/\varphi$ where $\varphi = 0.9$ Capacity Reduction Factor. If not, go to 6.
12. Find neutral axis depth and Moment of Inertia of section.
13. Calculate deflection δ.
14. Check $\delta \leqslant$ Span/360. If not go to 6.
15. Calculate cost of slab/ft^2.

Figure 10.3 Algorithm for minimum cost slab

10.2.3 Development of a subroutine library

In reviewing the various steps in the algorithm of Figure 10.3, it is possible to identify a number of basic computational tasks which may form the basis for a group of related subroutines, e.g.

Step 8   Calculate balanced steel ratio.
Step 10   Calculate ultimate load RM.

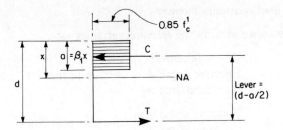

Figure 10.4 Assumed stress distribution at Ultimate Load

Step 12 Calculate neutral axis depth of singly reinforced section.
Step 12 Calculate Moment of Inertia of transformed section.

Each of these basic operations has been implemented in a simple subroutine, as outlined in the following sub-sections.

(1) Balanced steel ratio (see Figure 10.4 for notation)

$$\frac{x}{d-x} = \frac{\varepsilon_c}{\varepsilon_y} \text{ at balance} \tag{10.2}$$

i.e.

$$\frac{x}{d} = \frac{\varepsilon_c}{\varepsilon_c + \varepsilon_y} = \frac{0.003}{0.003 + f_y/E_s}. \tag{10.3}$$

Also $C = T$, therefore

$$0.85 f'_c \beta_1 x = A_s f_y$$

and substituting for x

$$\rho_b = \frac{A_{st}}{bd_b} = 0.85 \beta_1 \frac{\varepsilon_c}{\varepsilon_c + \varepsilon_y} \frac{f'_c}{f_y}. \tag{10.4}$$

Also

$$\beta = \phi(f'_c).$$

Thus relevant parameters are

$$\rho_b = \phi(f'_c, f_y) \quad \text{assuming } E_s = 29 \times 10^6 \text{ psi.}$$

The parameters selected include the proportion β_1 since this is frequently required in addition to ρ_b. Thus the routine specification is

SUBROUTINE ASTBAL (FC, FY, BETA1, RHOB)
                          Input─────┘   └──────Output

(2) Ultimate load resistance moment

In a similar way we identify the relevant variables as:

$$\left.\begin{array}{r}\text{Breadth } b \\ \text{Effective depth } d \\ \text{Steel area } A_{st}\end{array}\right\} \text{section properties}$$

$$\left.\begin{array}{r}\text{Concrete stress } f_c' \\ \text{Steel stress } f_y\end{array}\right\} \text{material properties.}$$

The routine specification is

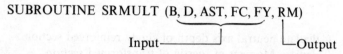

SUBROUTINE SRMULT (B, D, AST, FC, FY, RM)

Input ⸺⸺⸺⸺⸺⸺⸺⸺⸺ Output

Note: Although this could be written as a FUNCTION subroutine there is some advantage in using subroutines as standard — e.g. standard calling statement, type independent of name, etc.

(3) Neutral axis depth

The routine specification is

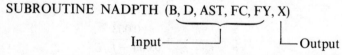

SUBROUTINE NADPTH (B, D, AST, FC, FY, X)

Input ⸺⸺⸺⸺⸺⸺⸺⸺⸺ Output

As before the input data comprises section and material properties.

(4) Moment of inertia

Several methods may be used to compute deflections. The method selected here employs the inertia of the transformed cracked section — i.e. ignoring concrete in the tension zone and transforming steel to an equivalent area of concrete by use of the ratio of the elastic moduli E_s/E_c.

Again, the required data comprises section and material properties

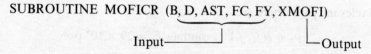

SUBROUTINE MOFICR (B, D, AST, FC, FY, XMOFI)

Input ⸺⸺⸺⸺⸺⸺⸺⸺⸺ Output

Relocatable libraries

When a number of subroutines have been developed which are likely to be of recurring use to one or more users in a group, it is advantageous to assemble these in a library of semi-compiled sub-programs.

10.2.4 Example 10.3: Minimum cost design of a floor

The following memorandum outlines a slightly more elaborate floor design which is amenable to computer simulation:

PERPLEX BUILDING SYSTEMS

Memorandum:
To: Slye Droole, Design Section.
From: Mark E. Ting, Buying Dept.
Re: *Modular Floor Systems.*

We are planning to market a simple but versatile floor system comprising precast concrete slab units simply supported on steel beams. In order to develop some typical costs, I require an automated design procedure which will assist in the selection of optimum values for the spacing of the steel beams, and the width, thickness, and reinforcement of the slab units. The floor areas to be processed are rectangular, a typical example being 150 ft long by 35 ft wide.

The following costs may be assumed:

(1) Concrete in slab units $2.50/cu. ft
(2) Formwork on underside and long
    edges of precast slabs $1.25/sq. ft
(3) Reinforcement $0.35/lb
(4) Erection of slab units where one
    unit has a weight of W lb $(W/300)^{1.5}$ per unit
(5) Steel beams (erected) $0.25/lb.

Live loads on the floors are typically in the range of 80–150 lb/sq. ft. Assume material stresses of 4000 psi in concrete and 50 000 psi in steel.

Due to recent publicity of the failure of grossly overloaded floors, I intend to stress in our advertising the high ductility of our floors, and the value of early distress signs. For that reason I suggest that you design the slab units with a low steel ratio — say 30% of the balanced design steel ratio.

The problem posed by this assignment may be tackled in the following manner.

1. Specify material properties and costs.
2. Specify overall floor dimensions, length, and span.
3. Select a trial value N for the number of panels.
4. Select a trial value for the slab unit breadth B.
5. Design slab unit.
6. Calculate reactions on external (end) beams.
7. Design end beams.
8. Calculate reactions on internal beams.
9. Design internal beams.

10. Determine costs for concrete, reinforcement, formwork.
11. Obtain weight of slab unit and calculate erection costs.
12. Obtain weight of steel and calculate cost of beams.
13. Output total floor cost as function of N and B.

Additional routines may be identified for use in this study, e.g.

(1) Design a singly reinforced section.   (subroutine SRSECT)
(2) Design a one-way spanning slab.   (subroutine SLAB1W)
(3) Calculate maximum BM in a continuous slab
 (although not used in this example).   (subroutine SLABBM)
(4) Select a steel beam section.   (subroutine IBEAM)

These routines have been included in Appendix A.

10.3 PROBABILISTIC MODELS

Symbolic models involving one or more variables subject to uncertainty or which incorporate relationships comprising random functions are termed *probabilistic models*. The fundamental distinction is that the outcome of an analysis for a given set of input variables is not uniquely determined — i.e. experimental results are, in general, not repeatable.

Computed results must be described in statistical terms — usually the *expected value* and some measure of variability (e.g. *variance*). Two types of probabilistic model may be identified.

(1) Static models

This type of model concerns an analysis, some of the inputs for which are subject to uncertainty. The outcome of the analysis is thus also uncertain.

(2) Dynamic models

With this type of model, the engineer attempts to duplicate the dynamic behaviour of certain characteristics of a system, process, or operation over a period of time.

When the processes to be analysed are complex, it is usually not possible to obtain the probability or other statistical measure of an event by analytical means and the use of numerical experiments becomes necessary. The method described in the following section is usually called a *Monte Carlo* procedure. This technique involves generating a series of numbers which serve as observations of one or more random variables. These random quantities provide the data for calculating a particular result or outcome. If a sufficiently large number of these numerical experiments are carried out, estimates can be made of averages, probabilities, or frequency distributions.

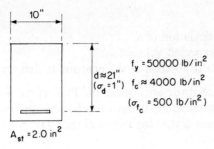

Figure 10.5   Singly reinforced section subject to some uncertainty

10.3.1 Static models

This type of model is now described with reference to the example illustrated in Figure 10.5.

Example 10.4: Probability of failure of a reinforced concrete beam

A singly reinforced concrete beam has a design cross-section as indicated in Figure 10.5. The effective depth is subject to some uncertainty and may be defined as being normally distributed about a mean of 21 in with a standard deviation of 1 in. Similarly, the concrete strength f_c has an expected value of 4000 lb/in^2 and a standard deviation of 500 lb/in^2. Estimate the probability that the resistance moment is less than 95% of the design capacity.

The probability density of the effective depth d may be illustrated graphically as shown in Figure 10.6.

Note: The notation $f_d(x)$ means '... the probability that the variable d has a value of x ...'.

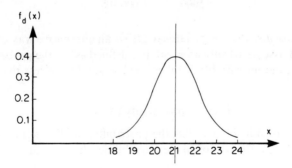

Figure 10.6   Probability density of effective depth d

The mathematical definition of the normal probability density is given by:

$$f_d(x) = \frac{1}{\sigma_d\sqrt{2\pi}} \exp\left[-\tfrac{1}{2}(x - m_d/\sigma_d)^2\right] \qquad (10.5)$$

where

σ_d is the standard deviation of d.
m_d is the mean or expected value of d.
x is any value on the abscissa of the probability density function (PDF).

It should be noted that the area under the PDF curve is 1.0.

The distribution of concrete strength may be described in a similar way (Figure 10.7) but it will be noted that the values of the PDF $f_{f_c}(x)$ are several orders of

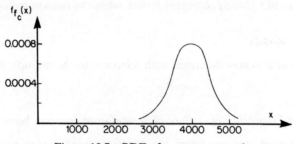

Figure 10.7   PDF of concrete strength

magnitude smaller on account of the relatively large values of x which must be used. The area under the curve is still 1.0, however.

x	$f_{f_c}(x)$
3000	0.000 108
3500	0.000 484
4000	0.000 798
4500	0.000 484
5000	0.000 108

The cumulative distribution function (CDF) is an alternative way of describing the uncertainty associated with an event. It is defined as '... the probability of the event that the random variable X takes a value equal to or less than argument x...', i.e.

$$F_X(x) = \text{probability } (X \leqslant x). \tag{10.6}$$

For the PDF shown in Figure 10.8, the probability of $(X \leqslant x_1)$ is given by the shaded area, which is the sum of all the probabilities that $X =$ any value less than or equal to x_1.

Then

$$F_X(x_1) = \int_{-\infty}^{x_1} f_X(x)\,dx. \tag{10.7}$$

For example the cumulative distribution function for the effective depth d will be as shown in Figure 10.9.

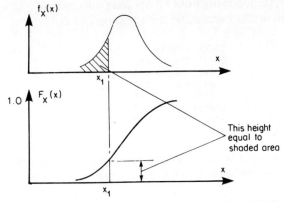

Figure 10.8 Cumulative distribution function (CDF)

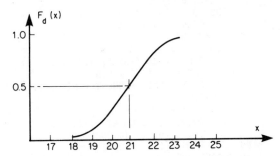

Figure 10.9 Cumulative distribution for effective depth d

Note that

(1) the probability ($d < 18$) is negligible,
(2) the probability ($d < 21$) is 0.5 (50% chance),
(3) the probability ($d < 24$) is almost 1.0 (virtually a certainty).

To model this situation a simple computer program is constructed which will generate *pseudorandom* numbers which correspond to the required distribution ('pseudorandom' because the series of random numbers produced artificially will be unique only for a finite length of series — after that the sequence may be repeated):

SUBROUTINE GAUSS (ISEED, STDEV, AVE, VAL)

                        Input   Output

The seed value (ISEED) is automatically replaced during execution of the routine. If subroutine GAUSS is used inside a loop, the value of ISEED should be initialized *once* only, outside the loop.

It should also be noted that random number generators are usually sensitive to the seed value used, and may generate a biased sequence.

For example, the following FORTRAN code will generate a series of 20 values of D with mean of 21 in and standard deviation of 1 in:

```
            ISEED = 2345
            DO 10 I = 1, 20
               CALL GAUSS (ISEED, 1.0, 21.0, D)
               WRITE (6, 20) D
     20        FORMAT (F12.3)
     10     CONTINUE
            STOP
            END
```

The distribution could be tested by generating a much larger number of values and counting the number of times the value falls within prescribed ranges:

```
            DIMENSION COUNT (30)
            ISEED = 2345
            DO 10 I = 1, 30
               COUNT (I) = 0.0
     10     CONTINUE
            DO 20 I = 1, 100
               CALL GAUSS (ISEED, 1.0, 21.0, D)
               INT = IFIX (D + 0.5)
               COUNT (INT) = COUNT (INT) + 1.0
     20     CONTINUE
            DO 30 I = 1, 30
               COUNT (I) = COUNT (I)/100.0
               WRITE (6.40) I, COUNT (I)
     40        FORMAT (15, F10.4)
     30     CONTINUE
            STOP
            END.
```

Such a program should give values close to the ideal normal distribution and might be run several times with different seed values to test sensitivity.

The reinforced concrete beam of Example 10.4 may be modelled in the following way:

(1) Calculate the design resistance moment (RMDES) for the design values of D and FC (using subroutine SRMULT).
(2) Store the test value (RMTST) = 0.95* RMDES.
(3) Generate random values for D and FC using subroutine GAUSS.
(4) Calculate a random resistance moment RM.
(5) Compare with RMTST — if RM < RMTST, count one failure.
(6) Repeat steps (3)–(5) (say) 100 times and count the number of failures.

Then

$$\text{probability (RM} < 0.95 \text{ RMDES)} \simeq \frac{\text{Number of failures}}{100}.$$

Typical code to do this is:

```
          B = 10.0
          AST = 2.0
          FY = 50000.0
          D = 21.0
          FC = 4000.0
          CALL SRMULT (B, D, AST, FC, FY, RMDES)
          RMTST = 0.95*RMDES
          ISEED = 2345
          COUNT = 0.0
          DO 10 I = 1, 100
              CALL GAUSS (ISEED, 1.0, 21.0, D)
              CALL GAUSS (ISEED, 500.0, 4000.0, FC)
              CALL SRMULT (B, D. AST, FC, FY, RM)
              IF (RM.LT.RMTST) COUNT = COUNT + 1.0
              WRITE (6, 20) I, RM, COUNT/FLOAT (I)
    20        FORMAT (I5, F15.1, F10.4)
    10    CONTINUE
          STOP
          END
```

It should be remembered, of course, that the above program must be run along with a library of routines (or equivalent) to satisfy the external requirements for routines SRMULT and GAUSS.

10.3.2 Modelling of arbitrary distributions

In the previous example it was assumed that both random variables could be modelled by a normal distribution. While this might be reasonable for the effective depth, the distribution of concrete strength is likely to be significantly different.

In many cases, the distribution of a random variable must be modelled to conform with experimental observations. Tests on 10 concrete specimens (cubes or cylinders) give the following results:

f_c = 2500–3000 psi   1 specimen
      3000–3500 psi   3 specimens
      3500–4000 psi   4 specimens
      4000–4500 psi   2 specimens
      Total no. of tests   10 specimens.

The probability density function of this experiment might be illustrated as shown in Figure 10.10. It should be noted that the ordinate is not strictly a probability, since the area under the curve is *not* unity. This can be corrected easily by dividing ordinate values by the area A where:

$$A = 500(0.1 + 0.3 + 0.4 + 0.2) = 500. \tag{10.8}$$

With reference to Figure 10.11 it is clear that A is equivalent to the stress interval used to categorize the test results, i.e.

$$f_{f_c}(x) = \frac{\text{'proportion of tests'}}{500}.$$

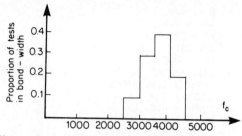

Figure 10.10 Histogram of experimental results

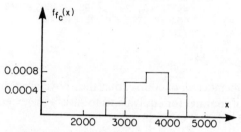

Figure 10.11 Probability density function (PDF) of concrete strength

The corresponding *cumulative distribution function* (CDF) is easily obtained as the integral of $f_{f_c}(x)$.

It will be seen that to obtain $F_{f_c}(x)$ it is necessary to *multiply* again by the stress interval 500. Thus the CDF could be obtained simply by accumulating the proportions or by accumulating the number of tests and dividing by the total number of tests (10).

10.3.3 Generation of random variates

Having obtained the CDF we can use this to generate random values of concrete strength f_c which are consistent with the distribution.

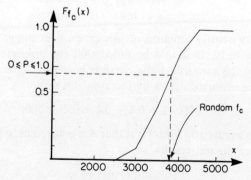

Figure 10.12 Cumulative distribution function for f_c

The procedure is to assign to $F_{f_c}(x)$ a random number P which is uniformly distributed in the interval $0.0 < P < 1.0$, and then find the corresponding value of x by interpolation. In Figure 10.12, for example, $P = 0.7$ yields $x = 3870$. A variety of techniques could be devised to carry out this inverse function evaluation. One approach is illustrated by the following code:

```
            ISEED = 12
            CALL URAND (ISEED, P)
            FC = 2750
            IF (P.LT.0.1) GOTO 90
            FC = 3250
            IF (P.LT.0.4) GOTO 90
            FC = 3750
            IF (P.LT.0.8) GOTO 90
            FC = 4250
         90 CONTINUE
            .
            .
            .
```

This method is rather crude and does not allow for values other than the averages of the stress intervals. Also, for many intervals the coding becomes unwieldy.

An improvement is illustrated in the following FORTRAN code:

```
            DIMENSION FCR(5), CDF(5)
            DATA FCR/2500., 3000., 3500., 4000., 4500./
            DATA CDF/0.0, 0.1, 0.4, 0.8, 1.0/
            ISEED = 12
            CALL URAND (ISEED, P)
            DO 10 I = 1, 4
            J = 1
            IF (P.LT.CDF(I+1)) GOTO 20
         10 CONTINUE
         20 CONTINUE
            FC = FCR(J) + (FCR(J+1) - FCR(J))*(P-CDF(J))/(CDF(J+1) - CDF(J))
```

This strategy employs the full range of values of FC and compresses the search coding into a single DO-loop.

An alternative procedure is to construct an auxiliary array of values of FC with (say) $10 + 1$ or 11 elements in it, each one corresponding to a cumulative probability value of $0.0, 0.1, 0.2, \ldots, 0.9, 1.0$.

The process may be visualized graphically, as illustrated in Figure 10.13, by subdividing the cumulative distribution function (CDF) uniformly.

Thus, for the data given earlier,

$F_{f_c}(x)$	x	$F_{f_c}(x)$	x
0.0	2500		
0.1	3000	0.6	3750
0.2	3167	0.7	3875
0.3	3333	0.8	4000
0.4	3500	0.9	4250
0.5	3625	1.0	4500

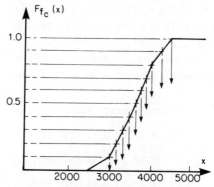

Figure 10.13   Uniform interpolation of CDF

The following coding can then be used:

```
      DIMENSION FCR (11)
      DATA FCR/2500., 3000., 3167., 3333., 3500., 3625.,
     +          3750., 3875., 4000., 4250., 4500./
      ISEED = 12
      CALL URAND (ISEED, P)
      P = 10.0*P
      IP = IFIX(P) + 1
      FRAC = P+1.0 - FLOAT(IP)
      FC = FCR(IP) + FRAC*(FCR(IP+1) - FCR(IP))
```

Obviously, the CDF need not always be sub-divided into 10 increments.

10.3.4   Generalized procedures

The method outlined above can be generalized so that the number of points describing the cumulative distribution function may be chosen arbitrarily. Two subroutines DISTN1 and DISTN2 which perform this function are included in the library of Appendix A. In addition, a routine for linear interpolation of a unimodal function (INTER1) is included since this is used by DISTN1.

If the earlier problem of Section 10.3.1 (i.e. probability that $RM < 0.95 RM_{des}$) is modified so that the concrete strength is described by the 10 specimen tests (instead of being assumed to be normally distributed) the program would take the following form:

```
        DIMENSION FCR(5), CYL(5), WK(5), CDF(21)
        DATA FCR/2500., 3000., 3500., 4000., 4500./
        DATA CYL/1.0, 3.0, 4.0, 2.0, 0.0/
        B = 10.0
        D = 21.0
        AST = 2.0
        FC = 4000.0
        FY = 50000.0
        CALL SRMULT (B, D, AST, FC, FY, RM)
        RMTST = 0.95*RM
```

```
      ISEED = 12345
      CALL DISTN1 (FRCR, CYL, 5, WK, CDF, 21)
      COUNT = 0.0
      DO 10 I = 1,100
         CALL GAUSS (P, 1.0, 21.0, D)
         CALL URAND (ISEED), P)
         CALL DISTN2 (CDF, 21, P, FC)
         CALL SRMULT (B, D, AST, FC, FY, RM)
         IF (RM. LT. RMTST) COUNT = COUNT + 1.0
         WRITE (6,20) I, RM, D, FC, COUNT/FLOAT (I)
20       FORMAT (15, F15.1, 2F10.3, F10.4)
10    CONTINUE
      STOP
      END
```

Note that the sequence of pseudorandom numbers may possess certain statistical properties which are sensitive to the initial seed value used. Any simulation model should be tested for such sensitivity. Another example of probabilistic modelling is given in Chapter 12.

10.4 DYNAMIC MODELS

Numerical representation of random variables may be easily extended from static situations as described above to dynamic situations in which certain properties of the system vary with time. The process involves the use of artificial statistical experiments to estimate unknown or dynamically varying quantities.

As was seen with the earlier static models, the results or outcomes approach the theoretical expected value as the number of experiments is increased. Thus simulation can be computationally expensive.

Since dynamic systems are by definition dependent on the time history of inputs or events, the frequency of occurrence of events (or the time between successive events) may be an important statistical property. We now consider a specific example in which this is the case.

10.4.1 Continuous variables

Ships arriving at an unloading dock have an average rate of arrival, but in reality the interval between arrivals may fluctuate about the mean. In many cases, these inter-arrival times can be represented by a negative exponential curve. Thus, if $f_T(x)$ is the probability that the inter-arrival time T is equal to x, then

$$f_T(x) = \lambda\, e^{-\lambda x} \qquad (10.9)$$

where the parameter λ is the average arrival frequency or $1/\lambda$ is the long-term inter-arrival time.

The cumulative distribution function $F_T(x)$ is obtained by integrating (10.9) to give

$$F_T(x) = 1 - e^{-\lambda x} \qquad (10.10)$$

where the constant of integration is obtained from the condition that $F_T(x = \infty) = 1.0$.

Generation of inter-arrival times can be done by the method described in Section 10.3.3 but, because of the simple form of (10.10), a direct solution is possible by inverting the function. Thus, if $F_T(x) = p$ where p is uniformly distributed in the interval $0 \leqslant p \leqslant 1$, then

$$T = -\frac{1}{\lambda} \ln(1 - p). \tag{10.11}$$

Equations (10.9) and (10.11) are illustrated graphically in Figure 10.14.

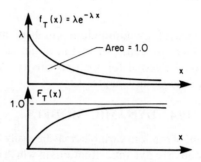

Figure 10.14 Distribution of inter-arrival times

10.4.2 Discrete (integer) variables

The number of ships arriving in a given time period (say per day or per week) should theoretically be a constant, but fluctuates in real life. The outcome must be an *integer* quantity. Such a situation may often be represented as a Poisson process.

The probability must be described in a discrete form as shown in (10.12) and Figure 10.15, in which the variable n must take integer values:

$$f_N(n) = \frac{\alpha^n e^{-\alpha}}{n!} \quad n = 0, 1, 2, 3, \ldots . \tag{10.12}$$

The cumulative distribution function is also discontinuous and the integration

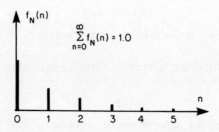

Figure 10.15 Probability distribution of the number of arrivals in a given time interval

process reduces to a summation. Thus

$$F_N(n) = \sum_0^n f_N(n) \quad n = 0, 1, 2, 3, \ldots. \tag{10.13}$$

For $\alpha = 0.2$ the discrete probabilities are tabulated as shown:

$n =$	0	1	2	3
$f_N(n) =$	0.8187	0.1637	0.0164	0.0011
$F_N(n) =$	0.8187	0.9825	0.9989	0.9999

The discontinuous CDF may be approximated by a continuous curve as indicated in Figure 10.16, as long as the real number obtained by interpolation on the curve is truncated to the nearest integer. In this way the routines DISTN1 and DISTN2 may be used for discrete distributions as well as continuous ones.

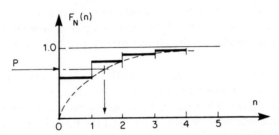

Figure 10.16 Approximation of a discrete (integer) CDF

10.4.3 Example 10.5: A maintenance problem

A production plant contains a large number of units which operate in parallel. The units break down at certain intervals and an analysis of past records shows that the number of units which break down on any one day can be represented by a Poisson process. Thus:

$$\text{Probability } (n \text{ units break down}) = \frac{a^n e^{-a}}{n!} \tag{10.14}$$

where the parameter $a = 0.2$.

Repair times appear to be normally distributed with an average of 5 days and a standard deviation of 1 day.

A single maintenance engineer is currently employed full-time and obviously can work on only one unit at any time. The daily salary for the engineer is $120 (irrespective of whether he is working or not). Cost of non-production of one unit is $500 per day.

The problem is to determine whether or not an additional maintenance engineer (or more) should be hired.

The problem can be tackled by developing a Monte Carlo simulation

procedure to estimate the long-term average daily cost for different numbers of maintenance engineers. The algorithm may be summarized briefly as follows:
1. Define costs, parameters, and number of engineers, M.
2. Set up a loop to simulate n days of operation (say $n = 100$).
3. For each day, generate a number of breakdowns.
4. For each breakdown, generate an estimate of the required repair times.
5. Assign available man-days of maintenance to reduce the outstanding repair time.
6. Keep track of the total 'down-time' T and thus a running total of cost.
7. After n days of simulation, estimate the average daily cost as

$$C = 120M + 500T/n. \tag{10.15}$$

No information is given as to the time within any particular day that the simulated breakdown occurs. A reasonable assumption is that these are uniformly distributed throughout the working day.

Clearly the value of n must be sufficiently large for reasonable estimates to be obtained. Also, for values of $M > 1$, the book-keeping is somewhat complicated. A digital computer solution is therefore desirable to allow sensitivity analysis to be carried out.

A possible solution is indicated in the FORTRAN coding of Figure 10.17. In this algorithm, the cumulative distribution function is set up by means of subroutine DISTN1, as in Section 10.3.3, using the probability density defined by (10.14). The value returned by DISTN2 within the main loop is truncated to an integer value, as illustrated in Figure 10.16. A running log of the breakdowns is maintained in the arrays TD(·) and REP(·) which hold respectively the time the unit went down and the anticipated repair time, both quantities measured in days.

The most complex part of the algorithm is the determination of the time the unit comes back 'up' on-line TU(·). When the number of maintenance engineers is 1, this is easily obtained as

$$TU_1 = TD_1 + REP_1$$
$$TU_j = \max\{(TD_j + REP_j), TU_{j-1} + REP_j\} \quad j = 2, 3, \ldots, n. \tag{10.16}$$

When more than one engineer is employed, the quantity TU_{j-1} in (10.16) must be replaced by the Mth largest value in the first $(j-1)$ elements of TU(·). A simple subroutine MTHVAL provides this value.

Once the array TU(·) has been computed, it is a simple matter to calculate the total down-time and hence the average daily cost by (10.15).

For the rates and parameters suggested in this example, the results obtained are summarized in Tables 10.1 and 10.2. Table 10.1 shows the details of the down-time, repair time, and up-time for each breakdown in a simulated sequence of 100 days. The initial seed is the same for each simulation involving a different number of maintenance engineers so only the TU( ) array changes.

Table 10.2 shows the summary results of the total production time lost and the

```
      DIMENSION PROB(10),UNIT(10),WK(10),CDF(21)
      DIMENSION TUC(100),REP(100),TD(100)
      LOGICAL PRINT
      PRINT=.FALSE.
      WRITE(6,5)
    5 FORMAT(*   DO YOU WANT DETAILED LISTING?...Y/N   *)
      READ(5,6)ANS
    6 FORMAT(A1)
      IF(ANS.EQ.1HY) PRINT=.TRUE.
C DEFINE NO. OF MAINTENANCE ENGINEERS
    1 WRITE(6,10)
   10 FORMAT(33H  SPECIFY NO. OF MTCE. ENGRS...I5)
      READ(5,20)M
   20 FORMAT(I5)
      IF(M.LT.1) STOP
C ZERO TIME ARRAY
      DO 30 IT=1,100
      TU(IT)=0.0
      TD(IT)=0.0
      REP(IT)=0.0
   30 CONTINUE
C SET PARAMETERS AND DAILY RATES
      AVE=5.0
      STDEV=1.0
      ALPHA=0.2
      SALARY=120.0
      PROD=500.0
C SET UP ARRAYS FOR ROUTINE DISTN1
      EXPALF=EXP(-ALPHA)
      DO 40 K=1,10
      UNIT(K)=FLOAT(K-1)
      PROB(K)=ALPHA**(K-1)*EXPALF/FACTRL(K-1)
   40 CONTINUE
C SET UP CUMULATIVE DISTRIBUTION FUNCTION
      CALL DISTN1(UNIT,PROB,10,WK,CDF,21)
C SET UP LOOP FOR NO. OF DAYS
      WRITE(6,50)
   50 FORMAT(26H  SPECIFY NO. OF DAYS...I5)
      READ(5,20)NDAYS
      ISEED=12345
      J1=0
      J2=0
      DO 100 IDAY=1,NDAYS
C GENERATE NO. OF BREAKDOWNS
          CALL URAND(ISEED,P)
          CALL DISTN2(CDF,21,P,BRK)
          IBRK=IFIX(BRK)
          IF(PRINT.AND.IBRK.GT.0) WRITE(6,203)IDAY,P,IBRK
  203     FORMAT(*  DAY*,I4,* (P=*,F7.4,*)  ...IBRK=*,I3)
          J1=J2+1
          J2=J1+IBRK-1
C GENERATE BREAKDOWN TIME AND ESTIMATE OF REPAIR TIME
          IF(IBRK.LE.0) GOTO 100
          DO 60 J=J1,J2
              CALL URAND(ISEED,P)
              TD(J)=FLOAT(IDAY-1) + P
              CALL GAUSS(ISEED,STDEV,AVE,REP(J))
              IF(PRINT) WRITE(6,204)J,TD(J),REP(J)
```

Figure 10.17   FORTRAN program for simulation of the maintenance problem

```
      204   FORMAT(32X,I3,2F7.3)
       60   CONTINUE
      100   CONTINUE
C     CALCULATE TU(), TIME WHEN UNIT RESUMES PRODUCTION
            DO 110 J=1,M
              TU(J)=TD(J) + REP(J)
      110   CONTINUE
            MP1=M+1
            DO 130 J=MP1,J2
              TU(J)=TD(J) + REP(J)
              IF(M.EQ.1) TUJM1=TU(J-1)
              IF(M.GT.1) CALL MTHVAL(TU,J-1,M,TUJM1)
              TUJ=TUJM1 + REP(J)
              IF(TUJ.GT.TU(J)) TU(J)=TUJ
      130   CONTINUE
C     CALCULATE TOTAL DOWN TIME
            DTIM=0.0
            DO 150 J=1,J2
              IF(PRINT) WRITE(6,201)J,TD(J),REP(J),TU(J)
      201     FORMAT(I5,3F10.3)
              DTIM=DTIM + TU(J) - TD(J)
      150   CONTINUE
C     CALCULATE TOTAL COST AND AVERAGE DAILY COST
            TOTCST=SALARY*FLOAT(M*NDAYS) + DTIM*PROD
            AVECST=TOTCST/FLOAT(NDAYS)
            WRITE(6,200)M,J2,DTIM,TOTCST,AVECST
      200   FORMAT(/,* FOR*,I4,* MAINTENANCE ENGINEERS*,/,
           +        *          NO. OF FAILURES  =    *,I4,/,
           +        *          TOTAL DOWN TIME  =    *,F8.3,/
           +        *          TOTAL COST       = $*,F10.2,/
           +        *          AVE. DAILY COST  = $*,F10.2)
            GOTO 1
            END
C
            SUBROUTINE MTHVAL(VAL,NV,M,VALM)
            DIMENSION VAL(NV)
            DO 10 I=1,M
              BIG=0.0
              DO 20 J=1,NV
                IF(VAL(J).LT.BIG) GOTO 20
                BIG=VAL(J)
                IBIG=J
       20     CONTINUE
              VAL(IBIG)=-VAL(IBIG)
       10   CONTINUE
            VALM=-VAL(IBIG)
            DO 30 J=1,NV
              VAL(J)=ABS(VAL(J))
       30   CONTINUE
            RETURN
            END
C
            FUNCTION FACTRL(N)
            FACTRL=1.0
            IF(N.EQ.0) RETURN
            DO 10 J=1,N
              FACTRL=FACTRL*FLOAT(J)
       10   CONTINUE
            RETURN
            END
```

Figure 10.17 — continued

Table 10.1 Detailed results of the simulation of the maintenance problem

Day	Random number	Number of breakdowns		Time down (days)	Repair time (days)	Time up (days) for $M =$		
						1	2	4
9	.8818	1	1	8.69	6.21	14.90	14.90	14.90
11	.8208	1	2	10.02	6.04	20.94	16.05	16.05
20	.8200	1	3	19.68	4.93	25.87	24.61	24.61
22	.9749	5	4	21.08	4.69	30.55	25.77	25.77
			5	21.95	4.80	35.35	29.41	26.75
			6	21.66	4.67	40.02	30.44	26.33
			7	21.79	3.81	43.83	33.22	28.42
			8	21.53	3.18	47.01	33.62	28.95
23	.8751	1	9	22.32	4.08	51.09	37.30	30.41
32	.9080	1	10	31.66	3.51	54.60	37.13	35.17
35	.8472	1	11	34.39	4.91	59.52	42.04	39.30
36	.8763	1	12	35.63	3.70	63.21	41.00	39.32
39	.8557	1	13	38.41	5.40	68.61	46.40	43.80
42	.9112	1	14	41.22	4.23	72.84	46.27	45.45
44	.8320	1	15	43.69	5.32	78.15	51.58	49.01
53	.8463	1	16	52.54	6.65	84.81	59.19	59.19
54	.9588	3	17	53.50	4.00	88.81	57.50	57.50
			18	53.35	5.62	94.43	63.12	58.97
			19	53.47	3.60	98.04	62.79	57.07
56	.9617	3	20	55.78	5.27	103.31	68.06	62.34
			21	55.48	6.83	110.14	69.96	64.33
			22	55.96	4.34	114.48	72.41	63.31
62	.9852	6	23	61.68	3.75	118.23	73.70	65.42
			24	61.32	4.63	122.86	77.03	66.97
			25	61.27	4.83	127.68	78.53	68.14
			26	61.35	4.76	132.44	81.79	69.09
			27	61.36	3.90	136.34	82.43	69.33
			28	61.63	5.27	141.61	87.06	72.24
68	.9379	1	29	67.30	4.22	145.83	86.65	72.36
82	.9814	6	30	81.86	5.01	150.85	91.67	86.87
			31	81.16	3.86	154.70	90.92	85.02
			32	81.60	6.51	161.21	97.42	88.11
			33	81.73	5.74	166.95	97.40	87.47
			34	81.83	4.38	171.33	101.78	89.40
			35	81.51	5.66	176.99	103.08	92.53
85	.9474	1	36	84.55	4.32	181.31	106.10	91.79
89	.9436	1	37	88.46	4.31	185.62	107.39	92.77

average total daily cost including both loss of production and salaries. The cost is a minimum with 4 maintenance men but this result may be sensitive to the initial seed value and the number of days simulated. On the more practical side, management would likely err on the side of too few men to reduce problems of boredom and embarrassment of lay-offs should the simulation turn out to be pessimistic.

Table 10.2 Summary of results for the maintenance problem

No. of days simulated = 100
Total no. of breakdowns = 37

Number of maintenance engineers	Total time lost (days)	Average loss of production ($/day)	Average total cost ($/day)
1	1802.05	9010	9130
2	453.32	2267	2507
3	277.48	1387	1747
4	222.06	1110	1590
5	198.42	992	1592
6	181.53	908	1628
7	176.92	885	1725
8	176.92	885	1845
9	176.92	885	1965

The sensitivity of these results to the number of days (n) and the initial seed value for the random number generator is left as an exercise for the interested reader. In this connection, a useful discussion is given by Au, Shane, and Hoel (Chapter 10, Simulation).[1] Also, it should be noted that the algorithm uses several assumptions which may not always be realistic. For example, it is assumed that an engineer continues to work on a breakdown until repair is completed, irrespective of any other minor breakdowns which might occur. Also, it is assumed that only one man can work on a repair job at any time. The inclusion of subtleties such as these in the simulation model presents a stimulating challenge to the systems analyst.

10.5 EXERCISES

10.5.1 A construction project involves five operations which are related by the prerequisites listed in the following table. The duration of each activity is assumed to be normally distributed with mean values and standard deviation as given. Determine the probability of the project completion time exceeding 20 months.

Activity	Prerequisites	Expected duration (months)	Standard deviation (months)
1	—	11	3
2	—	6	1
3	2	4	1
4	1, 3	9	2
5	2	10	2

10.5.2 Flows into a reservoir during equal time intervals are to be generated by the following equation:

$$V_{j+1} = \bar{V} + r(V_j - \bar{V}) + \Delta V(1 - r^2)^{1/2}$$

where

V_{j+1} = volume of inflow in period $j + 1$
V_j = volume of inflow in period j
$\bar{V}$ = average volume of inflow (30×10^6 m^3)
r = correlation coefficient (0.4)
ΔV = a normally distributed random component with a mean of zero and a standard deviation of 10×10^6 m^3.

If the withdrawal for each time period is 20×10^6 m^3, determine the reservoir storage required for a synthetic series of 50 time periods. Construct a curve relating the required storage volume to the withdrawal amount.

10.5.3 Ships carrying 10 000 tons of raw material supply a stockpile for a manufacturing process. Demand from the stockpile is initially 250 000 tons per year and is expected to increase uniformly to 8 times that figure over a 20-year period. The time taken by the ships from the source to the stockpile is observed to be variable with sample times for 20 trips as tabulated below. The shipping season is limited to 300 days per year.

Observed trip times (days)

12.0	9.5	10.0	15.0	11.5	13.5	15.5
12.5	12.0	10.5	9.0	16.0	14.0	13.5
11.0	13.5	14.0	12.0	13.0	14.0	

Unloading is done by means of a single conveyor belt and unloading time is estimated to be normally distributed with a mean of 3 days and a standard deviation of 0.6 days.

Ships forced to wait while another ship is unloading incur a cost of $2000/day. The cost of a second conveyor belt is $750 000. Ignoring inflation determine at what time in the 20-year life of the plant it is worth installing a second conveyor belt. Assume a discounting rate of 13%.

What would be the effect of inflation at 10% per annum on this decision?

10.6 REFERENCES

1. Au, T., Shane, R. M., and Hoel, L. A. (1972). *Fundamentals of Systems Engineering — Probabilistic Models*, Addison-Wesley, Reading, Mass.
2. de Neufville, R., and Stafford, J. H. (1971). *Systems Analysis for Engineers and Managers*, McGraw-Hill, New York.
3. de Neufville, R., and Marks, D. H. (Eds) (1974). *Systems Planning and Design*, Prentice-Hall, Englewood Cliffs, New Jersey.

Chapter 11
Decision Analysis

11.1 INTRODUCTION

This chapter is concerned with methods of dealing with problems which have a degree of uncertainty, and ultimately selecting attractive courses of action. The uncertainty in the problem can be reflected in the *choice* which a decision maker faces and the *outcome* that will probably result from making such a choice. If there is no choice available then by definition there is no decision to make.

In some cases, where given a choice the outcome is known with certainty, the type of decision problem posed is referred to as a *deterministic* problem. In other words, the outcome is completely fixed or determined once the choice has been made. If, however, the outcome from a given choice could still be influenced by chance or random events then this is referred to as a *stochastic* decision problem.

We can now identify the third component of a decision problem, i.e. the *criterion*. A course of action cannot be chosen until the decision maker has some form of rational criteria on which to base that choice. Some of these criteria are more conservative than others and the decision maker should apply the criterion that most closely suits the needs of his or her company. In this chapter several criteria will be examined, some in more detail than others.

11.2 THE MAXIMIN CRITERION

If a company is in poor financial shape and cannot afford to take extreme risks, then they might well wish to avoid the worst possible outcomes of various actions. This argument is best presented by an example.

The Red Dragon Construction Company have the choice of purchasing three different types of earth-moving equipment. If they buy type A vehicle then the worst that can happen is that maintenance could be high, whereas it is known from discussions with other companies that type B vehicles are prone to clutch trouble. The third vehicle, type C, is relatively trouble free for the first few years but replacement parts are expensive compared with the two others. The situation may be summarized by use of the *pay-off* table (Table 11.1).

The figures in the table are the net incomes (in $1000) from each type of vehicle if the maintenance turns out to be low (I), average (II), or high (III). These

Table 11.1 Red Dragon Construction Company net income (pay-off)

Action (Vehicle bought)	Outcome (Net profit from use of vehicles)		
	I Low maintenance cost	II Average maintenance cost	III High maintenance cost
A	45	25	15
B	75	65	−15
C	95	30	−20

numbers are referred to as conditional outcomes, and thus the net profit of using machine type A, if the maintenance is average, will be 25 ($25 000).

The *maximin* rule requires that the company will 'choose the action which maximizes the minimum pay-off which can be achieved'. Hence, the company's choice will be type A — providing that maintenance costs will be low or average — despite the more attractive returns that the purchase of vehicles B and C will bring. The maximin rule will therefore result in Table 11.2.

Table 11.2 Optimal action for maximin rule

Action	Outcome			Minimum value	Optimal action
	I	II	III		
A	45	25	15	15	*
B	75	65	−15	−15	
C	95	30	−20	−20	

The maximin criterion is extremely conservative and can only be justified under very exceptional circumstances. Therefore, it would appear that a more positive approach to the solution is required.

If, in the previous problem, it had been decided to buy vehicle A and outcome II had resulted, the company could have had a pay-off of $65 000 if vehicle B had been used. This implies that a loss of revenue of $40 000 would result and a good opportunity would be lost. Hence, if the company were more inclined to take risks and wished to minimize these lost opportunities then a new criterion would be appropriate.

A conditional opportunity loss (COL) table may now be determined from the pay-off table. This is found as follows. In column I, the maximum pay-off would be achieved by choosing vehicle type C, and hence actions B and A would have resulted in missing opportunities of (95 − 75), and (95 − 45) respectively. The complete table for all possible outcomes would be as shown in Table 11.3.

Table 11.3 Conditional opportunity loss (COL) table for Red Dragon Construction Company

Action	I	II	III
A	(95 − 45) 50	(65 − 25) 40	(15 − 15) 0
B	(95 − 75) 20	(65 − 65) 0	(15 − (−15)) 30
C	(95 − 95) 0	(65 − 30) 35	(15 − (−20)) 35

The zeros in the table imply that if the outcome had been known, then an optimal choice could have been made, i.e. no opportunities would have been lost. In this particular case all rows have zeros in them and a different action would be required for the three outcomes.

The new criterion would now be to minimize the maximum opportunity loss.

11.3 THE MINIMAX CRITERION

This rule requires that a maximum value be selected for each row in the COL table. The optimal action is that which has the smallest of these maximum values, and if applied to the COL table for the Red Dragon Construction Company, the results shown in Table 11.4 are obtained.

Table 11.4 Optimal action for minimax

Action	Outcome			Maximum value in row	Optimal action
	I	II	III		
A	50	40	0	50	
B	20	0	30	30	*
C	0	35	35	35	

In this case the minimax criterion leads us to choose action B, which is different from the choice made by using the maximin criterion. This is hardly surprising as the two criteria are fundamentally different. The maximin rule implies that a 'play safe attitude at all costs' be adhered to, whereas the minimax rule contrives to avoid lost opportunities.

In some instances it is quite possible that the two criteria would lead to the same optimal choice. If in the pay-off table, outcome III would have resulted in a net profit of 20 for vehicle B, then the action of choosing vehicle B would have been optimum for either criterion. This is no contradiction to the previous statements made about the minimax and maximin criteria, but only serves to show that the answers obtained depend entirely on the numbers in the problem. It is this fact which amplifies the basic weakness in both criteria, i.e. no account is taken of the probabilities of the different outcomes taking place. This is a great disadvantage in the majority of cases as the decision maker would require some kind of estimate, however crude, of what is likely to happen. Therefore, the estimate of an outcome occurring must somehow be brought into the analysis to extend the decision theory discussed so far. This could then be incorporated into the pay-off table to enable the decision maker to arrive at a more rational choice.

11.4 BASIC CONCEPTS OF PROBABILITY THEORY

As stated previously, it is now necessary to incorporate scientific prediction techniques via probability theory to enable the decision maker to determine an optimal choice. The application of probability theory, which was conceived by mathematicians and based on assumptions that might not be appropriate in certain situations, must, however, be applied carefully. It is not implied that probability theory should not be used, but merely that care should be taken to understand the assumptions underlying the chosen actions and check that they are applicable in any particular case.

In probability theory an *event* is the term used for something which may or may not happen. A decision analysis problem could well comprise many occurrences or events and the difficulty lies in determining the probability factor for each event.

11.4.1 Unconditional probability

A probability statement, e.g. 'the probability that it will freeze is 0.60', may be written as

$$P(A) = 0.60.$$

The symbol A represents the event 'it freezes', and the operator P stands for 'the probability of'. This, the simplest form of probability, is termed *unconditional probability* and is often referred to as a *simple* probability. It pays no regard for other events, e.g. if there are two events, A, 'it will freeze today', and B, 'it will snow today', and the occurrence of B is quite independent of A and vice versa, in the sense that freezing does not cause snowing and it can snow without it necessarily freezing, then the two events have unconditional probabilities of $P(A)$ and $P(B)$ respectively.

The essence of probability theory lies in the concept of complementary events, i.e. if one event does not occur then the alternative event must take place. In a very simplified manner the two events could be 'the sun shines' and, secondly, 'the sun does not shine'. It is obvious that if one occurs the other cannot and vice versa. Whichever event does occur, the sum of the probabilities of both events occurring must always be 1.00. Therefore, we can introduce a third event to incorporate the possible variations in the complementary events, i.e. 'the sun shines or the sun does not shine'.

If the probability of the sun shining is given by $P(A)$, of the sun not shining by $P(B)$, and the event 'the sun shines or the sun does not shine' by $P(C)$, then we can write that $P(C) = 1.00$. Also, the event C is really the net result of A or B and this is denoted by writing:

$$C = A \cup B$$

where the symbol $\cup$ is short-hand for *or*.

Now we have

$$P(C) = P(A \cup B) = 1.00 \tag{11.1}$$

and as event A or B must occur then

$$P(A \cup B) = P(A) + P(B) = 1.00. \tag{11.2}$$

The concept of complementarity can be extended to more than two events providing they are *collectively exhaustive and mutually exclusive*. This implies that the events must cover all possibilities and that the occurrence of one of them excludes the occurrence of the other. A collection of probabilities relating to a collectively exhaustive and mutually exclusive set of events is called a *probability distribution*. The sum of the individual probabilities must be 1.00. Therefore, if there were four collectively exhaustive and mutually exclusive events A, B, C, and D we would have

$$P(A) + P(B) + P(C) + P(D) = 1.00. \tag{11.3}$$

11.4.2 Conditional probability

In real-life situations, events do not usually occur in isolation but are strongly or weakly linked to other events. Therefore, if event A was 'the production of a poor batch of concrete' and event B was 'the delivery of several batches of inferior quality cement', then A might very well depend on B and so we would say that A was *conditional* upon B. This is an example of probability being dependent on having certain information, or on something else having happened before. This type of probability is often written as

$$P(A \mid B)$$

which means 'the probability of A given that (the vertical line) B has happened'.

For instance, suppose it was found that, in the quality control of concrete test specimens, of those which failed to reach a certain strength after seven days 70% had surface defects. Then, if A denotes the event 'failed to pass the seven-day strength test' and B is 'bad surface defects', $P(B \mid A) = 0.7$.

If, in the previous example, it was known that 20% of the concrete specimens would fail the strength test, then what is the probability that a test specimen will fail the test *and* have the surface defects? This is known as a *joint event*, i.e. 'event A happens *and* B happens', and is denoted by the following symbolic notation:

$$P(A \cap B) \tag{11.4}$$

where ∩ means 'and'.

It is known that $P(B|A) = 0.7$ and $P(A) = 0.2$, in other words, there is a 20% chance that A will take place and, after it has, there is a 70% chance that it will be followed by B. Thus, 70% of 20% go through A and B so that

$$P(A \cap B) = 0.7 \times 0.2 = 0.14.$$

Expressed symbolically in its most general form for any events A and B we can write:

$$P(A \cap B) = P(A) \times P(B|A). \tag{11.5}$$

The standard rules of algebraic manipulation may be applied to this expression, which makes it extremely useful in calculating probabilities which are unknown.

11.5 BAYES' STRATEGIES

It was stated at the beginning of this chapter that the objective of any decision analysis problem was to select the most favourable or optimal action taking into account the different possible outcomes. If probability theory is used in helping to make this choice then the most favourable strategy is known as Bayes' strategy. This makes use of the concept of *expected pay-off*, which is determined from a knowledge of the probabilities of certain events occurring. The idea is best demonstrated by the following example.

If the introduction of an action X has two possible outcomes 1 and 2, of which the pay-offs are respectively V_1 and V_2, then it is possible to estimate the expected pay-off if the probabilities of outcomes 1 and 2 occurring are known. If $P(1)$ and $P(2)$ are the respective probabilities of outcomes 1 and 2, then the expected pay-off of action X is

$$\text{Expected pay-off }(X) = P(1) \times V_1 + P(2) \times V_2. \tag{11.6}$$

The expected pay-off is often referred to as *expected value*, and for the action X this may be written as $EV(X)$.

If $V_1 = 80$, $V_2 = 10$, $P(1) = 0.70$, and $P(2) = 0.3$, then

$$EV(X) = 0.70 \times 80 + 0.3 \times 10 = 59.$$

This should not be interpreted as the real pay-off of action X as obviously we would either receive 80 or 10 depending on which outcome resulted. It is clearly impossible to obtain a pay-off of 59 on any one single action of choosing X, but if we did this consistently over a long series of choices then the *mean* value would be 59.

It is obvious that action X cannot be treated as one in a series of identical actions as real-life problems do not conform quite so obligingly to mathematical theories. However, if we are to apply the theory of probability to real-life problems then this assumption must be assumed to be valid. If it is obviously not valid then the theory will not be applicable and it would be unwise to draw too many conclusions from the answers obtained. The choice of whether or not to use a particular theory in real-world problems is as much a decision problem as any other.

Once the expected value of an action is calculated irrespective of the number of possible outcomes that might arise from it, we are in a position to make certain deductions (i.e. decisions). It must be borne in mind, however, that the sum of the probabilities must always equal 1.00, or in other words they must be collectively exhaustive and mutually exclusive. Also, the probability of an outcome occurring can never be less than zero.

If we now assign probabilities to the possible outcomes that the Red Dragon Construction Company expect from the purchase of the vehicles, then the pay-off table is as shown in Table 11.5. The highest EV is for action C and that is the optimal act or Bayes' strategy. Recall that by using maximin it was action A and the minimax criterion gave action B.

Table 11.5 Pay-off table based on Bayes' strategy

		Outcome			
		I	II	III	
Probability		0.55	0.25	0.20	EV of action
Action	A	45	25	15	34.00
	B	75	65	−15	54.50
	C	95	30	−20	55.75*

The same values of probabilities are applied to the opportunity loss table and used to calculate the expected opportunity losses (EOL). The optimal act, according to Bayes' strategy, will be the one with the *minimum EOL*. The result is given in Table 11.6.

The minimum EOL is now for action C and that is the Bayes' strategy for this criterion.

The prospective decision maker would by now be thoroughly confused and

Table 11.6 Expected opportunity loss based on Bayes' strategy

		Outcome			
		I	II	III	
Probability		0.55	0.25	0.20	EOL
Action	A	50	40	0	37.50
	B	20	0	30	17.00
	C	0	35	35	15.75*

quite rightly so. It would seem that his major decision would first be to choose the optimal method for determining his optimal action.

Four different criteria have chosen three different optimal actions, which may be summarized as:

Criterion	Indicated action
Maximin pay-off	A
Minimax opportunity loss	B
Maximum EV	C Bayes' strategy
Minimum EOL	C Bayes' strategy

The final two answers are the only ones that incorporate any form of probability. As already stated, it was no surprise that the first two criteria gave different answers as they were founded on completely different approaches. However, the last two criteria give the same result and this may or may not be coincidence. In fact, it can be proved mathematically that Bayes' strategies are always the same for maximum expected value and minimum expected opportunity loss. This only applies to Bayes' strategies in which probabilities are involved in the calculation. In reality, this implies that pay-off or opportunity loss could be used interchangeably provided that the probabilities are included in the relevant calculation. As the numbers in the COL table are always zero or positive it is usually preferable to use this method. It should be borne in mind that the recommended actions would result in nothing but losses and hence the problem has to be carefully and properly defined.

11.6 DECISION TREES

In any engineering project of appreciable magnitude the number of decisions to be made are numerous. The whole procedure is further complicated by the fact that the decisions are not isolated but are linked to one another, however tenuously. In many cases these decisions are sequential in nature and each step forward relies to a large extent on the decision made previously. For example, if a construction company is bidding for a contract to build a new bridge, their first action must be to assess the cost of the contract as accurately as possible. They should bear in mind that their competitors will be doing exactly the same thing, and all will have the intention of submitting the lowest tender. Hence, a choice has to be made initially on the time and money to be spent on the preparation of the tender. They could gamble and carry out a very superficial assessment, or else they could devote considerable effort to the tender. Having taken this decision there will be an outcome — their tender might be rejected completely, they could be asked to re-submit with further details, or else they could secure the contract. Each of these outcomes can be thought of as being poor, fair, or good, and given that they have a fair outcome, i.e. the authority in charge of the contract is requesting a more detailed cost analysis, then they must decide between re-analysing the cost and time estimates, or withdrawing altogether. This decision will obviously rely on the feedback obtained from the awarding authority and their likelihood of securing the contract. Given that decision, there will be an outcome which will lead to another decision, and so on.

There are two important facets to this type of decision problem — time and uncertainty. In real life both the actions and outcomes of any problem require a certain time period to develop. This brings into play the added complexity of the time-value of money, but this can usually be accounted for by discounting. However, the probabilities involved in the decision making series could well be influenced by time and also by added information being received. In a later section such influence of further information on the probabilities will be examined in more detail.

Already it is possible to observe that the degree of uncertainty inherent in such problems is further aggravated by being compounded into a chain of events. It is difficult enough to decide on the correct probability of a single action but much more so if the actions are linked in a chain in which each action could result in several outcomes. However, such probabilities are often assessed by bookmakers in quoting odds on sporting encounters and usually, much to the disgruntlement of the gambler, they turn out favourably. However, despite the difficulties involved, the decision maker has to solve such problems *in any case*. If he is consistently incorrect in his predictions of the outcomes then he will soon find himself in the ranks of the unemployed.

Now, the many interwoven and complex decisions that must be made in any sequential structure lend themselves favourably towards representation by means of diagrams. Such diagrams are called *decision trees* and we will now consider the methods by which they can be best utilized. An example will be used to demonstrate the salient features of the technique.

The Red Dragon Construction Company have developed a new rapid-hardening cement which shows exceptional promise. Initially, they have the alternatives of either test marketing it or abandoning it. If the result of the test market is negative the company may decide to abandon the product. In this case all they could hope to recoup would be $20 000 gained from the sale of their supplies for construction purposes where low grade concrete is required. However, if they decide to test market the cement, it will cost $150 000 and the response of the civil engineering industry could be either positive or negative with probabilities of 0.7 or 0.3. If it is positive they could either abandon the product or market it full scale. The result of such an all-out sales campaign could be a low, medium, or high demand with corresponding net pay-offs of $-400, 400$, or 2000 in units of $1000 (i.e. the outcome could range from a loss of $400 000 to a gain of $2 000 000, all financial values being discounted to present-day values).

As in many problems involving systems analysis, the description is already lengthy and the reader could easily be confused as to what steps the Red Dragon Construction Company should follow. Even so, the description is by no means complete and represents a drastically simplified version of the true situation. For example, it is hardly likely that the company would accept a net loss of $130 000 if

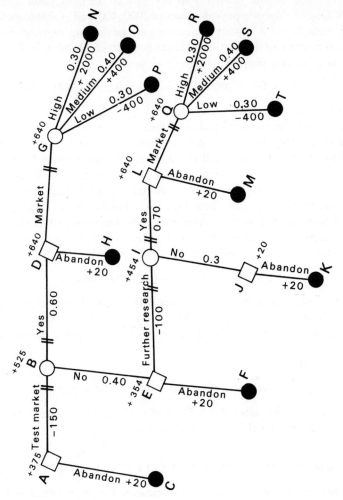

Figure 11.1 Decision tree for the Red Dragon cement marketing problem

the results of the test marketing were negative. It would be more likely that a further amount of research money would be ploughed into the development of the product in the expectation of improving the probabilities of success. Figure 11.1 depicts the tree network for the problem. The square nodal points represent points at which decisions have to be made, the white circles are the outcome results from the decisions, and the black circles are termination points. The branches between the nodes are either decisions or outcomes and each is labelled with an identification, a probability, or a net value where relevant. The numbers printed above the node points are expected values (EV) and the optimal paths are clearly marked (with a double bar). The analysis commences at the top right-hand side of the network and gradually works back to the start node A. Therefore, nodes N, O, and P having pay-off values of 2000, 400, and −400 respectively are reached from G via a probabilistic outcome, i.e. having reached node G the outcome is now out of the control of the decision maker. The EV for G is therefore calculated as shown previously, i.e.

$$EV(G) = 0.3 \times 2000 + 0.4 \times 400 + 0.3 \times (-400) = 640$$

and this value is inserted above node G.

Nodes H and G are reached from D, which is a decision point. If it is decided to move to H there will be a predetermined benefit of $20 000 (or +20), but if it is decided to move to G, there will be an expected value of 640. As the criterion being used is to maximize the expected value the decision should be to market the product, which is branch DG. This is marked the optimal act, and of course is conditional upon being at D. The value of D is therefore 640, i.e. the larger of 640 and 20. Before we can proceed further back to B along this path the EV of node E must be determined. This is achieved by a similar process to the one which has already been described. Thus EV(L) is +640 and hence EV(I) is given by

$$EV(I) = 0.7 \times 640 + 0.3 \times 20 = 454.$$

Since the cost of further research (link EI) is 100, EV(E) is the greater of (454 − 100) and 20, i.e. EV(E) = +354. The value of B is now found by taking the EV of the two outcomes 'yes' and 'no' which have led to the two separate sub-trees.

The EV of B is

$$EV(B) = 0.60 \times 640 + 0.40 \times 354 = 525.$$

The value of the branch AB is 525 − 150 = 375 and the value of AC is 20. The optimal initial act therefore is to 'test market'.

11.7 LIMITATIONS OF THE EXPECTED-VALUE TECHNIQUE

So far throughout this chapter, the discussion has been concerned with the concept of expected value (EV), or, if given its full title, expected monetary value (EMV). There are certain situations in which the use of EMV may lead to dire consequences for the potential decision maker. This is because there is no account taken of the risk potential involved in the project. In the previous example the Red Dragon Construction Company stand to make a profit of $375 000 from the use of maximum EMV, and on this basis they would proceed with the project. However, if in reality things did not work out too well, and a low outcome was the eventual result, then they would incur a loss of over $500 000 (e.g. route ABEILQT implies a loss of −150 − 100 − 400 = −650). If the company is a relatively small business and such a loss would mean disaster, then it is obviously preferable that it should not embark on the project. The measure of EMV indicates that the company should go ahead with the project but the circumstances may be such that this could be the wrong decision.

Suppose that the company concerned had the opportunity of embarking on a similar project but where the pay-off and losses were only a fifth of the previous project. In this instance a loss of this magnitude would not have the same 'life or death' significance that the previous project held and the company would be able to absorb such a deficit with no major difficulty. As the EMV would still be positive, then in this case the company should proceed with the project.

We now have the paradox of the EMV criterion sometimes giving us the correct answer and sometimes not. This obviously has to be resolved and the key to the answer lies in the magnitude of the losses involved. A company could be able to recover from small losses but not from a catastrophic loss. It is possible to avoid the whole situation by returning to the measurement of minimax loss which was discussed in Section 11.3. However, this criterion takes no account of probability so the situation would not be improved. The answer to the problem is to adapt the criterion of EMV so that it will reflect in the analysis the risk that a company can accept. This may be done by introducing the concept of utility.

11.8 CONCEPT OF UTILITY

The main ideas involved in utility are relatively simple. In the decision tree problem, an abbreviation of money was used in the sense that *one unit* represented $1000. This was done mainly to avoid the tediousness of having to write all the zeros involved with such large sums of money. In this case we could have defined one *monetary unit*, or one m.u., as being equal to $1000. It was implied in the context used that the conversion factor from monetary units to dollars was a constant. This is not necessarily always the case and it is quite possible that the conversion factor could vary in a non-linear manner. However, it is imperative that the variation of the conversion factor is always known so as to avoid any ambiguity.

Therefore, it is convenient to construct a chart which enables amounts to be converted from money to m.u. or back again, for any particular problem. For instance, if the project under consideration involved sums of money ranging between +$200 000 to −$200 000, we might wish to choose a convenient conversion factor of $2000 = 1 m.u., and so the amounts would be between 100 m.u. and −100 m.u. This linear relationship is shown in Figure 11.2.

If the negative part of the curve in Figure 11.2 is bent downwards more steeply as the losses increase there is no longer a linear variation between money and m.u. The graph shows a real money loss of −$200 000 to be equivalent to −500 m.u., and this reflects the Red Dragon Construction Company's true thoughts about going out of business. This is the essence of the concept of utility.

There is a danger of bending the curve too quickly and over-exaggerating the risk involved in the project. This could well lead to the situation where the company would never be prepared to take risks and therefore never stand to make any gains.

The utility curve need not be linear in its middle range and may take any form which reflects the views of the decision makers. The linear curve shown in Figure

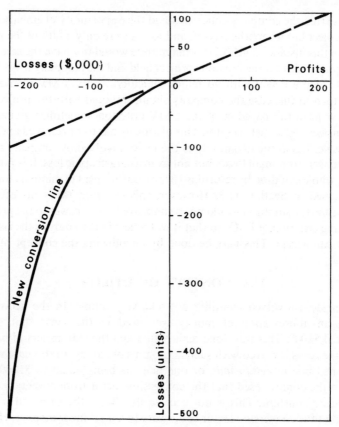

Figure 11.2 Non-linear exaggeration of losses

11.3 indicates an attitude of carefully playing the odds and is often termed as *risk neutral*. The convex curve (bending upwards) in Figure 11.3 is indicative of a decision maker who wishes to have one big pay-off rather than play carefully and obtain a number of much smaller pay-offs. The curve is referred to as being *risk prone* and would be unlikely to be accepted by most modern corporations. The concave curve of Figure 11.3 illustrates a typical *risk averse* utility curve. It must be stressed that the most suitable utility curve for a particular company could comprise a mixture of shapes of the three types shown in the diagram, and need not take the form of one particular type.

11.9 PRIOR AND POSTERIOR PROBABILITIES

The situation often arises that a forecast of probabilities made well in advance of an occurrence, such as the tendering process for a contract, requires re-assessment as later events shed further light on the matter. There is a prime example of this in weather forecasting which requires continuous updating as

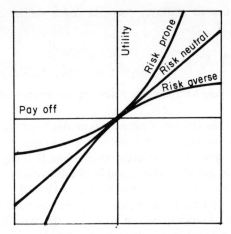

Figure 11.3 Risk-prone and risk-averse utility curves

weather stations, aircraft, and satellites keep transmitting the most recent reports. A weather report made early in the day could have changed drastically by evening. This is a result of the probabilities being revised from their initial values as other, but still indefinite, information becomes available.

An initial probability statement is known as a *prior* probability distribution and a *posterior* probability distribution is one which has been modified as a result of more recent information. In this instance the term distribution is used to indicate that the total probability of 1.00 is spread over a number of collectively exhaustive and mutually exclusive events. It is self-evident that what is posterior to one sequence of events becomes the prior event to future occurrences. The calculation of the inter-relationships between prior and posterior probabilities will be made in this section via the use of *Bayes' theorem*.

The theorem is a means of deriving probabilities and linking effects with their possible causes. To demonstrate the use of the theorem it is assumed that two causes, C_1 and C_2, have two possible effects E_1 and E_2. For example, the two causes C_1 and C_2 could be the delivery of a batch of concrete made by companies A and B respectively. The resulting effects might be that E_1 is a poor-quality concrete batch but E_2 is quite satisfactory. Also required are the prior probabilities $P(C_1)$ and $P(C_2)$ that a sample chosen at random came from A and B respectively. The two conditional probabilities that a certain cause would result in a given effect are also required. The conditional probability that a batch is faulty given that it comes from company A is written as $P(E_1|C_1)$. Similarly, $P(E_1|C_2)$ is the probability that a batch is faulty given that it comes from company B.

Therefore, it is now possible to calculate the probability that a batch comes from A, given that it was faulty, as

$$P(C_1|E_1) = \frac{P(E_1|C_1)P(C_1)}{P(E_1|C_1)P(C_1) + P(E_1|C_2)P(C_2)}. \tag{11.7}$$

This expression is known as Bayes' formula and a similar expression can be calculated for $P(C_2 | E_1)$:

$$P(C_2 | E_1) = \frac{P(E_1 | C_2)P(C_2)}{P(E_1 | C_2)P(C_2) + P(E_1 | C_1)P(C_1)}. \tag{11.8}$$

The application of these formulae may now be demonstrated by the following example.

The Red Dragon Construction Company is unable to produce sufficient ready mixed concrete and has decided, as a short-term 'stop gap', that it will purchase the quantity it needs from two rival companies A and B. It expects A to produce 30% and B 70% of this quantity. The reason for the difference in quantity is largely due to the fact that company A is much smaller than B and has insufficient capacity to fully meet the demands of Red Dragon Construction Company. This is regrettable since it is known from past experience that only approximately 15% of the concrete that A supplies is faulty, whereas 25% of company B's concrete is known to be sub-standard.

A new batch of inferior concrete has recently been supplied which, unfortunately, cannot be traced to its source. The manager of the Red Dragon Construction Company is getting very disillusioned with the situation and decides to estimate the probability that this concrete came from the rogue company B.

The vital statistics are therefore

$$P(C_1) = 0.30, \qquad P(C_2) = 0.70,$$
$$P(E_1 | C_1) = 0.15, \quad P(E_1 | C_2) = 0.25.$$

If we now substitute these values into Bayes' formula we get

$$P(C_1 | E_1) = \frac{0.15 \times 0.30}{0.15 \times 0.30 + 0.25 \times 0.70} = 0.205$$

and

$$P(C_2 | E_1) = \frac{0.25 \times 0.70}{0.25 \times 0.70 + 0.15 \times 0.30} = 0.795.$$

These two values now become the posterior probabilities of the distribution of C_1 and C_2.

To further demonstrate the application of Bayes' theorem let us assume that the untraceable batch of concrete was quite satisfactory. This now alters the problem quite dramatically as although we have the formula for calculating $P(C_1 | E_2)$ and $P(C_2 | E_2)$, all the probabilities are not known. However, use may be made of the fact that the sets of values of C and E must each form a collectively exhaustive and mutually exclusive group, i.e.

$$P(E_1 | C_1) + P(E_2 | C_1) = 1.00.$$

Therefore, as

$$P(E_1 | C_1) = 0.15, \quad \text{then } P(E_2 | C_1) = 0.85$$

and as
$$P(E_1 \mid C_2) = 0.25, \quad \text{then } P(E_2 \mid C_2) = 0.75.$$
Hence
$$P(C_1 \mid E_2) = \frac{P(E_2 \mid C_1)P(C_1)}{P(E_2 \mid C_1)P(C_1) + P(E_2 \mid C_2)P(C_2)} \tag{11.9}$$
$$= \frac{0.85 \times 0.30}{0.85 \times 0.30 + 0.75 \times 0.70} = 0.327.$$

Similarly,
$$P(C_2 \mid E_2) = 0.673.$$

In this case the values of the posterior and prior probabilities are much closer than previously. This is due to the fact that defective batches are much scarcer than satisfactory ones and so information that a batch is defective is much less common and therefore becomes more significant.

Until now we have dealt exclusively with problems involving two causes and two effects but Bayes' formula can be extended quite easily to situations involving numerous causes and effects. If, for example, there are two effects and three causes and E_1 has been observed, the posterior probability for the first cause would be

$$P(C_1 \mid E_1) = \frac{P(E_1 \mid C_1)P(C_1)}{P(E_1 \mid C_1)P(C_1) + P(E_1 \mid C_2)P(C_2) + P(E_1 \mid C_3)P(C_3)}. \tag{11.10}$$

The numerator in this expression is identical to the previous formula and the only addition to the denominator is the inclusion of the conditional probability for the first effect and the third cause, multiplied by the prior probability for the third cause. For additional causes the denominator is further extended by adding on the appropriate terms.

The problem used to demonstrate the previous calculations was simple and bears little relationship to real life. Hence, a much more complicated problem will now be examined which shows clearly how Bayes' theorem enables the decision maker to incorporate information into an analysis and how widely divergent initial options come closer together in the light of factual information.

11.10 PILE SELECTION EXAMPLE†

A site engineer is confronted with the dilemma of selecting the appropriate pile length in a situation where depth to bedrock is uncertain. However, it has been determined that the depth to bedrock is not less than 10 metres and not greater than 15 metres. The piles in question come in two standard lengths of 10 metres and 15 metres respectively. If the engineer makes an incorrect choice then costs will be incurred by having to cut the 15 metre piles if bedrock depth is 10 metres, and by having to splice sections onto the 10 metre piles if bedrock depth is 15 metres. The situation can be summed up quite simply by use of the decision

† Adapted from Benjamin and Cornell, p. 573.[1]

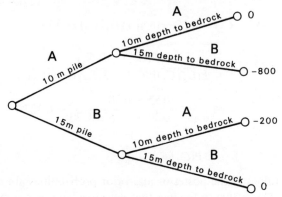

Figure 11.4 Simple decision tree for pile selection problem

Table 11.7

State of nature	Action	
	Drive 10 m pile	Drive 15 m pile
Depth to bedrock 10 m	No loss	5 m of pile cut-off (200 unit loss)
Depth to bedrock 15 m	Splicing of pile required (800 unit loss)	No loss

tree shown in Figure 11.4. The 'losses' concerned have been entered as negative values and represent the 'opportunity loss' associated with not choosing the optimal action. The problem may also be summarized by means of a pay-off table (Table 11.7).

The prior probabilities of state have been calculated by the engineer to be as follows:

$$P'(10) = 0.65$$
$$P'(15) = 0.35$$

where the single prime indicates 'prior'.

These values have been estimated from knowledge obtained from large-scale geological maps, and depths of piles which have been driven in the vicinity of the site. However, the engineer is of the opinion that further site investigation, in the form of sonic tests, is required to determine more reliable probability values for the depths to bedrock. Because of soil irregularities and measuring errors, the depths indicated by the sonic tests are not wholly reliable. The equipment available is capable of measuring to the nearest 2.5 m, and the outcome of the site investigation is given in Table 11.8.

Table 11.8 Sample probabilities

	True state	
Sample outcome	10 m depth	15 m depth
10 m indication	0.65	0.10
15 m indication	0.10	0.75
12.5 m indication	0.25	0.15

This table implies that, if the true depth is 15 m, the equipment is 75% reliable but has a 10% chance of being completely incorrect. Also, there is a 15% chance that it will register an ambiguous reading of 12.5 m. The total of all the probabilities must equal unity. It may be seen by comparing the two columns that the instrument is more likely to register a reading which is too deep rather than one which is too shallow. These sample likelihoods may be in part subjective as they could well depend on many factors, e.g. the design tolerances of the equipment and the judgement of the engineer.

Assume that the sonic tests are carried out and that the depth indicated by the instrument is 12.5 m. The posterior probabilities of state, given the observation of 12.5 m, are given by equations (11.7) and (11.8). Thus,

$$P''(10|12.5) = \frac{0.25 \times 0.65}{0.25 \times 0.65 + 0.15 \times 0.35} = 0.756 \qquad (11.11)$$

$$P''(15|12.5) = \frac{0.15 \times 0.35}{0.25 \times 0.65 + 0.15 \times 0.35} = 0.244 \qquad (11.12)$$

assuming that the state must be either 10 m or 15 m and that the posterior probabilities must therefore sum to unity.

It should be noticed that the ambiguity of the result has in fact altered the probabilities of state, caused by the difference in the error probabilities 0.25 and 0.15.

The expected utilities may now be determined by multiplication of the posterior probabilities by the corresponding utilities and summing the products, e.g. if a 10 m pile is driven, then

$$EV(10 \text{ m pile}) = -30 \times 0.756 + -830 \times 0.244 = -225.2$$

Similarly, the posterior expectation for a 15 m length pile given the sonic test results in an outcome which is the ambiguous result of 12.5 m is

$$EV(15 \text{ m pile}) = -230 \times 0.756 + -30 \times 0.244 = -181.2$$

Hence, the maximum value of these two, i.e. -181.2, is chosen as being the expected utility.

The same procedure is now repeated for the sonic test given outcomes of 10 m

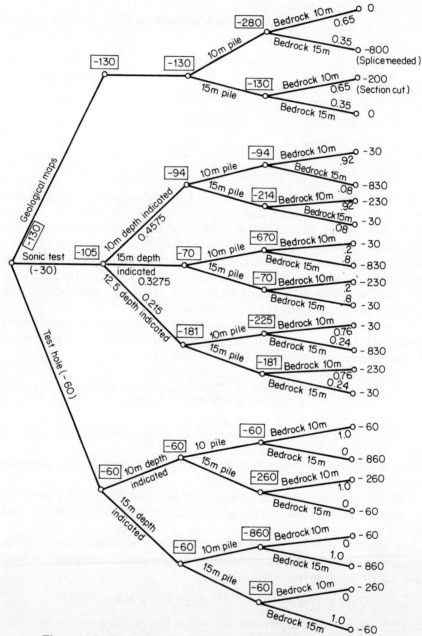

Figure 11.5 Expanded decision tree for the Pile Selection Problem

and 15 m depths respectively. These expected utilities will not be calculated in detail and the results are given at the appropriate nodes in Figure 11.5.

Now the expected value commensurate with relying on the sonic tests may be found by weighting the optimal utilities found previously by the probabilities that the appropriate depth will be the outcome and summing. The probabilities are determined as follows:

$$P[10\text{ m}|\text{sonic test}] = P[10\text{ m}|\text{sonic result of }10\text{ m}]P'[10\text{ m}]$$
$$+ P[10\text{ m}|\text{sonic result of }15\text{ m}]P'[15\text{ m}]$$
$$= 0.65 \times 0.65 + 0.1 \times 0.35$$
$$= 0.4575.$$

Similarly,

$$P[15\text{ m}|\text{sonic test}] = P[15\text{ m}|\text{sonic result of }15\text{ m}]P'[10\text{ m}]$$
$$+ P[15\text{ m}|\text{sonic result of }10\text{ m}]P'[15\text{ m}]$$
$$= (0.75 \times 0.35 + 0.1 \times 0.65)$$
$$= 0.3275.$$

Therefore,

$$P[12.5\text{ m}|\text{sonic test}] = 1 - 0.4575 - 0.3275 = 0.215.$$

The optimal action is then to choose the longer pile. Even though the engineer believes that the true state is more likely to be 10 m (i.e. $P'(10) = 0.65$), the extra cost of splicing compared with cutting (800 against 200) outweighs the relative confidence that the true state is 10 m (0.756 to 0.244).

If the sonic test indicated depths of 10 m or 15 m, the posterior probabilities would of course be different. Repeating the calculation as in (11.11) and (11.12) the possible result may be summarized as follows:

$$P''(10|10) = 0.923; \quad P''(15|10) = 0.077$$
$$P''(10|12.5) = 0.756; \quad P''(15|12.5) = 0.244$$
$$P''(10|15) = 0.198; \quad P''(15|15) = 0.802.$$

The final decision analysis may now utilize these posterior probabilities to construct a decision tree as described in Section 11.6 and using Bayes' strategy of equation (11.6).

If we now consider that the engineer is placed in a more realistic setting by being allowed to drill an exploratory hole to determine the true depth accurately, the decision tree is as given in Figure 11.5. Comparing the expected value of the three experiments, i.e. -130, -105, and -60, it is obvious that the site engineer should not make a decision before obtaining more information from one of the available experiments. Of the two experiments, the reliable but more expensive test drilling is preferable (in this case) to the cheaper but less reliable sonic experiments. The engineer should absorb the initial cost of the drilling and then order accordingly, rather than run the risk of making a wrong decision. This risk results from his present uncertainty, which is reduced but not eliminated by the less expensive sonic experiments.

11.11 EXERCISES

11.11.1 The quality of a concrete mix delivered to a site is variable. In order to decide upon a course of action with respect to acceptance of the delivered concrete it is classified, after compression testing, as either 2000, 3000, or 4000 psi grade. The classification is based on the closest measured strength. The differences in the strengths occur because of bad, adequate, or good operating conditions at the mixing plant, and are linked to these conditions by the following probabilities:

Operating conditions	Concrete strength		
	2000	3000	4000
Bad	0.7	0.3	0
Adequate	0.2	0.5	0.3
Good	0.1	0.6	0.3

The probabilities that the operating conditions are bad, adequate, or good are 0.6, 0.2, and 0.2, respectively. A delivered batch of concrete is now observed to be poor (2000 psi). What are the revised probabilities for the operating conditions?

11.11.2 A large engineering company manufactures fork lift trucks in one of their divisions. There are many minor competitors but only one major rival who has recently introduced a new type of truck which is apparently selling well. The first company has to decide whether to introduce a competing truck but are concerned over the likely reaction of the rival company. The rival company could expand production of the new truck, keep it at the same level, or cut it back. From past experience the chances of the three actions are respectively 50%, 30%, and 20%. The only information that the first company can obtain is from a study of the rival's advertisement in the technical press and they have assigned probabilities to the rival reducing, maintaining, and increasing their advertising. The probabilities are expressed as follows:

Cause			Effects Advertising will be		
			I Reduced	II Maintained	III Increased
Rival is going to	A	Cut back	0.60	0.30	0.10
	B	Keep at the same level	0.40	0.40	0.20
	C	Expand	0.20	0.30	0.50

It is observed that, during the following month, advertising is increased. What would then be the revised probabilities for the rival's course of action?

11.11.3 The Caredig River Authority is planning to build a dam costing £1.2 million. Based on historical records, it is estimated that there is a 0.25 probability that one or more serious floods would occur during the life of the dam and a 0.10 probability that one or more major floods would occur. The probabilities that the two possible spillway types will fail during these two levels of flooding are estimated as follows:

	Serious flood		Major flood	
	Fail	Safe	Fail	Safe
Large spillway	0.05	0.95	0.1	0.9
Small spillway	0.10	0.90	0.15	0.85

If a spillway fails to function properly during a major flood the dam will be destroyed. The replacement cost of the dam will be the same as the original cost. In addition to a total loss of the dam and its spillway, other property damages will be incurred. It is estimated that in case of failure during a serious flood, there is a 70 percent and 30 percent chance of other property losses amounting to £0.6 million and £1.8 million, respectively.

(a) Model this decision problem with a decision tree.
(b) What is the optimum decision based on the EMV criterion?
(c) How would you caution the Authority about basing its decision on the EMV criterion?

11.11.4 At a construction site, 15 760 cubic metres of soil are to be excavated using a single excavator with a capacity of 2.5 cubic metres per load and capable of 20 loads per hour. A maximum of four small bulldozers, each of capacity 15 cubic metres per hour, are to be used to push soil towards the excavator. The nature of the site is such that two bulldozers can be accommodated without loss of efficiency, but if three or more are used, the productivity is reduced by 20% due to interference. Given the following assumed data, draw a utility data curve for the excavation activity.

    Working time for a shift — 8 hours
    Excavator: ownership cost — $10 per hour
    Bulldozers: ownership cost — $5 per hour
    Operating cost — $15 per hour (for either machine)
        Second shift, $20 per hour
        Third shift, $27.50 per hour.

11.11.5 A civil engineering contracting firm is considering submitting a bid proposal for the supply and installation of the machinery system for the swing span of a bridge. The chief engineer of the company has estimated that a bid proposal can be prepared at a cost of $3000, but that such a bid will only have a 20% chance of being accepted. As an alternative, the company can invest $12 000 on an extensive research study of the project before preparing the bid proposal. Such a proposal will have a 30% chance of being accepted.

If the company does get the job, it can either expand its personnel and staff or sub-contract work to other companies. If a major portion of the work is to be performed by sub-contractors, it is estimated that there is a probability of 70%, 20%, and 10% for making a profit of $500 000, $750 000, and $1 000 000 respectively. If the work is to be accomplished mostly by expanding staff, there is a probability of 60%, 35%, and 5% for profits of $250 000, $1 000 000, and $1 500 000 respectively.

(a) Suppose that you were the company chairman and that your company has total assets of $5 000 000. Prepare a utility curve to reflect *your* risk characteristics.
(b) Based on the above utility curve, what is your optimum policy?
(c) What is the maximum monetary investment that you would be willing to provide for the preliminary proposed study?

11.11.6 An oil company has the option of exploring for oil at only one of three sites in the North Sea. The expected drilling costs for each site are $1 000 000, $200 000, and $2 000 000 respectively, and their probable well outputs are given in the following table:

		P(output), barrels per day		
Site	0	20 000	100 000	1 000 000
1	0.05	0.4	0.5	0.05
2	0.1	0.75	0.1	0.05
3	0.7	0.05	0.05	0.2

From past experience, it is known that a well in this area will net the company $25 for each barrel a day of production.

(a) At which site should the company drill?
(b) Suppose the company commissioned detailed seismic tests for $500 000 and now has a revised estimate of the outputs at each site (see following table). In retrospect, was it worthwhile to have the tests done? How have the possible decisions been altered?

| | P(output), barrels per day | | | |
Site	0	20 000	100 000	1 000 000
1	0	0.1	0.8	0.1
2	0	0.8	0.15	0.05
3	0.5	0.05	0.05	0.4

(c) If the company's geologist now estimates that the drilling costs for each site have the probabilities shown in the following table, where should the company drill?

| | P(drilling cost), millions of dollars | | | |
Site	0.2	1	2	4
1	0.1	0.8	0.1	0
2	0.8	0.2	0	0
3	0	0.5	0.3	0.2

11.12 REFERENCES

1. Benjamin, I. R., and Cornell, C. A. (1970). *Probability, Statistics and Decision for Civil Engineers*, McGraw-Hill, New York.
2. Coyle, R. G. (1972). *Decision Analysis*, Thomas Nelson & Sons Ltd., London.

Chapter 12
Additional Worked Examples

12.1 INTRODUCTION

The preceding chapters have each specialized in one method of optimization and each method has been illustrated with reference to examples. Unfortunately, the more difficult part of systems optimization is in the translation of a vaguely specified or poorly perceived problem into the form of a mathematical model. The task is then to identify the most suitable optimization methods for the more formally defined mathematical programming problems. Such skills can be hinted at in general terms but there is really no substitute for experience. The present chapter contains a number of hypothetical examples which will serve to illustrate a variety of approaches. In general, the problems are imprecisely defined and the explanation takes the reader through a typical sequence of steps leading to formal problem description and solution. Fully detailed solutions are not presented for reasons of space, but the algorithms are described in sufficient detail to allow a reader with sufficient experience in coding to prepare a computer program for the solution of the problems. In certain cases, typical results are given but the reader may find it instructive to vary certain system parameters to observe the sensitivity of the solution to these quantities. In the preparation of these programs, frequent reference may be made to subroutines listed in Appendix A.

Frequently, the choice of optimization algorithm may be obvious but in cases where there is either a choice or some uncertainty about the best method, the chart shown in Figure 12.1 may provide some guidance. This diagram is based on a suggestion by J. Windsor of the University of Durban.

12.2 THE FASSBUCK TRADING CO.

The memorandum describes a design situation which is reminiscent of the Thirstville project described in Section 1.4 with the difference that the discharge is continuous and no balancing reservoir is required. The economics of the project are significantly more complicated, principally due to the relatively short life of pump sets compared with the economic project life.

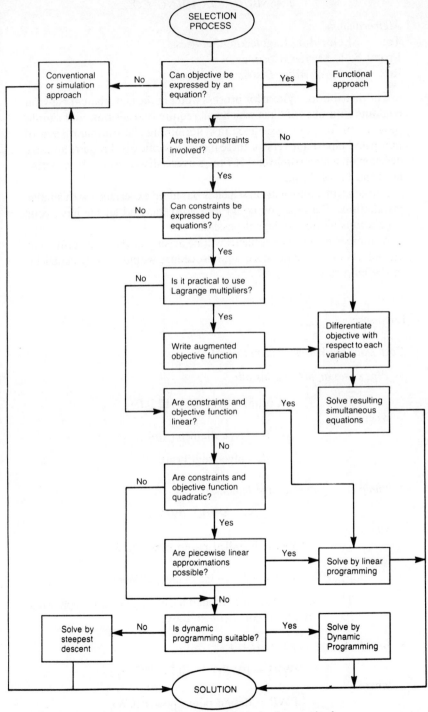

Figure 12.1 Selection of optimization method

FASSBUCK TRADING CO.

Memorandum
To: Al Gorithm, Engineering Services.
From: S. T. Mayter, Costing.
Re: *Sizing of Water Distribution Mains.*

The distribution system for process water in the new plant at Newton contains a number of links which will require booster pumps. I should be grateful for your assistance in determining the optimum sizes of pump and pipe diameters for these mains. Specifically, I require an early decision on a line to deliver 0.567 m^3/s over a distance of 1830 m with a net head lift of 7.6 m.

I attach some information which is based on experience with similar installations. Perhaps you could look into this and let me have your suggestions within 10 days if possible.

A number of similar problems will be arising in the near future and any efforts to automate this design procedure would be very beneficial in the long run.

Enclosure:

Cost Data for Water Distribution Mains

1. *Pipe cost* (*supply and install*)

$$\text{Cost(\$)} = DL(400 - 12\sqrt{D})$$

where

D = pipe diameter (m)

L = pipe length (m).

2. *Pump cost* (*pump set and building*)

$$\text{Cost(\$)} = \text{PWRI} \cdot \text{PWRcost}$$

where

PWRI = installed pump power (kW)

and

$$\log_{10}(\text{PWRcost}) = 4.387 - 0.574 \log_{10}(\text{PWRI}).$$

Note: The pump set has an estimated life of 12 years and represents 30% of the total cost defined above.

3. *Installed pump capacity* (*PWRI*)

$$\text{PWRI} = \text{PWR}(1.3 + 0.7\, e^{-\text{PWR}/150})$$

where

PWR = design pump power (kW).

4. *Design pump horse-power* (PWR)

$$PWR = 9.81QH/0.7 \quad (kW)$$

where

Q = discharge (m³/s)

H = total pump lift, including friction (m).

Note: The pump set efficiency of 70% includes both electrical and mechanical losses.

5. *Annual costs*

   (a) Pump power $100 per kW per year
   (b) Maintenance 1% of capital cost for pipeline and pump set (assumed constant).

6. *Hydraulic performance*
   The friction loss may be based on the following formulae:

$$h_f = \frac{fLV^2}{2gD}$$

$$\frac{1}{\sqrt{f}} = 2\log_{10}(3.7D/k)$$

where

h_f = head loss (m)
V = mean velocity (m/s)
L = pipe length (ft)
g = 9.81 m/s²
D = pipe diameter (m)
k = roughness height of 0.5 mm.

7. *Financing*
   Assume an economic life of 30 years for the system with a constant discount rate of 9%. Salvage value of the pump sets should be taken as 10%.

The physical arrangement of the system is shown in Figure 12.2 in which the relevant design variables are shown. System cost is dependent on the physical system parameters of pipe length L, diameter D, and installed pump-power PWRI. In addition, there is significant dependence on the pump life (PLIFE), the discounting rate, the method of determining salvage value, and possibly also inflation. The mathematical problem may be stated, in terms of physical parameters only, as follows:

$$\underset{h_f^*,\, PWRI^*,\, D^*}{\text{Minimize}} Z = \phi(L, H_{\text{lift}}, Q, k, h_f, PWRI, D) \qquad (12.1)$$

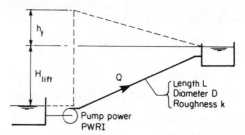

Figure 12.2 General arrangement of pump–pipeline system

subject to

$$g_1(L, h_f, Q, D, k) = 0$$
$$g_2(\text{PWRI}, h_f, H_{\text{lift}}, Q) = 0.$$

Note that, of the seven design variables, the first four are taken to be system constants, so that only h_f, PWRI, and D are variable. Furthermore, since there are two equality constraints, it is possible that by appropriate substitution the problem may be reduced to an unconstrained problem in a single variable. It seems reasonable that either diameter D or pump power PWRI should be selected as the design variable since these are likely to be discrete quantities. Since the pipeline is a larger capital cost than the pump, the diameter D is chosen as the design quantity.

The physical design procedure then takes the following form:

(i) Define system values for L, H_{lift}, Q, and k, and a set of diameters D_j ($j = 1, 2, \ldots$).
(ii) Select jth element of array of pipe diameters D_j.
(iii) Calculate friction coefficient as a function of the relative roughness D_j/k.
(iv) Calculate headloss h_f from Darcy–Weisbach equation.
(v) Obtain total pump-lift = $H_{\text{lift}} + h_f$.
(vi) Calculate design pump power PWR.
(vii) Calculate installed pump power PWRI.

The calculation of the costs involves a number of variables, only some of which are defined in the initial problem statement. No allowance is made for inflation and it would be prudent to make some provision for this in order to observe the sensitivity of the optimal policy to uncertainty in the inflation rate. Further uncertainty is introduced with respect to the salvage value since the method to be used is not specified (see Section 9.4). The declining-balance method will be adopted here.

The overall cash-flow is illustrated in Figure 12.3, in which:

P_1 = capital cost of pipeline, pump house, and first pump set
P_2 = capital cost (inflated) of second pump set
P_3 = capital cost (inflated) of third pump set
S_1 = salvage value of first pump set

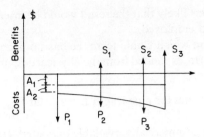

Figure 12.3   Cash flow for the Fassbuck Project

S_2 = salvage value of second pump set
S_3 = salvage value of third pump set
A_1 = annual maintenance of pipeline (constant)
A_2 = initial annual operating cost of pumps (inflating).

It should be noted that if the expected pump set life and the economic life of the project are subject to variation (for sensitivity analysis, say), it is possible that fewer or more than three pump sets may be required. This flexibility should be built-in to the program.

If the economic life of the project is NL i.e. 30 years and the pump set life is NP i.e. 12 years, then the number of pump sets required is given by

$$NS = INT(NL/NP) + 1 \tag{12.2}$$

The age of the last pump set at the end of the project is

$$NX = NL - NP.INT(NL/NP). \tag{12.3}$$

If the salvage value S is given as some fraction α of the initial cost of the pump set P (say $\alpha = 0.1$) then the depreciation rate d is implicitly defined by the relation

$$S = P(1 - d)^{NP}$$

or

$$d = 1 - \left(\frac{S}{P}\right)^{1/NP}. \tag{12.4}$$

Thus, using the values of the example,

$$d = 1 - \left(\frac{0.1P}{P}\right)^{1/12} = 0.1746 \quad \text{or} \quad 17.46\%.$$

The salvage value of the last pump set is then defined by

$$S_3 = P_3(1 - d)^{NX}$$
$$= P_3(1 - 0.1766)^6$$
$$= 0.316 P_3.$$

It will be found that after discounting to present value, the salvage value of the pump set is rather insignificant compared with the total cost, especially for the last pump set. It seems likely that the result would be insensitive to the choice of depreciation method employed.

A typical algorithm which would form the basis of a computer program may take the following form, expanded from the physical design procedure mentioned earlier:

(1) Define system values for pipe length L, static lift H_{lift}, discharge Q, and pipe roughness k.
(2) Define array of commercially available diameters D_j $(j = 1, 2, \ldots)$.
(3) Define economic parameters: interest rate i; inflation rate r; economic life NL; pump set life NP; salvage fraction $\alpha = S/P$.
(4) Calculate number of pump sets required NS and age of last pump set NX.
(5) Calculate depreciation rate d, assuming declining balance method.
(6) Select jth diameter D_j $(j = 1, 2, \ldots)$.
(7) Calculate relative roughness D_j/k and hence obtain friction coefficient f.
(8) Find head loss from Darcy–Weisbach equation.
(9) Get total pump lift $= H_{lift} + h_f$.
(10) Calculate design pump power PWR.
(11) Calculate installed pump power PWRI = PWRfact × PWR.
(12) Calculate capital cost of pipeline C_1.
(13) Calculate pipeline maintenance $C_2 = 0.01 C_1$.
(14) Calculate cost of pumphouse and first pump set C_3.
(15) Calculate value of first pump set $C_4 = 0.3 C_3$.
(16) Calculate fraction of pump set cost for design power $C_5 = C_4/\text{PWRfact}$.
(17) Calculate salvage value $S_1 = \alpha C_5$ and discount to present value C_6.
(18) Calculate equivalent present value C_7 of uniform series C_2.
(19) Calculate equivalent present value C_8 of geometric series for pump operation ($A_2 = \$100 \times$ PWR per year).
(20) Sum $(-C_1 - C_3 + C_6 - C_7 - C_8)$.
(21) Allow for second to penultimate pump sets.
(22) For $k = 2, 3, \ldots, (\text{NS} - 1)$, calculate inflated cost of kth pump set $C_9(k)$ where $C_9(k) = 0.3 \times$ Cost of pumphouse for design power PWR inflated for $k \times$ NP years and discounted to present value.
(23) Calculate salvage value $C_{10}(k) = \alpha C_9(k)$.
(24) Repeat steps (22) and (23) for $k = 2, 3, \ldots, (\text{NS} - 1)$ and accumulate costs C_{11}

$$C_{11} = +\sum C_9(k) - \sum C_{10}(k), \quad k = 2, 3, \ldots, (\text{NS} - 1).$$

(25) For last pump set, calculate inflated cost of NSth pump set C_{12} as in step (22) inflated for $(\text{NS} - 1) \times$ NP years and discounted to present value.
(26) Calculate salvage value C_{13} of last pump set with age NX years and discount to present value by NL years.
(27) Get total present value

$$(-C_1 - C_3 + C_6 - C_7 - C_8) - C_{11} - C_{12} + C_{13}.$$

(28) Print out diameter D_j, PWRI, and total present value.

(29) If $j <$ number of diameters, then increment $j = j + 1$ and go to step (6).
(30) Stop.

Typical results obtained from a program based on the above algorithm are shown in Figure 12.4.

DIAMETER (M)	POWER (KW)	PRESENT VALUE
0.300	4696.64	-7474115.37
0.350	2132.18	-3860833.87
0.400	1099.07	-2356161.56
0.500	422.45	-1311139.26
0.600	238.60	-1042950.95
0.700	168.89	-987780.08
0.750	150.77	-991953.61
0.850	130.08	-1029650.25
1.000	116.87	-1119865.99

Figure 12.4 Typical results for the Fassbuck Project

12.3 A PARKING-LOT PROBLEM

12.3.1 Background

The following hypothetical study is typical of many real-life instances involving allocation of a scarce resource (in this case, parking space) to a population of users with uncertain demand characteristics. The capacity for service is deliberately over-estimated in order to minimize or eliminate waste of a perishable resource. In this study, the numbers are reduced to facilitate presentation.

The following correspondence sets the scene.

UNIVERSITY OF ASHBY-DE-LA-ZOUCH

To: Rankine Justice, Parking
From: Dr N. A. Wiseman, Vice-President Academic.
Re: Parking.

I have recently received representations from the local committee of the Association of University Teachers concerning the fact that twenty-five parking stickers have been allocated for Parking Lot 'Q', where there are in fact spaces for only twenty cars. It is claimed that this is an extortionist practice by the University Administration and is resulting in considerable hardship and inconvenience to the faculty members concerned.

Please advise me as to the extent of inconvenience which is likely to result from this arrangement, giving a quantitative estimate of the number of days when overflow is likely to result.

N. A. Wiseman,
Vice-President Academic.

UNIVERSITY OF ASHBY-DE-LA-ZOUCH

To: Dr Hyam Hardup, Chairman, Civil Engineering.
From: Rankine Justice, Parking Superintendent.
Re: *Parking Lot 'Q'.*

I enclose a copy of Dr Wiseman's memorandum on the question of Parking Lot 'Q'. It was good of you to agree to undertake this job and, as discussed, I have made the necessary survey and attach the results for use in your simulation study.

I understand that you intend to simulate the behaviour of the twenty-five users for a period of, say, 100 days and from this obtain an indication of the number of days on which the maximum number of users at any time will exceed 20. It would be useful if your study could also provide information on the effect of increasing the number of spaces or decreasing the number of stickers issued.

                            R. Justice,
                            Parking Superintendent.

The survey referred to in the memo to Dr Hardup is described in Figure 12.5, in which the arrival and departure times are noted for the 25 users over a period of 5 days. These times were obtained by giving each car owner a survey slip upon arrival on which they were asked to note the arrival and departure times and then return the slip to the parking lot attendant.

12.3.2 Analysis of data

The first point to notice is that of $5 \times 25 = 125$ possible users, there are only 115 responses. The missing 10, or 8%, may be due to (i) users not bringing a car to work or (ii) users being unable to find space in Parking Lot 'Q'.

The second point is that, although the arrival times are scattered, the duration of parking is roughly inversely proportional to arrival time. This makes sense, since persons arriving around noon presumably do not wish to be late twice in one day and, consequently, leave around the normal time. Before embarking on a more detailed analysis, a simple plot such as Figure 12.6 shows a promising linear trend.

The following coding serves to analyse the data. The routine CURFIT (Appendix A) is designed to read data from a designated file with a prescribed format (2F10.3). Consequently, the raw data are processed by the program of Figure 12.7(a) to perform the following tasks:

(i)   remove the header and number of records,
(ii) convert from sexagesimal to decimal format,
(iii) construct a scratch (i.e. temporary) file with arrival time as the independent variable (x) and parking duration as the dependent quantity (y),

Arrival and Departure Times (Hr/Min) — 25 Cars over 5 Days

MBN633	13	1	16	30	MJJ009	14	0	15	48
MJJ009	14	2	15	50	TOR544	9	52	14	25
MKL001	7	52	14	14	MKL011	9	27	14	56
FUD809	12	52	16	1	BDS345	7	56	13	26
BDS345	8	32	13	31	EDS001	7	38	14	26
NXX275	8	38	13	57	MUD122	13	38	16	49
MGD354	8	20	13	32	FYU209	8	31	13	14
MJJ009	14	3	16	0	STELC1	8	27	13	57
JIM001	8	50	13	57	JIM001	8	40	14	14
MGD354	9	14	15	8	MBN633	10	57	15	42
STELC1	7	38	13	34	FUD809	13	13	16	24
FUD809	12	45	15	54	NBC788	8	22	13	31
MYC033	8	55	14	18	NXX275	8	34	13	54
NBC788	13	23	15	23	MCY033	9	18	14	44
CRR866	10	26	14	34	MBN633	11	56	15	48
NAS077	9	19	14	10	JIM001	7	45	13	15
TOR544	13	46	16	22	BDS345	9	31	14	7
MBN633	13	55	15	57	MKL011	15	43	16	45
MGD354	9	55	13	49	HYU643	10	26	14	32
MGD354	12	53	16	49	NBN788	15	53	17	27
JIM001	11	12	15	44	NAS077	12	9	15	21
MGD354	10	2	14	41	NAS077	8	24	14	18
MKL011	11	23	15	4	JIM001	13	42	16	17
MUD122	8	37	13	57	EDS001	9	8	13	56
FYU290	9	51	14	38	EDS001	9	59	14	36
BDS345	9	10	13	22	MDG354	10	28	14	0
HYU643	8	51	15	5	MUD122	13	11	15	44
NBC788	11	6	15	37	STELC1	8	37	13	58
MNN766	8	34	14	1	MYC033	8	22	13	39
NZE090	11	44	14	52	MKL011	12	46	15	40
DOF001	10	36	14	52	DPF001	9	20	13	51
NXX275	7	35	14	19	JIM001	8	44	14	36
MGD354	8	6	13	43	FUD809	8	37	13	57
NBC788	8	6	14	5	MGD354	8	58	13	47
MUD122	8	6	14	36	NYR500	13	17	16	9
MJJ009	9	45	15	22	EDS001	10	12	14	35
RUT622	8	35	14	20	RUT622	13	30	16	0
STELC1	13	18	16	45	NBC788	12	48	15	16
CRR866	9	26	14	7	MUD122	8	16	13	18
MYC033	12	47	16	31	STELC1	13	57	16	47
NBC788	9	52	14	38	FYU209	7	58	14	16
NYR500	15	4	17	41	MGD354	8	20	13	54
RUT622	12	55	15	52	DOF001	13	9	15	49
MYC033	13	16	16	51	FUD809	7	34	14	15
TOR544	13	47	16	33	MKL011	8	41	13	45
MYC033	13	28	15	59	NZE090	13	3	16	21
TOM002	12	50	16	0	MKL011	7	31	13	55
STELC1	12	47	16	12	FUD809	13	55	17	2
RUT622	9	42	14	13	HYU643	7	49	13	24
MGD354	8	46	14	20	MJJ009	9	12	13	52
TOR544	11	14	14	49	MJJ009	9	45	14	41
MNN776	9	0	13	17	TOR544	9	22	14	1
EDS001	13	8	16	4	JIM001	13	44	15	43
TOR544	7	33	13	31	NAS077	13	42	16	24
BDS345	9	50	14	25	FUD809	10	14	14	49
					MJJ009	12	45	15	47
					MKL011	9	23	14	44

Figure 12.5 Survey data for Parking lot "Q"

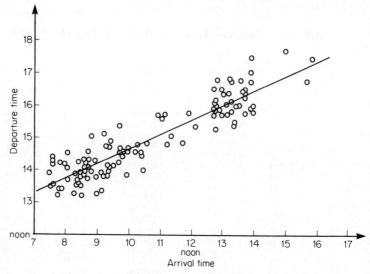

Figure 12.6 Departure time vs. Arrival time

(iv) read the arrival time and construct a histogram of arrival frequencies in half-hour increments,

(v) process the data on the scratch file by means of routine CURFIT to estimate the linear correlation between parking duration and arrival time, and also to obtain the standard error of estimate of the dependent variable.

Using the data of Figure 12.5, the results of this analysis are as follows. The correlation between parking duration DUR and arrival time TA is given by

$$DUR = 10.178 - 0.553TA \text{ h.} \tag{12.5}$$

Obviously, this could be transferred into an equation for departure time TD, i.e.

$$TD = 10.178 + 0.447TA \text{ h.} \tag{12.6}$$

Equation (12.6) is shown relative to the raw data in Figure 12.6.

The standard error of estimate of DUR on TA is found to be 0.456 hour. Finally, the distribution of arrival times is found to be as shown in the following groupings:

Time	Arrivals	Time	Arrivals	Time	Arrivals
7.5–8.0	11	10.5–11.0	2	13.5–16.0	11
8.0–8.5	10	11.0–11.5	4	16.0–16.5	3
8.5–9.0	17	11.5–12.0	2	16.5–15.0	0
9.0–9.5	12	12.0–12.5	1	15.0–15.5	1
9.5–10.0	10	12.5–13.0	10	15.5–16.0	2
10.0–10.5	6	13.0–13.5	13		

```
      PROGRAM PRKANAL
C     CREATE ARRAY TO STORE NO. OF ARRIVALS IN EACH
C     HALF-HOUR INTERVAL.
      DIMENSION FREQ(48)
      NDAT=1
      NSCR=2
      REWIND NDAT
      REWIND NSCR
C     DEFINE NO OF DATA RECORDS
      N=12
C     READ INPUT DATA AND TRANSFORM TO REQUIRED FORMAT ON
C     SCRATCH TAPE FOR USE BY ROUTINE CURFIT.
      DO 50 J=1,N
      READ(NDAT,60)ITAHR,ITAMN,ITDHR,ITDMN
60    FORMAT(10X,I6,I4,I6,I4)
      TA=FLOAT(ITAHR) + FLOAT(ITAMN)/60.0
      TD=FLOAT(ITDHR) + FLOAT(ITDMN)/60.0
      DUR=TD-TA
      WRITE(NSCR,65)TA,DUR
65    FORMAT(2F10.3)
C     COUNT NO OF ARRIVALS IN TIME INTERVAL
      K=IFIX(2.0*TA) + 1
      FREQ(K)=FREQ(K) + 1.0
50    CONTINUE
C     REWIND SCRATCH FILE TO BE READ BY ROUTINE CURFIT
      REWIND NSCR
      TYPE=3HLIN
      CALL CURFIT(NSCR,N,TYPE,A,B,R2,STERR)
      WRITE(6,100)A
      WRITE(6,101)B
      WRITE(6,102)R2
      WRITE(6,103)STERR
100   FORMAT(//,20H  Y-INTERCEPT      = ,F10.3)
101   FORMAT(   20H  GRADIENT Y/X     = ,F10.3)
102   FORMAT(   20H  CORLN. COEFFT.   = ,F10.3)
103   FORMAT(   20H  ST.ERR. OF EST.  = ,F10.3)
C     OUTPUT DISTRIBUTION OF ARRIVALS
      DO 70 J=1,48
      T1=0.5*FLOAT(J-1)
      T2=T1 + 0.5
      WRITE(6,75)T1,T2,FREQ(J)
75    FORMAT(3F10.1)
70    CONTINUE
      STOP
      END
```
Figure 12.7 (a) FORTRAN code to analyse the raw data

12.3. Simulation

Whether or not a five-day survey is typical of the behaviour of the users over a longer time period is open to question. However, within these limitations, the results of the analysis provide a basis for generating synthetic data with the same statistical properties, thus allowing the time period of study to be extended indefinitely.

```
      PROGRAM PRKSIM
C     THIS PROGRAM SIMULATES THE ARRIVAL AND DEPARTURE
C     OF PARKING LOT USERS OVER A PERIOD OF MANY DAYS.
C     THE AVAILABLE NO. OF SPACES IS 20 AND THE NO. OF
C     USERS MAY BE SPECIFIED BY THE USER (21 TO 25)
C     THE DISTRIBUTION OF ARRIVAL TIMES AND THE CORR-
C     ELATION OF DURATION AND ARRIVAL IS BUILT IN TO
C     THE PROGRAM.
C     USES ROUTINES GAUSS, DISTN1 AND DISTN2
C
      DIMENSION TIME(18),ARRVLS(18),WK(18)
      DIMENSION CDF(51),TA(25),TD(25),TAD(50)
      DIMENSION ITAG(50),NCRS(300)
C     DEFINE ARRIVALS IN HALF-HOUR INTERVALS FROM 7.30
      DATA ARRVLS/11.,10.,17.,12.,10., 6., 2., 4., 2., 1.,
     +            10.,13.,11., 3., 0., 1., 2., 0./
C     SPECIFY CORRELATION COEFFICIENTS FOR PARKING
C     DURATION AS A FUNCTION OF ARRIVAL TIME
      A=10.178
      B=-0.553
      STERR=0.46
      DISPLAY "   SUPPLY NO. OF USERS"
      ACCEPT NUSERS
      DISPLAY "   SUPPLY NO. OF DAYS FOR SIMULATION"
      ACCEPT NDAYS
      NCOMPS=0
      MAXCOM=0
C     DEFINE TIME INTERVALS
      DO 10 J=1,18
      TIME(J)=7.0 + 0.5*FLOAT(J)
   10 CONTINUE
C     INITIALIZE RANDOM VARIABLES
      DISPLAY "   SUPPLY SEED INTEGER FOR RANDOM NUMBER"
      ACCEPT IY
C     SET UP CUMULATIVE DISTRIBUTION FUNCTION
      CALL DISTN1(TIME,ARRVLS,18,WK,CDF,51)
C     START LOOP FOR NDAYS OF SIMULATION
      DO 20 IDAY=1,NDAYS
C     START LOOP FOR NO. OF USERS
      DO 30 IUSER=1,NUSERS
C     TEST IF USER IS A NON-STARTER (I.E. .LE.8 PERCENT)
      CALL URAND(IY,P)
      IF(P.GT.0.08) GOTO 40
C     ASSIGN ARBITRARY ARRIVAL AND DEPARTURE TIMES
      TA(IUSER)=23.999
      TD(IUSER)=24.0
      GOTO 30
   40 CONTINUE
C     GENERATE ARRIVAL TIMES FROM CDF.
      CALL URAND(IY,P)
      CALL DISTN2(CDF,51,P,TA(IUSER))
C     GET EXPECTED VALUE OF PARKING DURATION.
      DUR=A + B*TA(IUSER)
C     ADD NORMALLY DITRIBUTED RANDOM COMPONENT
      CALL GAUSS(IY,STERR,DUR,RANDUR)
      TD(IUSER)=TA(IUSER) + RANDUR
   30 CONTINUE
C     COMBINE ARRIVAL AND DEPARTURE TIMES IN ONE ARRAY
```

```
      C  FOR PROCESSING WITH ROUTINE TGSORT
            DO 50 J=1,NUSERS
               TAD(J)=TA(J)
               TAD(J+NUSERS)=TD(J)
         50 CONTINUE
            NUSER2=2*NUSERS
            CALL TGSORT(TAD,ITAG,NUSER2,-1)
            NCARS=0
            MAXCAR=0
      C  COUNT NO. OF CARS AND STORE MAXIMUM
            DO 60 J=1,NUSER2
            IF(ITAG(J).LE.NUSERS) NCARS=NCARS+1
            IF(ITAG(J).GT.NUSERS) NCARS=NCARS-1
            IF(NCARS.GT.MAXCAR) MAXCAR=NCARS
         60 CONTINUE
            NCRS(IDAY)=MAXCAR
      C  GET NO. OF USERS TURNED AWAY AND NOTE MAXIMUM
            ICOMPS=MAXCAR-20
            IF(ICOMPS.LT.0) ICOMPS=0
            NCOMPS=NCOMPS + ICOMPS
            IF(MAXCOM.LT.ICOMPS) MAXCOM=ICOMPS
         20 CONTINUE
      C
      C  OUTPUT RESULTS
            DISPLAY "  SIMULATION FOR",NDAYS," DAYS"
            DISPLAY "  WITH ",NUSERS," PERMITS ISSUED"
            DISPLAY "  DAILY USERS"
            WRITE(6,100)(NCRS(K),K=1,NDAYS)
        100 FORMAT(4X,10I4)
            DISPLAY "  TOTAL NO. OF COMPLAINTS=",NCOMPS
            DISPLAY "  MAX. COMPLAINTS IN 1 DAY=",MAXCOM
            STOP
            END
```

Figure 12.7 (b) FORTRAN code to simulate use of Parking Lot 'Q'

The aim of the simulation program will be to generate arrival and departure times for a population of 25 potential users on each of, say, 100 days and from this to estimate the number of days on which users are turned away. In addition, it will be useful to define some measure of inconvenience to estimate the sensitivity of this parameter to a reduction in the number of users allocated to that area.

Figure 12.7(b) shows typical FORTRAN coding to carry out this simulation. The principal steps in the program are as follows:

(i) Set up arrays to receive the generated arrival and departure times and also auxiliary arrays to define the experimentally determined distributions of arrival times for use in routine TGSORT.
(ii) Define the histogram information by means of a data statement. Note that routine DISTN1 requires array sizes one more than the number of intervals. Since the time intervals are uniformly distributed, these values can be more easily generated by a DO-loop.
(iii) Define the correlation parameters for ease of changing (in the event that further data are obtained) and initialize the seed value for the random number generator.

(iv) Set up two loops, one to define a specified number of days for simulation and an inner loop to define the number of users.
(v) The core of the simulation involves five steps:
(1) Generate a random number $P(0.0 < P < 1.0)$ to test if the user is a 'non-starter', i.e. if $P \leqslant 0.08$, set arbitrary values for TA( ), TD( ) = 23.99 and 24.
(2) Generate a second random number $P(0.0 < P < 1.0)$ for use in routine DISTN2, which generates a synthetic arrival time consistent with the observed distribution.
(3) Obtain the expected duration of parking as a deterministic function of the arrival time.
(4) Generate a normally distributed duration of parking, using the expected value and the standard deviation found from the sample.
(5) Obtain the departure time.
(vi) For each simulated day, the total number of users requiring a parking space can be determined in various ways. The method employed here is to use routine TGSORT to generate an integer array containing subscripts of the arrival *and* departure times in ascending order. A simple scan of this integer array allows the maximum number of potential users to be determined. A more sensitive measure is the total 'user-days' of inconvenience which is summed over the entire 100-day period of simulation.

12.3.4 Results

Figure 12.8 shows typical results from one simulation of 100 days. Thus, the probability of not finding a space is given approximately by

$$P(\text{no space}) = 31/(0.92 \times 2500) = 1.35\%.$$

```
SUPPLY NO. OF USERS ?  25

SUPPLY NO. OF DAYS FOR SIMULATION ?  100

SUPPLY SEED FOR RANDOM NUMBER ?  987654321

SIMULATION FOR    100   DAYS
WITH      25  PERMITS ISSUED
DAILY USERS
    19   21   20   21   17   22   19   17   16   19
    18   19   20   19   16   15   17   20   21   19
    16   19   18   17   18   18   18   15   19
    20   22   18   19   22   20   21   19   17   18
    19   21   20   20   19   21   18   16   19   19
    19   17   20   18   20   18   18   21   18   20
    19   17   22   19   16   20   21   20   21   23
    22   20   16   21   22   20   20   17   20   20
    20   18   18   21   19   22   18   20   20   18
    22   21   19   17   16   19   18   19   16   16
TOTAL NO. OF COMPLAINTS=        31
MAX. COMPLAINTS IN 1 DAY=        3
```

Figure 12.8 Typical results of Parking Lot simulation

This preliminary analysis may be extended to refine the model and to study the sensitivity of the results to changes in certain system parameters.

Clearly, the study is only as reliable as the data on which it is based. However, before any sensitivity analysis can be attempted with respect to the system parameters (e.g. the correlation coefficients, standard error of estimate, and the distribution), it is desirable to test the reliability of the results subject to changes in the initial seed value and the length of simulation.

With some minor changes in the coding, it is possible to run the simulation for a much longer period (say, 1000 days) and to print out at regular intervals updated statistics concerning the probability of being turned away. By observing the way in which these statistics stabilize around a long-term average, some feel may be obtained for the appropriate length of simulation period.

Sensitivity to the initial seed value should reduce with increase in the simulation period but much may depend on the random number generator and the machine word size.

A further refinement might be introduced by varying dynamically the percentage of 'no-shows'. For example, the initial estimate of 8% may include the percentage of users turned away. Some care is required, however, since a finite percentage of rejects will reduce the proportion of 'no-shows', which, in turn, will tend to increase the number of rejects. Thus, for a system in which this percentage is significant, the variation of the percentage could lead to a form of instability.

12.4 DESIGN OF AN INSULATED PRESSURE VESSEL

12.4.1 Preamble

The background to this problem is presented in the following two memoranda.

To: Dr Knightly-Chivers.
From: I. C. Hands, Llaregyb Sub-Office.
Re: *New Pressure Vessel Design.*

You will no doubt have heard of our innovative scheme to provide heated comfort stations for our Arctic workers. I am concerned that the preliminary estimate for the capital and operating cost of the hot water pressure vessel is much higher than anticipated. The vessel must be of cylindrical shape with hemispherical ends and is required to hold 15 cubic metres at a working pressure of 1500 kPa.

We are proposing to provide outside thermal insulation to reduce heat loss and thus operating cost. Could you arrange for one of your design team to prepare a preliminary design for minimum cost. I believe you have on file the necessary cost data, etc.

To: Newly Startit, Design Office.
From: Dr Knightly-Chivers.

Please find a copy of correspondence regarding the hot water pressure vessel for the Baffin Island group. Also noted below are relevant cost and design data. Please let me have a reply drafted for my signature within two weeks. Failure to meet this deadline may result in your having to complete the design in person at the Pangirtung office.

Steel working stress (use constant plate thickness based on cylindrical hoop tension only)	75 N/mm²
Cost of steel plate (cylindrical)	$0.03/cm³
Cost of steel plate (spherical)	$0.06/cm³
Full penetration weld of hemispherical ends	$70.00/m
Insulation	$22.00/m²/cm thick
Losses (assumed to be inversely proportional to thickness)	$28.00/m²/yr for 1 cm of insulation

Discount operating cost at 8.5% over 20 years.

12.4.2 The Mathematical model

The general arrangement of the pressure vessel, together with the relevant variables, is shown in Figure 12.9. The system parameters for the problem are as follows:

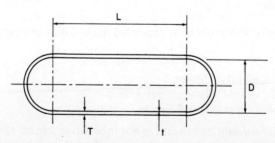

Figure 12.9 General arrangement of pressure vessel

Required volume V = 15 m³

Required pressure p = 1500 kPa

Working stress (hoop tension) f = 75 mPa.

The design variables to be determined are as follows:

Diameter	D (m)
Length	L (m)
Steel thickness	t (mm)
Insulation thickness	T (mm).

The objective function is a combination of the capital cost for fabricating the vessel and the operating cost arising from the heat loss. Thus:

Total cost $z = $ C1 (cost of steel in cylinder)
$+ C2$ (cost of steel in hemispheres)
$+ C3$ (cost of welding)
$+ C4$ (cost of insulation)
$+ C5$ (present value of annual heat losses).

The surface area and contained volume of the cylindrical and hemispherical parts of the vessel can be expressed as simple functions of the diameter D and length L. Thus:

$$\text{Area of cylinder} = \pi D L$$
$$\text{Area of sphere} = \pi D^2$$
$$\text{Volume of cylinder} = \frac{\pi}{4} D^2 L$$
$$\text{Volume of sphere} = \frac{\pi}{6} D^3.$$

The mathematical model may now be stated:

$$\text{Minimize } z = \phi(D, L, t, T) \tag{12.7}$$

subject to
$$g_1(p, D, t) = f \tag{12.8}$$
$$g_2(D, L) = V. \tag{12.9}$$

Equations (12.8) and (12.9) represent the working stress and required volume constraints respectively.

Since the insulation thickness T does not appear in the equality constraints, and since the costs for insulation (C4) and discounted losses (C5) can be expressed per unit surface area, the value of T may be obtained by *sub-optimization* (i.e. without reference to the other variables D, L, and t).

With reference to the cost figures in the memo,

$$C(T) = C4 + C5 = 22.00T + \frac{28.00}{T} \frac{P}{A} (8.5\%, 20) \tag{12.10}$$

where

$$\frac{P}{A}(8.5\%, 20) = \frac{1.085^{20} - 1}{0.085 \times 1.085^{20}}$$

$$= 9.4633 \quad \text{(see Section 9.2.5)}$$

$$\therefore C(T) = 22T + \frac{264.97}{T}. \tag{12.11}$$

For $C(T)$ = minimum set $dC(T)/dT = 0$, i.e.

$$22 - \frac{264.97}{T^2} = 0 \tag{12.12}$$

$$\therefore T^* = 3.47 \text{ cm}.$$

Assuming that insulation can be installed in multiples of 1 cm, a simple comparison between 3 cm and 4 cm by means of (12.11) yields

$$C(3 \text{ cm}) = \$154.32/\text{m}^2$$

$$C(4 \text{ cm}) = \$154.24/\text{m}^2.$$

A value of $T = 4$ cm will be used, not simply because of the saving of 8 cents, but because future fuel costs (and, thus, costs due to heat loss) are likely to increase, which will move the optimum value of T towards the larger value.

12.4.3 Treatment of constraints

With the sub-optimization of T, the problem reduces to one of minimization of a function of three variables subject to two equality constraints. It would be possible to use Lagrangian multipliers (Section 2.5) to solve this problem, but it is always worth examining the constraints to see if they are of sufficiently simple form to allow direct substitution for one of the design variables in the objective function.

The stress constraint (12.8) may be written as

$$2ft = pD$$

or

$$2 \times 75 \times 10^6 \times \frac{t}{1000} = 1500 \times 10^3 \times D$$

where t is in mm and D is in m. That is,

$$t \text{ (mm)} = 10D \text{ (m)}. \tag{12.13}$$

Obviously, direct substitution is possible.

The volume constraint (12.9) may also be reduced. Thus:

$$\frac{\pi}{4} D^2 L + \frac{\pi}{6} D^3 = V = 15 \text{ m}^3. \tag{12.14}$$

Since L appears in only one term, it is relatively simple to substitute for L in terms of the required volume V and the single remaining variable D. Thus:

$$L = \frac{15 - (\pi/6)D^3}{(\pi/4)D^2} = \frac{60}{\pi D^2} - \frac{2}{3}D. \qquad (12.15)$$

12.4.4 The unconstrained model

Substituting (12.11) and (12.15) in the objective function and evaluating the various terms $C1$ to $C5$ results in the following unconstrained model:

$$\underset{D^*}{\text{Minimize}} \; z = C1 + C2 + C3 + C4 + C5 \qquad (12.16)$$

where

$$C1 = \$0.03 \times 10^3 \times \pi D L t$$
$$= 300\pi D^2 \left(\frac{60}{\pi D^2} - \frac{2}{3}D \right)$$
$$= 1800 - 200\pi D^3$$
$$C2 = \$0.06 \times 10^3 \times \pi D^2 t$$
$$= 600\pi D^3$$
$$C3 = \$70.00(2\pi D + L)$$
$$= 70\left(2\pi D + \frac{60}{\pi D^2} - \frac{2}{3}D\right)$$
$$= 140\pi D - \frac{140}{3}D + \frac{4200}{\pi D^2}$$
$$C4 + C5 = \$154.24(\pi D L + \pi D^2)$$
$$= 154.24\pi D \left(\frac{60}{\pi D^2} - \tfrac{2}{3}D + D \right)$$
$$= 154.24 \left(\frac{60}{D} + \frac{\pi}{3}D^2 \right).$$

It will be noted that one longitudinal weld and two circumferential welds are included in the calculation of $C3$. However, in the estimation of $C4$ and $C5$, the area has been calculated in terms of the internal dimensions D and L, with no allowance for the thicknesses t and T.

The optimum value of D can now be obtained by setting $dz/dD = 0$, i.e.

$$\frac{dz}{dD} = 600\pi D^2 + 1800\pi D^2 + 140\pi - \frac{140}{3} - \frac{8400}{\pi D^3} \qquad (2.17)$$
$$- 154.24 \frac{60}{D^2} + 154.24 \frac{2\pi D}{3}.$$

By trial and error, the optimum diameter D^* is found and the corresponding values of L^* and t^* are obtained by substitution. Thus

$$D^* = 1.277 \text{ m}$$
$$L^* = 10.860 \text{ m}$$
$$t^* = 12.77 \text{ mm}$$
$$T^* = 40 \text{ mm}.$$

The total cost for the optimal design is made up as follows:

Cost of steel (cylinder)	$16 691.56
Cost of steel (hemisphere)	$3925.31
Cost of welding	$1321.88
Cost of insulation	$4284.97
Discounted losses	$3225.56
Total	$29 449.28.

This design is not practically feasible, since steel plate is available only in discrete thicknesses. If the value of plate thickness t is rounded up to 13 mm, some advantage may be taken of the increased hoop tension strength to increase the diameter and thus reduce the surface area of the vessel.

From (12.13), if $t = 13$ mm, the maximum diameter is 1.3 m. Then by the volume constraint of (12.15), the length L is 10.434 m. The modified design is then

$$D = 1.3 \text{ m}$$
$$L = 10.434 \text{ m}$$
$$t = 13 \text{ mm}$$
$$T = 40 \text{ mm}$$

and the total cost is increased to $29 454.11. This last calculation suggests an alternative approach to the problem. Since T is obtained by sub-optimization and the two equality constraints are relatively simple, it would be possible to develop an algorithm in which the plate thickness t (mm) is the only independent variable. From t, the value of D is found by (12.13) and, hence, L from (12.15). All quantities and costs are then calculable.

12.5 A NEW WATER SUPPLY

12.5.1 Background

The problem described in this section is somewhat similar to the Thirstville 'case-study' of Section 1.4. The supply is assumed to be by gravity main, but the problem is complicated by the irregular demand pattern and a more detailed design of the balancing tank. The scene is set by the following memorandum, together with the typical cross-section of Figure 12.10.

To: Dr S. T. Mater, Civil Engineering New Works.
From: Ms Chris Talgazing, Planning Department.
Re: *New Water Supply.*

I have now received from the Process Planning Section an estimate of the water supply which will be required by the new plant. The demand, averaged over four-hour intervals, is given for a one-week cycle in the attached table, from which you will note that considerable fluctuation in demand is to be expected.

The local authority has assured me that the town supply main can provide in excess of 0.1 m^3/s and I understand that the main passes within 3.2 kilometres of the demand point. The pressure elevation at the take-off point on the town main should be about 30 m above ground elevation at the plant.

I should like you to examine the costs of providing a water supply, including, if necessary, a reinforced concrete balancing tank. A typical cross-section of a similar reservoir is shown in the accompanying sketch; the same criteria for earth cover, ground slope, freeboard, etc., should be used in your estimate. I should point out, however, that the only ground available for such a tank is a long, level strip only 25 m wide.

I realize that there may be further information necessary before your study can be completed, and that estimates of construction costs are approximate. However, I hope your report will help to identify and define the main factors to be considered and show whether or not a balancing tank is justified.

Expected Water Demand
(flows in cubic metres per second, averaged over four-hour periods)

Time	12–4 a.m.	4–8 a.m.	8–12 noon	12–4 p.m.	4–8 p.m.	8–12 midnight
Sun.	0.012	0.020	0.037	0.031	0.020	0.017
Mon.	0.012	0.034	0.083	0.068	0.057	0.034
Tues.	0.023	0.040	0.068	0.062	0.045	0.028
Wed.	0.021	0.045	0.051	0.034	0.034	0.016
Thurs.	0.023	0.014	0.034	0.048	0.054	0.040
Fri.	0.021	0.014	0.034	0.048	0.054	0.040
Sat.	0.018	0.026	0.040	0.060	0.045	0.026

12.5.2 Problem formulation

The first step is to prepare a sketch of the system showing the relevant components, system parameters, and design variables. Figure 12.11 shows such a diagram, and the following system parameters and design variables are identified:

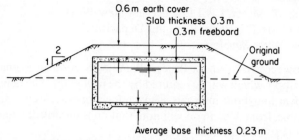

Figure 12.10 Typical section of in-ground tank

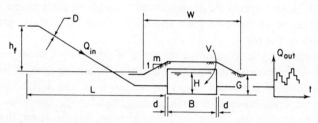

Figure 12.11 Diagrammatic sketch of the system (New Water Supply Problem — Section 12.5)

System parameters:
  Pipeline length L (m)   3200 m
  Available pressure head h_f (m)   30 m
  Available ground width W (m)   25 m
  Outflow (demand) Q_{out} (m³/s)   (see table)
  Embankment slope M 2:1
  Reinforced concrete bending modulus $\dfrac{RM}{bd^2}$ (N/mm)² 0.4 N/mm²

Design variables:
  Inflow (supply) Q_{in} (m³/s)
  Pipe diameter D (m)
  Inside tank breadth B (m)
  Inside tank length XL (m)
  Water depth H (m)
  Depth in ground G (m)
  Concrete wall thickness d (m)
  Tank volume V (m³)

The objective function which is to be minimized comprises only capital expenditures. No continuing costs are included.

Objective function $z = \quad C1 \quad$ (Excavation) \hfill (12.18)
$\qquad\qquad\qquad\quad + C2 \quad$ (Embankment)
$\qquad\qquad\qquad\quad + C3 \quad$ (Import fill/export surplus)

+C4 (Reinforced concrete)
+C5 (Formwork — outside, inside, and roof slab)
+C6 (Pipeline).

The following costs are assumed for the purpose of the analysis.

Excavation	$5.00/m^3	(12.19)
Form embankment	$2.00/m^3	
Import fill or dispose of surplus	$2.50/m^3	
Outside formwork	$12.00/m^2	
Inside vertical formwork	$18.00/m^2	
Suspended slab formwork	$25.00/m^2	
Reinforced concrete	$100.00/m^3	
Pipeline	$C6 = LD(390 - 11.5\sqrt{D})$.	

12.5.3 Identifying constraints

The eight design variables defined in Section 12.5.2 are not independent and the next step is to determine the interactions which exist between these variables and to define the relevant constraints.

Pipeline capacity

The inflow Q_{in} and pipe diameter D must be related to the available piezometric gradient, which in turn is defined by the pressure head h_f and the length L. A decision is needed as to the flow resistance law and the relevant friction loss parameters. The Strickler equation will be used here with an equivalent roughness height of $k = 0.3$ mm. (This is another system parameter, omitted from the list of Section 12.5.2.) Thus,

$$Q_{in} = \frac{8.41\sqrt{g}}{k^{1/6}} \frac{\pi}{4} D^2 \left(\frac{D}{4}\right)^{2/3} \left(\frac{h_f}{L}\right)^{1/2}. \qquad (12.20)$$

Balancing tank volume

Within certain limits, the required storage volume V will be dependent on the value of the inflow Q_{in} and the specified demand pattern Q_{out} (Figure 12.12).

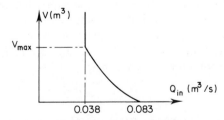

Figure 12.12 Storage volume as a function of inflow

Inspection of the table of Q_{out} values shows that if $Q_{in} \geqslant 0.083$ m³/s, there will be no need for a balancing tank. This, however, might require a large and expensive pipeline. At the other extreme, the value of Q_{in} must not be less than the average demand. This is given by

$$(Q_{out})_{ave} = \frac{1}{42} \sum_{i=1}^{42} (Q_{out})_i = 0.038 \text{ m}^3/\text{s}.$$

When $Q_{in} = 0.038$ m³/s, the balancing storage required will be a maximum. For intermediate values of Q_{in} (i.e. $0.038 \leqslant Q_{in} \leqslant 0.083$), some form of calculation or interpolation will be required.

Tank dimensions

The design variables include all three internal dimensions for the storage tank, as well as the required volume. Obviously, there is a simple relation between these quantities, i.e.

$$B \times XL \times H = V. \tag{12.21}$$

Wall thickness

For this example, it will be assumed that the reinforced concrete wall of the tank will experience the greatest bending moment when the tank is hydraulically tested before backfilling on the outside and before completion of the roof slab. Figure 12.13 shows this condition. The thickness of the wall will be based on the

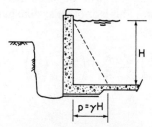

Figure 12.13 Hydrostatic loading of an unpropped cantilever

value of the bending moment on a simple, unsupported cantilever subject to hydrostatic loading. The analysis is more complex than this, but the assumption is probably adequate for the purpose of proportioning the tank.

For a fluid specific weight of $\gamma = 9810$ N/m³ and depth H, the pressure at the base is $p = \gamma H$. The total force is given by

$$P = Hp/2 = \gamma H^2/2 \tag{12.22}$$

Thus,

$$\text{BM} = PH/3 = \gamma H^3/6. \tag{12.23}$$

The wall thickness can now be determined by the relation

$$BM = RM = Kbd^2 \qquad (12.24)$$

in which the flexural strength factor K is given a low value of 0.4 N/mm^2 in order to reduce the risk of the concrete cracking on the wet, tension side. The quantity b in (12.24) represents the breadth of the reinforced concrete section, but in this case the wall may be designed for a unit width of 1 m so that $b = 1$.

Width constraint

If a balancing tank is to be constructed, the total width between the toes of the embankment on each side must be less than 25 m. Clearly, this distance will depend on the tank dimensions H and B, the wall thickness d, and the depth G to which the tank is sunk in the ground. Figure 12.14 shows the geometry of the cross-section.

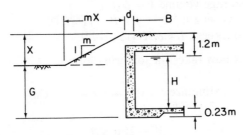

Figure 12.14 Relation between total width W and other variables

The embankment height is given by

$$X = H + 1.43 - G \qquad (12.5)$$

in which the number 1.43 is the sum of the fixed quantities shown in Figure 12.10. The total width W is then found as

$$W = B + 2d + 2mX \qquad (12.26)$$

and the necessary constraint takes the form

$$W - 25.0 \leqslant 0.0. \qquad (12.27)$$

12.5.4 Solving the mathematical model

From the preceding sections, the mathematical model may be set up.

$$\underset{Q^*, D^*, B^*, XL^*, H^*, G^*, d^*, V^*}{\text{Minimize } z} = C1 + C2 + \ldots + C6 \qquad \text{(refer 12.18)}$$

subject to

$$g_1(Q_{\text{in}}, D) = 0 \qquad \text{(refer 12.20)}$$

$$g_2(Q_{in}, Q_{out}, V) = 0$$
$$g_3(B, XL, H, V) = 0 \quad \text{(refer 12.21)}$$
$$g_4(d, H) = 0 \quad \text{(refer 12.22-24)}$$
$$(W - 25) \leq 0. \quad \text{(refer 12.25-27)}$$

This is a non-linear problem involving eight design variables, four equality constraints, and one inequality constraint. The problem can be greatly simplified if a sub-set of the design variables is chosen so as to allow the equality constraints to be substituted in the objective function, thus reducing the complexity of the model.

If the selected independent design variables are diameter D, tank breadth B, depth H, and in-ground depth G, the equality constraints can all be incorporated into the objective function as follows:

1. Calculate $Q_{in} = \phi(D)$ by (12.20).
2. Find required storage volume $V = \phi(Q_{in})$.
3. Obtain inside tank length $XL = \phi(V, B, H)$ (12.21).
4. Calculate wall thickness $d = \phi(H)$ (12.22-24).

The reduced model now takes the following form:

$$\underset{D^*, B^*, H^*, G^*}{\text{Minimize}} z = (C1 + C2 + \ldots + C6) \quad (12.28)$$

subject to

$$W - 25.0 \leq 0.$$

This model can be further reduced to an unconstrained model by incorporating a penalty term to ensure that the inequality constraint is satisfied, i.e.

$$\text{Minimize } z' = (C1 + C2 + \ldots + C6) + FAC(W - 25)\delta \quad (12.29)$$

where $\delta = 1$ if $W - 25 > 0$ and $\delta = 0$ if $W - 25 \leq 0$.

The multiplier FAC in (12.29) should be large enough to ensure that the penalty term is significant in comparison with the real objective function z. Equation (12.29) may now be optimized by a non-linear algorithm such as the Hooke and Jeeves pattern search (i.e. subroutine HJMIN).

12.5.5 Calculating balancing storage

As discussed in Section 12.5.3, a method is required to compute the necessary balancing storage as a function of the inflow Q_{in}. A suitable subroutine BALNCE is illustrated in Figure 12.15 which should be self-explanatory.

It would be possible to include a call of this routine within the cost routine used by HJMIN, but this would be rather inefficient. A better arrangement would be to include in the driving program a series of calculations which would determine corresponding values of inflow Q_{in} and storage volume V, which could then be

```
      SUBROUTINE BALNCE(QOUT,NQOUT,QIN,VOLUME)
C **************************************************************
C  THE ROUTINE OPERATES ON AN ARRAY OF REQUIRED OUTFLOWS
C  TO DETERMINE THE NECESSARY BALANCING STORAGE VOLUME WHICH
C  IS REQUIRED IF THE SPECIFIED INFLOW VOLUME IS SUPPLIED.
C  IF THE INFLOW IS LESS THAN THE AVERAGE DEMAND A VERY
C  LARGE VOLUME (E.G. 10.0E20) IS RETURNED.
C     QOUT   = ARRAY OF SIZE (NQOUT) CONTAINING THE REQUIRED
C              TIME SERIES OF OUTFLOWS.
C     NQOUT  = NO. OF OUTFLOWS.
C     QIN    = SPECIFIED AVAILABLE INFLOW.
C     VOLUME = COMPUTED BALANCING STORAGE VOLUME REQUIRED.
C  THE UNITS USED MUST BE CONSISTENT THROUGHOUT.  THUS
C  THE VOLUME IS DEFINED IN TERMS OF THE TIME INCREMENT
C  USED TO DEFINE THE OUTFLOW TIME SERIES.
C **************************************************************
      DIMENSION QOUT(NQOUT)
      SUMQ=0.0
      VOL=0.0
      VMIN=0.0
      DO 10 I=1,NQOUT
      DV=QIN-QOUT(I)
      SUMQ=SUMQ + QOUT(I)
C  TEST IF TANK IS FULL AND OUTFLOW .LE. INFLOW
      IF(DV.GE.0.0.AND.VOL.GE.0.0) GOTO 10
      VOL=VOL + DV
      IF(VOL.GT.0.0) VOL=0.0
      IF(VOL.LT.VMIN) VMIN=VOL
   10 CONTINUE
      VOLUME=-VMIN
C  CHECK THAT AVERAGE DEMAND IS AVAILABLE AT LEAST.
      QAVE=SUMQ/FLOAT(NQOUT)
      IF(QIN.LT.QAVE) VOLUME=10.0E20
      RETURN
      END
```

Figure 12.15   FORTRAN subroutine BALNCE

transferred to the cost routine for interpolation in much the same fashion as illustrated in Figure 12.12. Two points are worth noting:

(1) If the inflow Q_{in} is less than the average demand, the routine BALNCE automatically sets the required volume to an arbitrarily high value. This is equivalent to adding a penalty term if the constraint $Q_{in} \geq 0.038$ is violated.
(2) In calculating the cost of the balancing tank, a check should be made that the required volume is finite. If $Q_{in} > 0.083 \text{ m}^3/\text{s}$, then $V = 0.0$ and all the calculations associated with the tank can be skipped.

12.5.6   Typical solution

A typical solution using routine HJMIN is presented in this section. As described in Section 5.7.1, the method requires a main driving program and an objective function subroutine and must be executed in conjunction with the routine HJMIN as listed in Appendix A. The two subprograms will be discussed separately.

The dimension statements define the various arrays required. Two of these hold the values of the design variables and the corresponding incremental values to be used in the local search procedure of HJMIN. These appear in the calling statement. The other arrays are needed to store the outflow time history and the computed values of Q_{in} and V used to define the curve of Figure 12.12. Other design variables, system parameters, design quantities, and rates are transferred between the main program and the cost subroutine by means of labelled COMMON blocks. Those variables which constitute input to the routine COST are initialized either by a DATA statement, by simple assignments, by input from the keyboard, or by calculation. The inflow–storage function is defined by a set of 11 coordinate pairs which are evaluated in a DO-loop. Note that the minimum inflow is set slightly below the average demand to ensure that an arbitrarily high storage quantity is assigned, thus serving as a penalty term.

After the call of HJMIN, the optimal values of the design variables, together with other relevant information, are output. Some of the output is conditional on the balancing tank being of finite size. It is convenient to introduce a loop in the main program to allow alternative starting values to be defined. This helps to confirm the existence of a global minimum.

The objective function subroutine COST contains identical COMMON blocks and relevant dimension statements, as in the main program. It is convenient (and marginally more efficient), to re-assign the design variables as simple variables, rather than elements of an array. The first step is to calculate the pipeline cost and calculate by interpolation the storage volume required for the pipeline capacity. If no storage is required, the objective function calculation ends here. However, for finite storage volumes, the design of the tank makes up the bulk of the coding. The details of the design and the calculation of quantities should be fairly obvious from the coding and comment statements.

When the total real cost is calculated, penalty terms are added which correspond to the remaining constraints on the solution. The principal one is the available width of ground, but other (perhaps superfluous) non-negativity constraints have been added to keep the tank dimensions positive. It is easy to overlook the fact that the extrapolation step of the algorithm might produce a negative value of a variable which in turn generates a 'negative cost'.

The solution given by the program of Figure 12.16 is summarized in the output shown in Figure 12.17.

12.5.7 Allowance for discrete variables

The solution developed in the previous section may be impractical since the pipe diameter is assumed to be a continuous variable. A more realistic solution would be to remove the diameter from the array of design variables and introduce a loop in the main program to allow a series of discrete, commercially available pipe sizes to be defined. For this diameter, the inflow capacity and thus the storage volume would be fixed and transferred through COMMON block to the subroutine. The program of Figure 12.16 could be forced to operate in this way by

```
      PROGRAM TANKEX
      DIMENSION VAR(4),DVAR(4)
C SEE ROUTINE COST FOR DEFINITION OF DESIGN VARIABLES
      DIMENSION QOUT(42),QAR(11),VOLAR(11)
C THESE ARRAYS USED TO DEFINE DEMAND AND INFLOW/STORAGE
C RELATIONSHIP
      COMMON /DESIGN/QAR,VOLAR,L,HF,K,M,XL,VOL,W,D,EMBHT,
     +                QIN
      COMMON /RATES/CEXC,CEMB,CDIF,CCONC,CFORMO,CFORMI,CFORMS
      REAL K,L,M
      EXTERNAL COST
C DEFINE DEMAND TIME HISTORY
      DATA QOUT/0.012,0.020,0.037,0.031,0.020,0.017,
     +          0.012,0.034,0.083,0.068,0.057,0.034,
     +          0.023,0.040,0.068,0.062,0.045,0.028,
     +          0.021,0.045,0.051,0.034,0.031,0.016,
     +          0.023,0.040,0.062,0.071,0.051,0.034,
     +          0.021,0.014,0.034,0.048,0.054,0.040,
     +          0.018,0.026,0.040,0.060,0.045,0.026/
C RESTART WITH NEW INITIAL VALUES
    1 CONTINUE
C DEFINE INITIAL VALUES FOR DESIGN VARIABLES AND INCREMENTS
      DISPLAY "SUPPLY INITIAL VALUES FOR B, H, G, DIA"
      ACCEPT VAR(1),VAR(2),VAR(3),VAR(4)
      IF(VAR(1).LE.0.0) STOP
      DO 5 J=1,4
    5     DVAR(J)=VAR(J)/25.0
C DEFINE SYSTEM PARAMETERS
      L=3200.0
      HF=30.0
      K=0.0003
      M=2.0
C DEFINE COST RATES
      CEXC=5.00
      CEMB=2.00
      CDIF=2.50
      CCONC=100.0
      CFORMO=12.00
      CFORMI=18.00
      CFORMS=25.00
C GET AVERAGE AND MAXIMUM DEMAND FLOWS.
      QAVE=0.0
      QMAX=0.0
      DO 10 J=1,42
         QAVE=QAVE + QOUT(J)
         IF(QOUT(J).GT.QMAX) QMAX=QOUT(J)
   10 CONTINUE
      QAVE=QAVE/42.0
C COMPUTE POINTS ON INFLOW/STORAGE CURVE
C SET MIN. FLOW JUST BELOW QAVE TO ENSURE PENALTY
      QMIN=QAVE-0.001
      DO 20 J=1,11
         QIN=QMIN + (QMAX-QMIN)*FLOAT(J-1)/10.0
         CALL BALNCE(QOUT,42,QIN,VOL)
         QAR(J)=QIN
         VOLAR(J)=VOL*4.0*3600.0
   20 CONTINUE
```

Figure 12.16 (a) Main FORTRAN program for problem of New Water Supply

```
      C DEFINE PARAMETERS FOR ROUTINE HJMIN
        DATA RHO,EPS,NW/0.5,0.01,0/
        NMAX=1000
      C
        CALL HJMIN(VAR,DVAR,4,ANS,RHO,EPS,COST,NW,NMAX)
      C
      C NOW OUTPUT RESULTS
        WRITE(6,100)NMAX
        WRITE(6,101)ANS
        WRITE(6,102)VAR(4)
          WRITE(6,110)QIN
  100   FORMAT(18H SOLUTION FOUND IN,I5,11H ITERATIONS)
  101   FORMAT(15H MINIMUM COST=$,10X,F10.2)
  102   FORMAT(18H PIPE DIAMETER(M)=,7X,F10.3)
  110   FORMAT(16H AVERAGE INFLOW=,9X,F10.3)
      C SKIP REMAINING OUTPUT IF NO TANK REQUIRED.
        IF(VOL.GT.0.0) GOTO 30
        WRITE(6,103)
  103   FORMAT(25H NO STORAGE TANK REQUIRED)
        GOTO 1
   30   CONTINUE
        WRITE(6,104)VAR(1),XL,VAR(2)
        WRITE(6,105)VOL
        WRITE(6,106)VAR(3)
        WRITE(6,107)D
        WRITE(6,108)W
        WRITE(6,109)EMBHT
  104   FORMAT(19H TANK DIMENSIONS(M),6X,3F10.3)
  105   FORMAT(19H TANK VOLUME(CUB.M),6X,F10.3)
  106   FORMAT(21H TANK DEPTH IN GROUND,4X,F10.3)
  107   FORMAT(19H TANK WALL THKSS(M),6X,F10.3)
  108   FORMAT(15H WIDTH USED (M),10X,F10.3)
  109   FORMAT(18H EMBANKMENT HT.(M),7X,F10.3)
        GOTO 1
        END
```

Figure 12.16 (a) — continued

```
        SUBROUTINE COST(X,CST)
C****************************************************
C THIS ROUTINE DETERMINES AN ARTIFICIAL OBJ. FUN.
C FOR THE TANK + PIPELINE WITH A PENALTY TERM FOR
C THE AVAILABLE WIDTH CONSTRAINT.
C    X(1)    = INSIDE TANK BREADTH (B)
C    X(2)    = INSIDE DEPTH OF WATER (H)
C    X(3)    = DEPTH OF TANK IN GROUND (G)
C    X(4)    = PIPE DIAMETER. (DIA)
C USES COMMON BLOCKS /DESIGN/ AND /RATES/
C****************************************************
        DIMENSION X(4)
        DIMENSION QAR(11),VOLAR(11)
        COMMON /DESIGN/QAR,VOLAR,L,HF,K,M,XL,VOL,W,D,EMBHT,
       +               QIN
        COMMON /RATES/CEXC,CEMB,CDIF,CCONC,CFORMO,CFORMI,CFORMS
        REAL K,L,M
C REDEFINE DESIGN VARIABLES FOR CONVENIENCE
        B=X(1)
        H=X(2)
        G=X(3)
        DIA=X(4)
```

Figure 12.16 (b) Objective function subroutine

```
      C   CHECK D>0 AND GET PIPE COST
              IF(DIA.LT.0.0) DIA=0.0
              CPIPE=L*DIA*(390.0 - 11.5*SQRT(DIA))
              CST=CPIPE
      C   CALC QIN BY STRICKLER EQN.
              CONST=8.41*SQRT(9.81)/(K**0.1667)
              QIN=CONST*0.785*DIA*DIA*(DIA/4.0)**0.667*SQRT(HF/L)
      C   FIND HOW MUCH BALANCING STORAGE NEEDED WITH THIS FLOW.
              CALL INTER1(VOLAR,QAR,11,QIN,VOL)
      C   IF ZERO VOLUME IGNORE TANK CALCS
              IF(VOL.LE.0.0) RETURN
      C   GET TANK LENGTH AND DESIGN WALL THICKNESS
              XL=VOL/(B*H)
              BM=9810.0*H*H*H/6.0
              D=SQRT(BM/400000.0)
      C   GET OVERALL TANK DIMENSIONS
              HOA=H + 0.83
              BOA=B + 2.0*D
              XLOA=XL + 2.0*D
      C   GET QUANTITIES OF EXCAVATION AND EMBANKMENT
              VOLEXC=BOA*XLOA*G
              EMBHT=HOA+0.6-G
              IF(EMBHT.LT.0.0) EMBHT=0.0
              A1=BOA*XLOA
              A2=(BOA+M*EMBHT)*(XLOA+M*EMBHT)
              A3=(BOA+2.0*M*EMBHT)*(XLOA+2.0*M*EMBHT)
              VOLEMB=(EMBHT/6.0)*(A1 + 4.0*A2 + A3)
              HAG=HOA - G
              IF(HAG.LT.0.0) HAG=0.0
              VOLEMB=VOLEMB - HAG*BOA*XLOA
      C   FIND DIFFERENCE BETWEEN EXC AND EMB VOLUME.
      C   IGNORE BULKING
                  VOLDIF=ABS(VOLEXC-VOLEMB)
      C   GET CONCRETE VOLUME
              CONC=2.0*(BOA+XL)*HOA*D + 0.53*BOA*XLOA
      C   NOW GET OUTSIDE, INSIDE AND SLAB FORMWORK
              FORMO=2.0*(BOA+XLOA)*HOA
              FORMI=2.0*(B + XL)*(H + 0.3)
              FORMS=B*XL
      C   NOW CALCULATE COSTS FOR TANK CONSTRUCTION
              C1=VOLEXC*CEXC
              C2=VOLEMB*CEMB
              C3=CONC*CCONC
              C4=FORMO*CFORMO + FORMI*CFORMI + FORMS*CFORMS
              C5=VOLDIF*CDIF
              CST=CPIPE+C1+C2+C3+C4+C5
      C   CHECK WIDTH OF GROUND USED AND ADD PENALTY TERM
              W=BOA + 2.0*M*EMBHT
              PENW=1.0E07*(W-25.0)
              IF(PENW.LT.0.0) PENW=0.0
      C   ADD PENALTY TERMS FOR NON-NEGATIVE VARIABLES
              PENG=1.0E6*(-G)
              PENH=1.0E6*(-H)
              PENB=1.0E6*(-B)
              IF(PENG.LT.0.0) PENG=0.0
              IF(PENH.LT.0.0) PENH=0.0
              IF(PENB.LT.0.0) PENB=0.0
              CST=CST + PENW + PENG + PENH + PENB
              RETURN
              END
```

Figure 12.16 (b) — continued

```
SUPPLY INITIAL VALUES FOR B, H, G, DIA ? 15.0, 2.5, 2.5, 0.2
SOLUTION FOUND IN 381 ITERATIONS
MINIMUM COST=$              305617.00
PIPE DIAMETER(M)=                 .207
AVERAGE INFLOW=                   .046
TANK DIMENSION(M)              15.010        23.278       2.860
TANK VOLUME(CUB.M)            999.126
TANK DEPTH IN GROUND            1.947
TANK WALL THKSS(M)               .309
WIDTH USED (M)                 24.999
EMBANKMENT HT.(M)               2.343
```

Figure 12.17   Results from program of Figure 12.16

setting $DVAR(4) = 0.0$ after the four quantities are initialized. Although computationally inefficient, this small change would cause the diameter to remain constant at the value input by the user. Typical results are shown in Figure 12.18.

```
SUPPLY INITIAL VALUES FOR B, H, G, DIA ?15.0, 2.0, 2.0, 0.2
SOLUTION FOUND IN   251 ITERATIONS
MINIMUM COST=$              315963.19
PIPE DIAMETER(M)=                 .200
AVERAGE INFLOW=                   .042
TANK DIMENSIONS(M)             13.650         3.784       3.406
TANK VOLUME(CUB.M)           1477.795
TANK DEPTH IN GROUND            2.200
TANK WALL THKSS(M)               .402
WIDTH USED (M)                 24.999
EMBANKMENT HT.(M)               2.636
SUPPLY INITIAL VALUES FOR B, H, G, DIA ?15.0, 2.0, 2.0, 0.21
SOLUTION FOUND IN   207 ITERATIONS
MINIMUM COST=$              305653.56
PIPE DIAMETER(M)=                 .210
AVERAGE INFLOW=                   .048
TANK DIMENSIONS(M)             14.555        20.091       3.179
TANK VOLUME(CUB.M)            929.732
TANK DEPTH IN GROUND            2.180
TANK WALL THKSS(M)               .362
WIDTH USED (M)                 24.997
EMBANKMENT HT. (M)              2.429
SUPPLY INITIAL VALUES FOR B, H, G, DIA ?15.0, 2.0, 2.0, 0.22
SOLUTION FOUND IN   203 ITERATIONS
MINIMUM COST=$              306607.25
PIPE DIAMETER(M)=                 .220
AVERAGE INFLOW=                   .054
TANK DIMENSIONS(M)             14.775        14.774       3.009
TANK VOLUME(CUB.M)            656.897
TANK DEPTH IN GROUND            2.180
TANK WALL THKSS(M)               .334
WIDTH USED (M)                 24.480
EMBANKMENT HT.(M)               2.259
```

Figure 12.18   Results from program of Figure 12.16 modified for constant diameter

12.6 MINIMUM WEIGHT OF A PORTAL FRAME

12.6.1 Introduction

In Chapter 3 the minimum weight design of a rectangular portal frame was considered. In this section the problem is re-examined in a more general way with particular attention given to the following aspects of the problem:

(i) The dimensions of the frame should be variable and the type and intensity of the loading should be generalized.
(ii) The significance of the assumed linear relationship between the weight per unit length and fully developed plastic moment should be examined.
(iii) The effect on the optimal design of only discrete members being available should be studied.

The following memorandum provides background to the project:

Memorandum
To: Ben Tover, Drawing Office.
From: Willi Bendit, Fabricating Shop.
Re: *Minimum Weight Frames.*

We are anticipating enquiries regarding the supply of a number of rectangular portal frames. At the moment it is not clear what the dimensions will be, nor do we know the exact nature and intensity of the loading to be carried.

We have in stock a good selection of beam and column sections† and I should like to be in a position to respond quickly to any requests received.

I recall that on a previous occasion you developed a minimum weight design for a specific job subject only to concentrated loads, although I seem to remember that it was based on an assumption of linear relationship between section weight and plastic moment about which I had some doubts. Would you look into the possibility of preparing a computer program which would enable us to develop similar minimum weight designs for a variety of conditions?

12.6.2 Re-statement of the problem

For convenience the problem is re-stated as developed previously in Section 3.14. With reference to the frame of Figure 12.20(c) the printed problem may be written as follows for columns and beams of weight per unit length W_c and W_b respectively.

$$\text{Minimize } z = 2L_c W_c + L_b W_b \quad (12.30)$$

† Figure 12.19 shows the properties (mass per metre and plastic modulus) for the beam and column sections in stock.

Serial size	Mass per metre	Plastic modulus axis x–x	Serial size	Mass per metre	Plastic modulus axis x–x	Serial size	Mass per metre	Plastic modulus axis x–x
mm	kg	cm³	mm	kg	cm³	mm	kg	cm³
356 × 406	634	14247	914 × 419	389	17628	457 × 152	82	1797
	551	12078		343	15445		74	1620
	467	10009	914 × 305	289	12566		67	1439
	393	8229		253	10930		60	1284
	340	6994		224	9505		52	1094
	287	5818		201	8345	406 × 178	74	1502
	235	4689	838 × 292	226	9144		67	1343
Column core				194	7635		60	1195
	477	9700		176	6795		54	1046
356 × 368	202	3976	762 × 267	197	7156	406 × 152	74	1486
	177	3457		173	6186		67	1323
	153	2964		147	5163		60	1158
	129	2482	686 × 254	170	5616	406 × 140	46	886.3
305 × 305	283	5101		152	4989		39	718.7
	240	4245		140	4552	381 × 152	67	1254
	198	3436		125	3987		60	1106
	158	2680	610 × 305	238	7447		52	959.0
	137	2298		179	5512	356 × 171	67	1210
	118	1953		149	4562		57	1007
	97	1589	610 × 229	140	4141		51	892.9
254 × 254	167	2417		125	3672		45	771.7
	132	1861		113	3283	356 × 127	39	651.8
	107	1485		101	2877		33	537.9
	89	1228	610 × 178	91	2484	305 × 185	54	843.4
	73	988.5		82	2194		46	721.3
203 × 203	86	978.8	533 × 330	212	5849		40	623.1
	71	802.4		189	5212	305 × 127	48	704.9
	60	652.0		167	4560		42	609.2
	52	568.1	533 × 210	122	3198		37	539.3
	46	497.4		109	2820	305 × 102	33	479.6
152 × 152	37	310.1		101	2616		28	406.9
	30	247.1		92	2362		25	337.5
	23	184.3		82	2051	254 × 146	43	567.4
			533 × 165	73	1776		37	484.5
				66	1562		31	394.8
			457 × 191	98	2229	254 × 102	28	353.1
				89	2012		25	305.3
				82	1830		22	261.5
				74	1654	203 × 133	30	312.6
				67	1469		25	259.1

Figure 12.19 Mass per metre and plastic modulus for standard column and beam sections

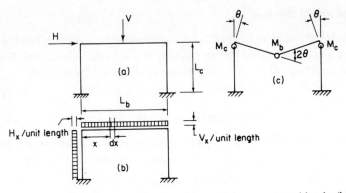

Figure 12.20 Rectangular portal frame subject to (a) concentrated loads; (b) distributed loads; (c) one of six possible collapse mechanisms

subject to

$$\begin{bmatrix} 0 & 4 \\ 2 & 2 \\ 2 & 2 \\ 4 & 0 \\ 2 & 4 \\ 4 & 2 \end{bmatrix} \begin{bmatrix} M_c \\ M_b \end{bmatrix} \geqslant \begin{bmatrix} VL_b/2 \\ VL_b/2 \\ HL_c \\ HL_c \\ HL_c + VL_b/2 \\ HL_c + VL_b/2 \end{bmatrix}$$

The six constraints correspond to the six collapse mechanisms shown in Figure 3.22.

Perhaps the first step towards generalization of the problem is to consider the effect of distributed loads as shown in Figure 12.20(b). The collapse mechanisms for both systems of loading will be the same; thus the energy absorbed will be the same function of the fully plastic moments M_b and M_c. The work done by the active loads may be expressed as:

$$\text{WD (concentrated load)} = \theta V L_b/2 \qquad (12.31)$$

$$\text{WD (distributed load)} = 2 \int_0^{L_b/2} V_x \, dx \times \theta$$

$$= \theta V_x L_b^2/4. \qquad (12.32)$$

Comparison of (12.31) and (12.32) confirms the intuitive suspicion that $V_x L_b \equiv V/2$. Similarly, a distributed horizontal load H_x may be represented by an equivalent concentrated load $H = H_x L_c/2$. This provides a simple way to handle distributed loads or to combine a distributed load (e.g. self-weight of the beam) with a concentrated applied load. Extension of this approach to cover unsymmetrically disposed concentrated loads or partially distributed loads appears to be possible but is not examined further here.

12.6.3 The weight function

In order to solve the model of (12.30) it is necessary to assume a relationship between the weight (or mass) per unit length (w_c, w_b) and fully plastic moment (M_c, M_b) of a section. The following two possibilities are considered:

Linear: $\quad w = aM + b$ (12.33)

Exponential: $\quad w = aM^b$. (12.34)

The fully plastic moment M can in turn be expressed in terms of the plastic modulus of the section Z_p and the yield stress of the material σ_y. Thus:

$$M = M_p = Z_p \sigma_y.$$ (12.35)

In Chapter 3 a linear relationship was assumed and in most related work on minimum weight design it is stated that the index of (12.34) has little effect on the minimum weight. Various approximations of the form (12.34) have been suggested in the literature.[1] For example, Heyman[2] has suggested an equation of the form

$$w \text{ (lb/ft)} = 3.4(Z_p - \text{in}^3)^{0.6}.$$ (12.36)

This translates into

$$w \text{ (kg/m)} = 0.947(Z_p - \text{cm}^3)^{0.6}.$$ (12.37)

For this example weight functions will be determined which approximate the tables of section properties accompanying Mr Bendit's memorandum. (These are taken from reference (3) for a grade 43 steel which has a yield stress of 245 N/mm².) It will be assumed that data files have been prepared in a format which can be accessed directly by routine CURFIT in order to determine the values of coefficients a, b in (12.33) or (12.34). For column sections the data could be arranged as follows:

14 247.0	634.0	35 6406
12 078.0	551.0	35 6406
10 009.0	467.0	35 6406
⋮	⋮	⋮
310.1	37.0	15 2152
247.1	30.0	15 2152
184.3	23.0	15 2152

where columns (1–10), (11–20), and (21–30) of the data file contain respectively the plastic modulus about the X–X axis (cm³), the mass per unit length (kg/m), and the section serial number which is a concatenation of the section breadth and depth in mm. Later it will be found convenient if the order of the sections is varied from that shown in the original table so that the mass/metre ratio increases monotonically.

A typical calling program is shown in Figure 12.21. Two points are significant. Firstly, it should be noted that the input file is not rewound by the routine

```
      NREAD=1
C  DATAFILE ATTACHED AS LOGICAL UNIT 1.
      REWIND NREAD
      PRINT (6,*)"ENTER FIRST AND LAST RECORDS TO BE READ...2I5"
      READ(5,*)N1,N2
      IF(N1.LE.1) GOTO 40
      N1M1=N1-1
      DO 25 JSKIP=1,N1M1
         READ(NREAD,*)DUMMY
   25 CONTINUE
   40 CONTINUE
C  SET NO. RECORDS TO BE READ
      NREC=(N2-N1+1)
      PRINT(6,*)"SPECIFY TYPE;'LIN','EXP','LOG'OR'PWR'"
      READ(6,60)TYPE
   60 FORMAT(A3)
      CALL CURFIT(NREAD,NREC,TYPE,A,B,R2,STERR)
      WRITE(6,70)A,B,R2
   70 FORMAT(8H A,B,R2=, 3F12.4)
      END
```

Figure 12.21 FORTRAN program for curve fitting of weight function

CURFIT prior to processing. This allows the user the flexibility to specify a subset of the data file by skipping any number of records and setting the number of data pairs to be read. This may be useful in attempting to fit a straight line to the smaller sections where the relationship is more non-linear.

Typical results for the section properties tabulated in the memo are shown in Figure 12.22(a) and (b).

12.6.4 A linear continuous solution

As a first step a solution is attempted using the linear approximation for the weight function (12.33) and assuming that an infinite variety of section sizes are available — i.e. that section properties are continuous variables.

The problem may be expressed in terms of the plastic modulus for beams and columns (Z_b and Z_c respectively) by substituting (12.33) and (12.35) in (12.30). Thus:

$$\text{Minimize } z = (2L_c a_c/\sigma_y)Z_c + 2L_c b_c + (L_b a_b/\sigma_y)Z_b + L_b b_b \qquad (12.38)$$

subject to

$$\begin{bmatrix} 0 & 4 \\ 2 & 2 \\ 2 & 2 \\ 4 & 0 \\ 2 & 4 \\ 4 & 2 \end{bmatrix} \begin{bmatrix} Z_c \\ Z_b \end{bmatrix} \geqslant \begin{bmatrix} VL_b/2\sigma_y \\ VL_b/2\sigma_y \\ HL_c/\sigma_y \\ HL_c/\sigma_y \\ HL_c/\sigma_y + VL_b/2\sigma_y \\ HL_c/\sigma_y + VL_b/2\sigma_y \end{bmatrix}$$

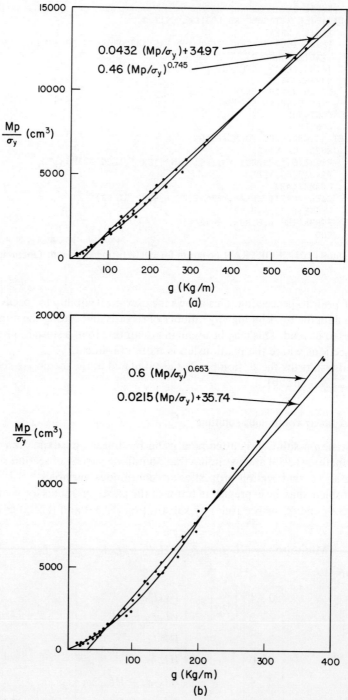

Figure 12.22 (a) Typical weight functions for columns. (b) Typical weight functions for beams

In Section 3.14 a simple version of the problem was solved numerically and also by graphical means since there are only two variables in the primal problem. The graphical method is attractive, especially if it can be generalized to some extent. The model of (12.38) can be written in non-dimensional form as follows:

$$\text{Minimize } z = L_b a_b [SZ_c + Z_b] \tag{12.39}$$

subject to

$$\begin{bmatrix} 0 & 4 \\ 2 & 2 \\ 2 & 2 \\ 4 & 0 \\ 2 & 4 \\ 4 & 2 \end{bmatrix} \begin{bmatrix} Z_c \\ Z_b/C \end{bmatrix} \geqslant \begin{bmatrix} 1 \\ 1 \\ R \\ R \\ 1+R \\ 1+R \end{bmatrix} \begin{matrix} \text{(a)} \\ \text{(b)} \\ \text{(c)} \\ \text{(d)} \\ \text{(e)} \\ \text{(f)} \end{matrix}$$

where

$$S = \frac{2L_c a_c}{L_b a_b}$$

$$C = VL_b/2\sigma_y$$

$$R = \frac{2HL_c}{VL_b}.$$

In the objective function of (12.39) the constant terms in the weight function have been dropped. This will not affect the optimal policy but must be corrected to obtain the correct minimum weight. Also it is seen that the coefficient S represents the negative slope of the iso-cost lines dZ_c/dZ_b. The six constraints are non-dimensional in terms of the ratio R and a normalizing plastic moment C.

Figure 12.23 shows a family of feasible space boundaries for different values of the ratio R. In every case it will be noted that only three possible vertices have to be considered and the optimum vertex will be decided by the four conditions:

$$S \leqslant 0.5$$

$$0.5 \leqslant S \leqslant 1.0$$

$$1.0 \leqslant S \leqslant 2.0$$

$$2.0 \leqslant S.$$

Consideration of the equations defining the locus of each vertex leads to a simple method of defining the active constraints for any case and thus the optimum solution. The solution can be summarized in the diagram of Figure 12.24 in which for any values of R and S the optimal values of Z_c/C and Z_b/C are defined. There are seven zones in the diagram representing the different types of optimal vertex

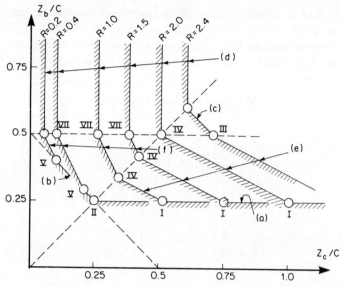

Figure 12.23 Feasible boundaries for model (12.38)

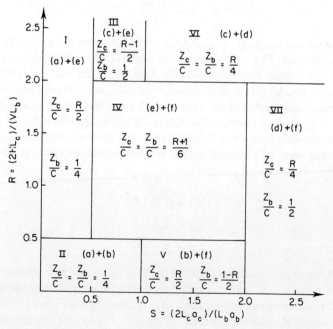

Figure 12.24 Failure mechanisms for minimum weight design

which may occur. In Figure 12.24 the vertex (or failure) types are distinguished by the Roman numerals I, II, ..., VII and the corresponding vertices in Figure 12.23 are similarly labelled. Also in each of the zones of Figure 12.24 the active constraints of (12.39) (i.e. (a), (b), ..., (f)) are shown, together with the corresponding optimal values for the dimensionless plastic moduli Z_c/C and Z_b/C for the column and beam respectively.

12.6.5 A non-linear continuous model

The graphical analysis of the previous section provides a convenient method of solving the problem without computer analysis provided that the linear weight function reasonably approximates the data. This may not be true, especially for small sections. Two approaches are possible to improve the reliability of the solution. If the approximate size of the column and beam sections is known from a preliminary analysis, the straight line of (12.33) may be adjusted to fit the data in the neighbourhood of the expected solution. This might be done by eye, by re-analysing the data for a limited set of sections, or by differentiating (12.34). Thus for beam sections (Figure 12.22(b))

$$w_b = 0.6 Z_b^{0.653}.$$

At $Z_b \simeq 750 \text{ cm}^3$

$$\frac{dw_b}{dZ_b} = 0.6 \times 0.653 \times Z_b^{-0.347}$$

$$= 0.0394$$

(compare with $w_b = 0.0215 Z_b + 35.74$).

Thus one way of taking account of the non-linear weight function is to use an iterative linear solution until the values of a_c and a_b as defined by dw_c/dZ_c and dw_b/dZ_b are consistent with the computed plastic moduli Z_c and Z_b. Clearly in some of the zones of Figure 12.24 the value of S may not be critical in determining the appropriate formulae for Z_c and Z_b.

A different approach to the problem may be developed by examining the constraints of the model (12.38). From (12.39(d)) it is clear that

$$Z_c \geqslant \frac{HL_c}{4\sigma}$$

or

$$M_c \geqslant \frac{HL_c}{4}.$$

If this value of M_c is substituted into the other five constraints, the value of M_b is defined as:

$$M_b = \max \begin{cases} VL_b/8 & \text{(a)} \\ VL_b/4 - M_c & \text{(b)} \\ HL_c/2 - M_c & \text{(c)} \\ HL_c/4 + VL_b/8 - M_c/2 & \text{(d)} \\ HL_c/2 + VL_b/4 - 2M_c & \text{(e)} \end{cases}$$

and thus $Z_b = M_b/\sigma_y$.

Using the non-linear weight functions (12.34) the total weight of the frame is given as

$$z = 2L_c w_c + L_b w_b$$
$$= 2L_c a_c Z_b^{bc} + L_b a_b Z_b^{bb}.$$

Thus the total weight is a function of a single variable Z_c, which can be minimized by trial or by a routine such as GOLDEN (Section 5.4.3). Figure 12.25 shows a typical FORTRAN program and subroutine for use by GOLDEN, together with typical results.

12.6.6 A non-linear discrete model

Having developed the non-linear model of the previous section it is a relatively obvious step to by-pass the curve-fitting process completely and instead select sections from the data files. The search of the data files can be done efficiently if the data are sorted such that section weight increases monotonically from the start of the file. The sorting could be done by machine but it is probably easier to do the job manually prior to constructing the data files.

A typical algorithm to obtain a discrete, non-linear solution might take the following form. Details of the coding are left as an exercise for the interested reader.

1. Read frame data, H, V, L_c, L_b, σ_y. Rewind column and beam data files.
2. Calculate $(Z_c)_{min} = HL_c/4$.
3. Read column section — get Z_c, w_c, number (if EOF stop).
4. If $Z_c \leqslant (Z_c)_{min}$ go to step 3.
5. Calculate $M = Z_c \sigma_y$.
6. Calculate $M_b = \max \begin{cases} VL_b/8 \\ VL_b/4 - M_c \\ HL_c/2 - M_c \\ HL_c/4 + VL_b/8 - M_c/2 \\ HL_c/2 + VL_b/4 - 2M_c \end{cases}$
7. Get $(Z_b)_{min} = M_b/\sigma_y$; if $(Z_b)_{min} < 0$ stop.
8. Rewind beam file.

9. Read a beam section; get Z_b, w_b, number (if EOF stop).
10. If $Z_b \leqslant (Z_b)_{min}$ go to step 9.
11. Calculate total weight from w_b and w_c.
12. Output results.
13. Go to step 3.

```
      PROGRAM TST(INPUT,OUTPUT,TAPE5=INPUT,TAPE6=OUTPUT)
      REAL LC,LB,M,MC,MB
C  DEFINE LABELLED COMMON TO TRANSFER PARAMETERS AND RESULTS
      COMMON /PLASTC/MC,MB,ZB,FY,H,V,LC,LB,AC,BC,AB,BB
      EXTERNAL WEIGHT
C  SPECIFY FRAME PARAMETERS
C  FRAME DIMENSIONS IN M.
      DATA V,H,LC,LB,FY/150000.0,100000.0,7.5,6.25,245.0/
      DATA AC,BC,AB,BB/0.46,0.745,0.6,0.653/
      ZCMIN=H*LC/(4.0*FY)
      DZC=ZCMIN/100.0
      EPS=DZC/10.0
      NW=6
      CALL GOLDEN(WEIGHT,ZCMIN,DZC,EPS,NW,ZC,WTMIN)
      WRITE(6,10)WTMIN,ZB,ZC
   10 FORMAT(* MINIMUM WEIGHT OF FRAME =*,F12.3,* OBTAINED WITH*,/,
     +       *  BEAM MODULUS OF*,F10.2,* CM^3*,/,
     +       *  COLUMN MODULUS OF*,F10.2,* CM^3*)
      END
      SUBROUTINE WEIGHT(ZC,WT)
      REAL LC,LB,M,MC,MB
      COMMON /PLASTC/MC,MB,ZB,FY,H,V,LC,LB,AC,BC,AB,BB
C  GET PLASTIC MOMENT IN COLUMN FOR TRIAL ZC
      MC=ZC*FY
C  COMPUTE MOMENT IN BEAM AS MAXIMUM OF FIVE VALUES
      MB=V*LB/8.0
      M=2.0*MB-MC
      IF(M.GT.MB) MB=M
      M=H*LC/2.0 - MC
      IF(M.GT.MB) MB=M
      M=M/2.0 + V*LB/8.0
      IF(M.GT.MB) MB=M
      M=2.0*M - MC
      IF(M.GT.MB) MB=M
      ZB=MB/FY
C  COMPUTE TOTAL WEIGHT FROM NONLINEAR WEIGHT FUNCTIONS
      WT=2.0*LC*AC*ZC**BC + LB*AB*ZB**BB
      RETURN
      END
 IN GOLDEN         765.3061        1302.6751
 IN GOLDEN         772.9592        1306.4283
 IN GOLDEN         785.3421        1312.4209
 IN GOLDEN         777.6890        1308.7290
 IN GOLDEN         770.0360        1304.9992
 IN GOLDEN         768.2293        1304.1132
 IN GOLDEN         767.1128        1303.5645
 IN GOLDEN         766.4227        1303.2251
   MINIMUM WEIGHT OF FRAME =    1303.225 OBTAINED WITH
   BEAM MODULUS OF       954.40 CM^3
   COLUMN MODULUS OF     766.42 CM^3
```

Figure 12.25 FORTRAN program to solve the non-linear continuous frame problem

A program based on the above algorithm will select and display a set of feasible solutions in which the column section increases from the minimum feasible value and the corresponding beam section reduces. The minimum weight design may then be obtained; alternatively, a design may be selected which is not optimal but which employs currently available sections.

12.7 REFERENCES

1. Moy, S. J. (1981). *Plastic Methods for Steel and Concrete Structures*, Macmillan, London.
2. Heyman, J. (1952). 'Plastic analysis and design of steel-framed structures', *Preliminary publication 4th Congress, I.A.B.S.E., Cambridge.* p. 95.
3. Constrado (1972). *Steel Designer's Manual*, C.L.S., London.

Appendix A
FORTRAN Subroutines

The subroutines described in the following pages are intended primarily for teaching purposes and as an aid in solving meaningful examples in optimization and simulation. As such, the authors have taken care to ensure the correctness of the code, but cannot be held responsible for any errors in the subroutines or for errors arising from their application.

Several of the programs may prove to be of use in the solution of practical problems or optimization. It should be recognized, however, that more efficient algorithms may be found in the literature. In particular the reader is referred to the texts by Siddall and Forsythe, Malcolm, and Moler.†

INDEX OF SUBROUTINES

ASTBAL	Finds the steel ratio for balanced design in a rectangular, singly reinforced concrete section.
BISECT	Finds a root of a specified function of a single variable by interval halving.
COMBIN	Generates in successive calls all possible combinations of a set of N integers.
CPM	Solves critical path network.
CURFIT	Least squares curve fitting of four functions.
DISTN1	Generates a vector of variable values having a uniformly distributed probability.
DISTN2	Operates on the vector generated by DISTN1 to produce random variates for an arbitrary distribution.
DYNAM	Generates tables of optimal values corresponding to the stages of a dynamic programming problem.
DYNSOL	Uses the tables generated by DYNAM to obtain the optimal solution and policy of a problem in dynamic programming.
FACTRL	Computes the factorial of a positive integer.
GAUSS	Generates random variates for a normal distribution.
GOLDEN	Finds the minimum of a function of a single variable by Fibonacci search.
HJMIN	Finds the minimum of an unconstrained function of several variables by the pattern search method of Hooke and Jeeves.
IBEAM	Selects a steel beam section (from a somewhat limited table) for a specified imposed bending moment.
INTER1	Finds by linear interpolation the value of a unimodal function for a specified argument.

† Siddall, J. (1972). *Analytical Decision-Making in Engineering Design*, Prentice-Hall, Englewood Cliffs, New Jersey.
Forsythe, G. E., Malcolm, M. A., and Moler, C. B. (1977). *Computer Methods for Mathematical Computation*, Prentice-Hall, Englewood Cliffs, New Jersey.

JSTATE   Finds the element of an array containing a specified value. (Used by DYNAM and DYNSOL.)
LPDATA   Facilitates input of data defining a linear programming problem. May be used prior to SIMPLEX.
MOFICR   Finds the moment of inertia of a cracked, transformed, singly reinforced, rectangular concrete section.
NADPTH   Finds the neutral axis depth of a rectangular, singly reinforced concrete section.
SIMPLEX   Solves (i.e. minimizes) a linear programming problem by the standard simplex method, generating the tableaux if desired.
SLABBM   Calculates the maximum bending moment in one or more equal continuous spans.
SLAB1W   Designs a one-way spanning slab for flexural strength only.
SRMULT   Calculates the resistance moment of a rectangular, singly reinforced concrete section by ultimate load theory.
SRSECT   Proportions a singly reinforced, rectangular concrete section for a specified bending moment.
TAB1
TAB2   Routines to generate the simplex tableaux. (Used by SIMPLEX.)
TAB3
TGSORT   Generates an integer array defining the ascending or descending order of the elements of a specified real array.
TREE   Finds the minimum cost spanning tree for any origin of a complete planar network.
URAND   Generates a uniformly distributed random number in the interval 0.0 to 1.0. (After Forsythe, Malcolm, and Moler.)

```
      SUBROUTINE ASTBAL(FC,FY,BETA1,RHOB)
C ***************************************************************
C  THE ROUTINE FINDS THE STEEL RATIO FOR BALANCED
C  DESIGN IN A SINGLY REINFORCED RECTANGULAR CONCRETE
C  BEAM USING ULTIMATE LOAD THEORY.
C     FC    = CONCRETE COMPRESSIVE STRENGTH PSI
C     FY    = YIELD STRENGTH OF STEEL PSI
C     BETA1 = RATIO OF STRESS BLOCK TO NEUTRAL AXIS
C             DEPTH  (COMPUTED)
C     RHOB  = COMPUTED VALUE OF STEEL RATIO FOR
C             BALANCED DESIGN
C ***************************************************************
      BETA1=0.85
      IF(FC.LE.4000.0) GOTO 10
      BETA1=0.85 - 0.05*(FC/1000.0 - 4.0)
   10 CONTINUE
      C1=0.85*BETA1*FC/FY
      RHOB=C1*87000.0/(87000.0+FY)
      RETURN
      END
```

```
      SUBROUTINE BISECT(FUN,XSTART,DX,EPS,NW,XROOT)
C *************************************************************
C    THE ROUTINE DETERMINES BY THE METHOD OF INTERVAL
C    HALVING THE ROOT OF AN EQUATION OF A SINGLE VARIABLE.
C      FUN    = NAME OF SUBROUTINE SUPPLIED BY THE USER
C                TO DEFINE THE EQUATION.
C                FOR EXAMPLE:
C                SUBROUTINE EQUN(X,F)
C                F=X*X-6.0*X+4.0
C                RETURN
C                END
C      XSTART = START OF COARSE SEARCH RANGE
C      DX     = INITIAL INCREMENT FOR COARSE SEARCH
C      EPS    = MINIMUM ACCEPTABLE INTERVAL OF UNCERTAINTY
C                FOR CONVERGENCE
C      NW     = OUTPUT CHANNEL FOR INTERMEDIATE PRINTOUT.
C                SET NW=0 TO SUPPRESS ALL PRINTOUT.
C      XROOT  = COMPUTED VALUE OF ROOT
C    A COARSE SEARCH IS CARRIED OUT TO FIND THE INITIAL
C    INTERVAL OF UNCERTAINTY.  IF NO ROOT IS FOUND WITHIN
C    10000*DX SEARCH IS ABANDONED.
C    NOTE THAT TWO ROOTS CLOSE TOGETHER MAY BE MISSED
C    BY THE COARSE SEARCH.
C    A DISTINCTION IS MADE BETWEEN ROOTS AND POSSIBLE
C    DISCONTINUITIES AND A WARNING PRINTED.
C    THE NAME OF THE ACTUAL SUBROUTINE CORRESPONDING TO FUN
C    MUST APPEAR IN AN EXTERNAL STATEMENT IN THE DRIVING
C    PROGRAM.
C *************************************************************
      X1=XSTART
      XROOT=XSTART
      DX1=DX
C    START COARSE SEARCH
      CALL FUN(X1,F1)
      IF(NW.GT.0) WRITE(NW,50)X1,F1
      FAC=1.0
    5 CONTINUE
      X2=X1+DX1
      CALL FUN(X2,F2)
      IF(NW.GT.0) WRITE(NW,50)X2,F2
C    TEST FOR CHANGE OF SIGN INFUNCTION
      IF(F1*F2.LT.0.0) GOTO 20
C    NO CHANGE OF SIGN
      FAC=2.0*FAC
      DX1=FAC*DX
      F1=F2
      X1=X2
```

```
      C  SET LIMIT ON EXTENT OF SEARCH
            IF(FAC.LT.10000.0) GOTO 5
            IF(NW.GT.0)WRITE(NW,10)XSTART,X2
       10   FORMAT(24H IN BISECT NO ROOT FOUND,
           +        8H BETWEEN,F12.3,4H AND,F12.3)
            STOP
       20   CONTINUE
      C  INTERVAL OF UNCERTAINTY DEFINED BY X1,X2
      C  NOW FIND SLOPE OF FUNCTION IN THIS INTERVAL
            SIGN=1.0
            IF(F2.LT.0.0) SIGN=-1.0
      C  SET VARIABLE TO TEST FOR DISCONTINUITY
            FM1=F1
            FM=F1
      C  MAIN ITERATION LOOP STARTS HERE
       30   CONTINUE
            NDISC=0
            XM=0.5*(X1+X2)
            IF(SIGN*FM.LT.SIGN*FM1) NDISC=1
            CALL FUN(XM,FM)
      C  OUTPUT ITERATION VALUES IF REQUIRED
            IF(NW.GT.0) WRITE(NW,50)XM,FM
       50   FORMAT(10H IN BISECT,2F16.4)
            IF(SIGN*FM.LT.0.0) GOTO 60
            GOTO 70
       60   CONTINUE
            X1=XM
            GOTO 80
       70   CONTINUE
            X2=XM
       80   CONTINUE
      C  TEST FOR CONVERGENCE
            IF(ABS(X2-X1).GT.EPS) GOTO 30
            XROOT=XM
            IF(NDISC.EQ.0) RETURN
            IF(NW.GT.0) WRITE(NW,90)
       90   FORMAT(39H IN BISECT POSSIBLE DISCONTINUITY FOUND)
            RETURN
            END
```

```
      SUBROUTINE COMBIN(LIST,N,M,KOUNT,FINISH)
C ****************************************************************
C THE ROUTINE IS INTENDED TO BE USED INSIDE A LOOP IN WHICH
C IT IS REQUIRED TO GENERATE ALL THE POSSIBLE COMBINATIONS
C OF A SET OF  N  INTEGERS.  AT EACH RETURN THE ROUTINE
C PROVIDES A VECTOR OF INTEGERS AND A LOGICAL VARIABLE TO
C INDICATE WHETHER OR NOT THE COMBINATIONS HAVE BEEN EXHAUSTED.
C WITHIN THE LOOP, THE VECTOR MAY BE USED TO CALCULATE SOME
C OBJECTIVE FUNCTION BASED ON THE PERMUTATED INTEGERS.
C    LIST    = INTEGER ARRAY (N) CONTAINING ON OUTPUT THE
C              COMPUTED VECTOR OF PERMUTATED INTEGERS.
C    N       = SIZE OF SET FOR WHICH ALL POSSIBLE COMBINATIONS
C              ARE REQUIRED.
C    M       = WORK STORE INTEGER CONTAINING THE CURRENT NUMBER
C              OF DIGITS IN THE COMBINATION.  M INCREASES FROM
C              1 TO N WITH SUCCESSIVE CALLS OF THE ROUTINE.
C    KOUNT   = COUNTER OF CURRENT NUMBER OF COMBINATIONS WHICH
C              HAVE BEEN RETURNED.
C    FINISH  = LOGICAL VARIABLE SET TRUE ON EXIT IF ALL POSSIBLE
C              COMBINATIONS HAVE BEEN RETURNED.
C IF A LOOP CONTAINING A SEQUENCE OF CALLS TO COMBIN IS TO BE
C EXECUTED MORE THAN ONCE IN A PROGRAM THE PARAMETER M SHOULD
C BE SET TO ZERO.  THIS IS BECAUSE M IS SET WITHIN THE ROUTINE
C BY A DATA STATEMENT WHICH WILL ASSIGNED ONLY AT COMPILATION
C TIME.
C ****************************************************************
      LOGICAL FINISH
      DIMENSION LIST(N)
      IF(M.GT.0) GOTO 100
C INITIALIZE INTEGER ARRAY TO ZERO ON FIRST CALL
      DO 10 J=1,N
      LIST(J)=0
 10   CONTINUE
C INITIALIZE PARAMETERS ON FIRST CALL.
      FINISH=.FALSE.
      K=0
      KOUNT=0
      M=1
C M  IS NUMBER OF DIGITS CURRENTLY IN COMBINATION.
 100  CONTINUE
      KOUNT=KOUNT+1
      IF(LIST(M).EQ.N) GOTO 200
C INCREMENT RIGHT HAND DIGIT.
      LIST(M)=LIST(M) + 1
      RETURN
 200  CONTINUE
      K=K+1
      MK=M-K
```

```
C     CHECK IF M HAS TO BE INCREASED.
      IF(MK.EQ.0) GOTO 300
      IF(LIST(MK).EQ.(N-K)) GOTO 200
      LISTMK=LIST(MK)
      K1=K+1
      DO 210 J=1,K1
      LIST(MK+J-1)=LISTMK+J
210   CONTINUE
      K=0
      RETURN
300   CONTINUE
      M=M+1
      K=0
C     START NEW STRING OF  M  DIGITS.
      DO 310 J=1,M
      LIST(J)=J
310   CONTINUE
C     CHECK IF COMBINATION CONTAINS ALL N DIGITS.
      IF(M.EQ.N) FINISH=.TRUE.
      RETURN
      END
```

```
      SUBROUTINE CPM(NUS,NDS,DUR,NACT,NNODE,STMIN,STMAX,
     1                FTMIN,FTMAX,EVTMIN,EVTMAX,FF,TF,NW)
C *******************************************************************
C   THE ROUTINE OPERATES ON AN ACTIVITY NETWORK TO DETERMINE
C   THE FREE FLOAT AND TOTAL FLOAT FOR EACH ACTIVITY THUS
C   ALLOWING IDENTIFICATION OF THE CRITICAL PATH.
C      NUS    = INTEGER ARRAY OF SIZE (NACT) CONTAINING THE NODE
C               NUMBERS AT THE START OF EACH ACTIVITY.
C      NDS    = INTEGER ARRAY OF SIZE (NACT) CONTAINING THE NODE
C               NUMBERS AT THE END OF EACH ACTIVITY.
C      DUR    = ARRAY OF SIZE (NACT) CONTAINING THE DURATION OF
C               EACH ACTIVITY.  NOTE THAT DUMMY LINKS MUST BE
C               INCLUDED AND GIVEN ZERO DURATION.
C      NACT   = NUMBER OF ACTIVITIES (INCLUDING DUMMY ACTIVITIES)
C      NNODE  = NUMBER OF NODES.
C      STMIN  = ARRAY OF SIZE (NACT) FOR COMPUTED VALUES OF EARLIEST
C               START TIME FOR EACH ACTIVITY.
C      STMAX  = ARRAY OF SIZE (NACT) FOR COMPUTED VALUES OF LATEST
C               START TIME FOR EACH ACTIVITY.
C      FTMIN  = ARRAY OF SIZE (NACT) FOR COMPUTED VALUES OF EARLIEST
C               FINISH TIME FOR EACH ACTIVITY.
C      FTMAX  = ARRAY OF SIZE (NACT) FOR COMPUTED VALUES OF LATEST
C               FINISH TIME FOR EACH ACTIVITY.
C      EVTMIN = ARRAY OF SIZE (NNODE) FOR COMPUTED VALUES OF
C               EARLIEST POSSIBLE EVENT TIME FOR EACH NODE.
C      EVTMAX = ARRAY OF SIZE (NNODE) FOR COMPUTED VALUES OF LATEST
C               POSSIBLE EVENT TIMES FOR EACH NODE.
C      FF     = ARRAY OF SIZE (NACT) FOR COMPUTED VALUES OF FREE FLOAT
C               FOR EACH ACTIVITY.
C      TF     = ARRAY OF SIZE (NACT) FOR COMPUTED VALUES OF TOTAL
C               FLOAT FOR EACH ACTIVITY.
C      NW     = OUTPUT CHANNEL FOR PRINTOUT OF TABLE
C               (SET ZERO TO SUPPRESS PRINTOUT)
C *******************************************************************
      DIMENSION NUS(NACT),NDS(NACT),DUR(NACT)
      DIMENSION STMIN(NACT),STMAX(NACT),FTMIN(NACT),FTMAX(NACT)
      DIMENSION EVTMIN(NNODE),EVIMAX(NNODE),FF(NACT),TF(NACT)
      REAL LPOMIN
C  SET FTMIN(), STMAX(), EVTMIN() AND EVTMAX() TO 1.0
      DO 10 IACT=1,NACT
      FTMIN(IACT)=-1.0
      STMAX(IACT)=-1.0
10    CONTINUE
      DO 20 INODE=1,NNODE
      EVTMIN(INODE)=-1.0
      EVTMAX(INODE)=-1.0
20    CONTINUE
      EVTMIN(1)=0.0
C  START OF ITERATION LOOP.
30    CONTINUE
      DO 40 IACT=1,NACT
      I=NUS(IACT)
      IF(EVTMIN(I).LT.0.0) GOTO 40
      FTMIN(IACT)=EVTMIN(I) + DUR(IACT)
```

```
40        CONTINUE
          DO 50 INODE=2,NNODE
          EPOMAX=0.0
          DO 60 IACT=1,NACT
          IF(NDS(IACT).NE.INODE) GOTO 60
          IF(FTMIN(IACT).GT.EPOMAX) EPOMAX=FTMIN(IACT)
60        CONTINUE
          EVTMIN(INODE)=EPOMAX
50        CONTINUE
          IF(EVTMIN(NNODE).LT.0.0) GOTO 30
          DO 70 IACT=1,NACT
          FF(IACT)-EVTMIN(NDS(IACT)) - FTMIN(IACT)
70        CONTINUE
C     NOW CALCULATE EARLIEST START TIMES STMIN().
          DO 80 IACT=1,NACT
          STMIN(IACT)=EVTMIN(NUS(IACT))
80        CONTINUE
C     NOW WORK BACKWARDS FROM LAST NODE TO GET LATEST POSSIBLE
C     EVENT TIMES EVTMAX() AT EACH NODE.
          EVTMAX(NNODE)=EVTMIN(NNODE)
C     START OF ITERATION LOOP.
130       CONTINUE
          DO 140 IACT=1,NACT
          J=NDS(IACT)
          IF(EVTMAX(J).LT.0.0) GOTO 140
          STMAX(IACT)=EVTMAX(J) - DUR(IACT)
140       CONTINUE
          DO 150 K=2,NNODE
          INODE=K-1
          LPOMIN=1.0E06
          DO 160 IACT=1,NACT
          IF(NUS(IACT).NE.INODE) GOTO 160
          IF(STMAX(IACT).LT.0.0) GOTO 150
          IF(STMAX(IACT).LT.LOPMIN) LPOMIN=STMAX(IACT)
160       CONTINUE
          EVTMAX(INODE)=LPOMIN
150       CONTINUE
          IF(EVTMAX(1).LT.0.0) GOTO 130
          DO 170 IACT=1,NACT
          J=NDS(IACT)
          TF(IACT)=FF(IACT) + EVTMAX(J) - EVTMIN(J)
          FTMAX(IACT)=EVTMAX(J)
170       CONTINUE
          IF(NW.EQ.0) RETURN
          WRITE(NW,100)
100       FORMAT(7X,35H NUS NDS    DURATN    STMIN     FTMIN ,
         1 35H    STMAX     FTMAX     FF()      TF(),/)
          DO 188 I=1,NACT
          WRITE(NW,200)I,NUS(I),NDS(I),DUR(I),STMIN(I),FTMIN(I),
         1 STMAX(I),FTMAX(I),FF(I),TF(I)
200       FORMAT(2X,3I4,7F9.1)
180       CONTINUE
          END
```

```
      SUBROUTINE CURFIT(NREAD,NO,TYPE,A,B,R2,STERR)
C ****************************************************************
C     THE ROUTINE READS DATA FROM A FILE AND ATTEMPTS TO FIT A
C     BEST LINE OR CURVE THROUGH THE DATA TO MINIMIZE THE ERROR
C     OF   Y  ON  X.  (I.E. LEAST SQUARES ANALYSIS)
C     THE EQUATION MAY BE OF THE FOLLOWING FORMS:
C
C     LINEAR        Y = A + B*X
C     EXPONENTIAL   Y = A*EXP(B*X)    A.GT.0.0 AND ALL X    .GT.0
C     LOGARITHMIC   Y = A + B*LN(X)            ALL Y       .GT.0
C     POWER         Y = A*X**B        A.GT.0.0 AND ALL X, Y .GT.0
C
C      NREAD   = NO. OF INPUT DATA FILE CONTAINING RECORDS OF
C                X AND Y IN FORMAT (2F10.3).
C                NREAD=5 USUALLY IMPLIES KEYBOARD INPUT OF DATA.
C      NO      = NUMBER OF DATA PAIRS TO BE READ.
C      TYPE    = HOLLERITH OR STRING CONSTANT DEFINING THE TYPE
C                OF EQUATION TO WHICH THE DATA IS TO BE FITTED.
C                E.G. 3HLIN, 3HEXP, 3HLOG OR 3HPWR
C      NO      = NUMBER OF DATA PAIRS TO BE READ.
C      A       = COMPUTED IY-INTERCEPT.
C      B       = COMPUTED SLOPE OF  Y  ON  X.
C      R2      = COMPUTED SQUARE OF COEFFICIENT OF CORRELATION.
C      STERR   = COMPUTED VALUE OF THE STANDARD ERROR OF ESTIMATE
C                OF ORDINATE  Y.
C     NOTE THAT THE ROUTINE ALSO PROVIDES THE SUMS OF VARIABLES,
C     (OR THE NATURAL LOG TRANSFORMS WHERE APPROPRIATE),
C     SQUARES, AND CROSS-PRODUCTS IN LABELLED COMMON BLOCK:-
C         COMMON /SUMS/SX,SY,SXX,SYY,SXY
C
C     *** N.B. *** THE INPUT DATA FILE "NREAD" IS NOT REWOUND BY
C     THE ROUTINE CURFIT BEFORE PROCESSING.   THIS IS TO ALLOW THE
C     USER TO SKIP RECORDS IN THE CALLING PROGRAM THUS RESTRICTING
C     THE INPUT DATA TO A SUBSET OF THE FILE.
C ****************************************************************
      COMMON /SUMS/SX,SY,SXX,SYY,SXY
      ITYPE=0
      IF(TYPE.EQ.3HLIN) ITYPE=1
      IF(TYPE.EQ.3HEXP) ITYPE=2
      IF(TYPE.EQ.3HLOG) ITYPE=3
      IF(TYPE.EQ.3HPWR) ITYPE=4
      IF(ITYPE.GT.0) GOTO 5
      WRITE(6,6)TYPE
6     FORMAT(12H  IN CURFIT ,A3,15H NOT RECOGNIZED)
      STOP
5     CONTINUE
C     ZERO SUMMATION VARIABLES.
      SX=0.0
      SY=0.0
      SXX=0.0
      SYY=0.0
      SXY=0.0
```

```
C     READ AND PROCESS DATA
      DO 10 J=1,NO
      READ(NREAD,20)X,Y
20    FORMAT(2F10.3)
      GOTO (34,32,33,32) ITYPE
32    CONTINUE
      IF(Y.GT.0.0) GOTO 35
      WRITE(6,36)TYPE,Y
36    FORMAT(22H   IN CURFIT WITH TYPE=,A3,E12.4,12H NOT ALLOWED)
      STOP
35    CONTINUE
      Y=ALOG(Y)
      IF(ITYPE.EQ.2) GOTO 34
33    CONTINUE
      IF(X.GT.0.0) GOTO 37
      WRITE(6,36)TYPE,X
      STOP
37    CONTINUE
      X=ALOG(X)
34    CONTINUE
C     NOW ACCUMULATE SUMS
      SX=SX + X
      SY=SY + Y
      SXX=SXX + X*X
      SYY=SYY + Y*Y
      SXY=SXY + X*Y
10    CONTINUE
C     NOW COMPUTE EQUATION OF LINE
      AN=FLOAT(NO)
      C1=AN*SXY - SX*SY
      B=C1/(AN*SXX - SX*SX)
      A=(SY - B*SX)/AN
C     GET R-SQUARED.
      R2=B*C1/(AN*SYY - SY*SY)
      SDYONX=SQRT((SYY - SY*SY/AN)/(AN-1.0))
      STERR=SDYONX*SQRT(1.0-R2)
      IF(ITYPE.EQ.1.OR.ITYPE.EQ.3) RETURN
      A=EXP(A)
      RETURN
      END
```

```
      SUBROUTINE DISTN1(VAL,FREQ,NVAL,WK,VAR,NVAR)
C ***************************************************************
C THE ROUTINE OPERATES ON STATISTICAL DATA DEFINING
C THE FREQUENCY OF EVENTS WITHIN EACH OF A SERIES OF
C BANDWIDTHS, TO PRODUCE AN ARRAY OF CUMMULATIVE
C DISTRIBUTION VALUES OF THE VARIABLE. THIS MAY
C THEN BE USED BY ROUTINE DISTN2 TO GENERATE
C RANDOM VARIATES OF THE BISTRIBUTION.
C     VAL    = ARRAY OF SIZE (NVAL) CONTAINING THE
C              VALUES OF THE VARIABLE DEFINING THE
C              (NVAL-1) BANDWIDTHS.
C     FREQ   = ARRAY OF SIZE (NVAL) CONTAINING THE
C              FREQUENCY OF EVENTS IN EACH BANDWIDTH
C              E.G. FREQ(I) IS THE FREQUENCY OF EVENTS IN THE
C              BAND VAL(I) TO VAL(I+1). THE LAST ELEMENT
C              FREQ(NVAL) IS NOT USED.
C     NVAL   = INTEGER NO. OF ELEMENTS IN VAL() & NO().
C     WK     = WORK ARRAY OF SIZE (NVAL)
C     VAR    = ARRAY OF SIZE (NVAR) CONTAINING ON EXIT
C              THE COMPUTED VALUES OF THE VARIABLE
C              CORRESPONDING TO A UNIFORMLY DISTRIBUTED
C              CUMMULATIVE PROBABILITY FUNCTION.
C                  (J/NVAR)....J=1,2....NVAR
C     NVAR   = NO. OF POINTS REQUIRED IN THE COMPUTED
C              CUMMULATIVE PROB. DISTRIBUTION FUNCTION.
C              MUST BE SPECIFIED AT ENTRY)
C  E.G. FOR NVAL=5
C    IF VAL(5)  = 2500    3000    3750    4000    4500
C    AND FREQ(5)= 1.0     4.0     6.0     2.0     0.0
C    THEN WK(K) = 0.0     0.08    0.58    0.83    1.0
C    AND THE COMPUTED VALUES OF VAR(11) WOULD BE
C       VAR(11) = 2500    3025    3175    3325    3475
C                 3625    3767    3867    3967    4200
C                 4500
C  USES ROUTINE INTER1.
C ***************************************************************
      DIMENSION VAL(NVAL),FREQ(NVAL),WK(NVAL)
      DIMENSION VAR(NVAR)
      WK(1)=0.0
      DO 10 IVAL=2,NVAL
      WK(IVAL)=WK(IVAL-1)+FREQ(IVAL-1)*(VAL(IVAL)-VAL(IVAL-1))
   10 CONTINUE
      DO 20 IVAL=1,NVAL
      WK(IVAL)=WK(IVAL)/WK(NVAL)
   20 CONTINUE
      DO 30 IVAR=1,NVAR
      P=FLOAT(IVAR-1)/FLOAT(NVAR-1)
      CALL INTER1(VAL,WK,NVAL,P,VAR(IVAR))
   30 CONTINUE
      RETURN
      END
```

```
      SUBROUTINE DISTN2(VAR,NVAR,P,ANS)
C *****************************************************
C  THE ROUTINE OPERATES ON A CUMMULATIVE PROBABILITY
C  DISTRIBUTION FUNCTION DEFINED BY A SERIES OF
C  POINTS AND GENERATES A RANDOM VARIATE OF THE
C  DISTRIBUTION FOR A SPECIFIED RANDOM NUMBER
C  UNIFORMLY DISTRIBUTED IN THE INTERVAL 0.0 TO 1.0
C    VAR    = ARRAY OF SIZE (NVAR) CONTAINING
C             VALUES OF THE VARIABLE CORRESPONDING
C             TO CUMMULATIVE PROBABILITIES UNIFORMLY
C             SPACED IN THE INTERVAL 0.0 TO 1.0.
C             E.G.   PROB(VARIABLE.LE.VAR(J))
C                  = (J-1)/(NVAR-1)
C    NVAR   = NO. OF POINTS USED TO DESCRIBE THE
C             CUMMULATIVE PROBABILITY DISTRIBUTION
C             FUNCTION.
C    P      = SPECIFIED RANDOM NUMBER IN THE
C             INTERVAL  0.0.LE.P.LE.1.0
C    ANS    = COMPUTED VALUE OF VARIATE CORRES-
C             PONDING TO P.
C  THE ARRAY VAR(NVAR) MAY BE GENERATED BY MEANS
C  OF ROUTINE DISTN1.
C *****************************************************
      DIMENSION VAR(NVAR)
      PARG=P*FLOAT(NVAR-1)
      IP=IFIX(PARG)+1
      FRAC=PARG+1.0-FLOAT(IP)
      ANS=VAR(IP)+(VAR(IP+1)-VAR(IP))*FRAC
      RETURN
      END
```

```
      SUBROUTINE DYNAM(STATE,D,NSTAGE,NSTATE,ND,SIGN,DOPT,FOPT,
     1                 FEASBL,TRANSF,RETFUN)
C ****************************************************************
C THE ROUTINE IS INTENDED FOR THE PARTIAL SOLUTION OF DYNAMIC
C PROGRAMMING PROBLEMS INVOLVING THE ALLOCATION OF A LIMITED
C RESOURCE AMONG A NUMBER OF COMPETING ACTIVITIES.
C S/R DYNAM GENERATES TABLES CONTAINING THE OPTIMUM
C ALLOCATION AND SUB-OPTIMIZED RETURN FOR A NUMBER OF DISCRETE
C LEVELS OF RESOURCE AT EACH STAGE OF THE SERIAL DECISION
C MAKING PROCESS.    THE STAGES ARE CONSIDERED IN REVERSE ORDER.
C    STATE  = ARRAY (NSTATE) CONTAINING THE POSSIBLE VALUES OF
C             DECISION VARIABLES FOR WHICH THE OPTIMUM TABLES
C             ARE TO BE GENERATED.
C    D      = ARRAY (ND) CONTAINING THE DISCRETE DECISION
C             VARIABLES TO BE USED.
C    NSTAGE = NO. OF STAGES OR ACTIVITIES.
C    NSTATE = NO. OF LEVELS FOR WHICH THE STATE VARIABLE
C             IS DEFINED.
C    ND     = NO. OF DECISION VARIABLES TO BE USED.
C    SIGN   = SET +1.0 FOR MAXIMIZATION
C                 -1.0 FOR MINIMIZATION.
C    DOPT   = ARRAY (NSTAGE,NSTATE) CONTAINING ON EXIT COMPUTED
C             VALUES OF THE OPTIMUM DECISION VARIABLES FOR
C             FOR EACH STAGE AS A FUNCTION OF THE INPUT STATE
C             VARIABLE STATE(J).....  (1.LE.J.LE.NSTATE)
C    FOPT   = ARRAY (NSTAGE,NSTATE) CONTAINING ON EXIT COMPUTED
C             VALUES OF THE OPTIMUM RETURN FROM ALL STAGES
C             DOWNSTREAM OF STAGE (I) (NSTAGE.GE.I.GE.0)  AS
C             A FUNCTION OF THE INPUT STATE VARIABLE STATE(J)
C    FEASBL = USER DEFINED ROUTINE TO CHECK FEASIBILITY.
C    E.G.
C    S/R FEASBL(STATE,D,NSTAGE,NSTATE,ND,I,J,K,PEN)
C    ...IF STAGE(I), STATE(J) AND DECISION D(K) ARE
C       FEASIBLE SET PEN=0.0, OTHERWISE LEAVE PEN
C       UNCHANGED.
C    TRANSF = USER DEFINED ROUTINE TO DESCRIBE THE STATE
C             TRANSFORMATION FUNCTION.
C    E.G.
C    S/R TRANSF(STATE,D,NSTAGE,NSTATE,ND,I,J,K,JDS)
C    ...COMPUTES THE OUTPUT STATE LEVEL JDS FOR THE CURRENTLY
C       DEFINED STAGE(I), INPUT STATE(J) AND DECISION D(K).
C    RETFUN = USER DEFINED ROUTINE TO COMPUTE RETURN
C    E.G.
C    S/R RETFUN(STATE,D,NSTAGE,NSTATE,ND,I,J,K,RTN)
C    ...COMPUTES RETURN FROM STAGE(I) UNDER INPUT STATE (J)
C       AND DECISION D(K).
C    N.B.   ROUTINES FEASBL, TRANSF AND RETFUN MUST APPEAR IN
C    AN EXTERNAL STATEMENT IN THE MAIN PROGRAM.
C    THE COMPUTED TABLES MAY BE PROCESSED BY ROUTINE DYNSOL
C    TO IDENTIFY THE OPTIMUM POLICY AND RETURN.
C ****************************************************************
```

```fortran
      DIMENSION DOPT(NSTAGE,NSTATE),FOPT(NSTAGE,NSTATE)
      DIMENSION STATE(NSTATE),D(ND)
      COMMON /ACC/EPS
      EPS=0.001*ABS(STATE(NSTATE)-STATE(1))
      DO 10 N=1,NSTAGE
      DO 20 J=1,NSTATE
      FMAX=-SIGN*1.0E10
      DOPT(N,J)=0.0
      FOPT(N,J)=0.0
      DO 30 K=1,ND
C     IF STATE AND DECISION FEASIBLE SET PENALTY = 0.0
      PEN=-SIGN*1.0E10
      CALL FEASBL(STATE,D,NSTAGE,NSTATE,ND,N,J,K,PEN)
C     GET RETURN FUNCTION
      CALL RETFUN(STATE,D,NSTAGE,NSTATE,ND,N,J,K,RTN)
      FD=RTN + PEN
C     IF N=1 IGNORE DOWNSTREAM EFFECTS.
      IF(N.EQ.1) GOTO 40
C     IF CURRENT DECISION INFEASIBLE IGNORE DOWNSTREAM EFFECTS
      IF(ABS(PEN).GT.EPS) GOTO 40
C     GET DOWNSTREAM STATE
      CALL TRANSF(STATE,D,NSTAGE,NSTATE,ND,N,J,K,JDS)
      FD = FD + FOPT(N-1,JDS)
40    CONTINUE
      IF(SIGN*FD.LE.SIGN*FMAX) GOTO 30
C     UPDATE BEST SOLUTION
      FMAX=FD
      DOPT(N,J) = D(K)
30    CONTINUE
      FOPT(N,J) = FMAX
20    CONTINUE
10    CONTINUE
      RETURN
      END
```

```
      SUBROUTINE DYNSOL(STATE,D,NSTAGE,NSTATE,ND,SIGN,DOPT,FOPT,
     1                          TRANSF,NPRINT,DSOL,ANS)
C ***************************************************************
C OPERATES ON THE TABLES DOPT, FOPT GENERATED BY ROUTINE
C DYNAM TO OBTAIN THE OPTIMUM POLICY AND MAX. OR MIN. RETURN.
C    STATE  = ARRAY (NSTATE) CONTAINING THE POSSIBLE VALUES
C             OF THE VARIABLES.
C    D      = ARRAY (ND) CONTAINING THE DISCRETE DECISION
C             VARIABLES TO BE USED.
C    NSTAGE = NO. OF STAGES OR ACTIVITIES.
C    NSTATE = NO. OF LEVELS FOR WHICH THE STATE VARIABLE
C             IS DEFINED.
C    ND     = NO. OF DECISION VARIABLES TO BE USED.
C    SIGN   = SET +1.0 FOR MAXIMIZATION
C                 -1.0 FOR MINIMIZATION.
C    DOPT   = ARRAY (NSTAGE,NSTATE) CONTAINING THE OPTIMUM
C             DECISION VARIABLE FOR EACH STAGE AS A FUNCTION
C             OF THE INPUT STATE VARIABLE.
C    FOPT   = ARRAY (NSTAGE,NSTATE) CONTAINING THE OPTIMUM
C             RETURN FROM ALL STAGES DOWNSTREAM OF THE
C             STAGE (I) (NSTAGE.GE.I.GE.1) AS A FUNCTION
C             OF THE INPUT STATE VARIABLE.
C    NPRINT = INTEGER SET ZERO TO SUPPRESS PRINTOUT OF TABLES
C             AND RESULT.
C    DSOL   = ARRAY (NSTAGE) CONTAINING ON EXIT THE COMPUTED
C             VALUES OF THE OPTIMUM POLICY.
C    ANS    = COMPUTED MAXIMUM OR MINIMUM COST OR RETURN.
C
C SEE COMMENT IN ROUTINE DYNAM FOR DESCRIPTION OF THE
C NECESSARY USER DEFINED ROUTINES FEASBL, TRANSF AND
C RETFUN.
C THE CALL OF THIS ROUTINE MUST BE PRECEDED BY A CALL OF
C ROUTINE DYNAM.
C USES ROUTINES JSTATE, TRANSF.
C ***************************************************************
      DIMENSION DOPT(NSTAGE,NSTATE),FOPT(NSTAGE,NSTATE)
      DIMENSION STATE(NSTATE),D(ND),DSOL(NSTAGE)
      COMMON /ACC/EPS
C FIND MAX. OR MIN. RETURN AND CORRESPONDING DECISON AND
C INITIAL STATE FOR FURTHEST UPSTREAM STAGE.
      FMAX = -SIGN*1.0E10
      DO 10 J=1,NSTATE
      IF(SIGN*FOPT(NSTAGE,J).LT.SIGN*FMAX) GOTO 10
      FMAX=FOPT(NSTAGE,J)
      DSOL(NSTAGE) = DOPT(NSTAGE,J)
      STUS = STATE(J)
      JOPT = J
   10 CONTINUE
      ANS = FMAX
      DEC=DSOL(NSTAGE)
```

```
C     FIND DOWNSTREAM STATE VARIABLE.
      CALL JSTATE(D,ND,DEC,K)
      CALL TRANSF(STATE,D,NSTAGE,NSTATE,ND,NSTAGE,JOPT,K,JDS)
      DO 20 I1=2,NSTAGE
      I=NSTAGE + 1 - I1
C     LOOK UP OPTIMUM TABLE FOR ITH. STAGE.
      DSOL(I) = DOPT(I,JDS)
      DEC=DSOL(I)
      CALL JSTATE(D,ND,DEC,K)
      JUS=JDS
C     GET DOWNSTREAM STATE VARIABLE SUBSCRIPT JDS
      CALL TRANSF(STATE,D,NSTAGE,NSTATE,ND,I,JUS,K,JDS)
20    CONTINUE
      IF(NPRINT.EQ.0) RETURN
C     PRINT OUT SOLUTION AND TABLES.
      NW=6
      WRITE(NW,35)ANS
35    FORMAT(26H    OPTIMUM COST OR RETURN=,E14.5)
      DO 36 I=1,NSTAGE
      WRITE(NW,37)I,DSOL(I)
37    FORMAT(9H    XOPT(,I2,3H) =,E14.5)
36    CONTINUE
      DO 100 I=1,NSTAGE
      WRITE(NW,101)I
101   FORMAT(//,12H     AT STAGE,I4,/,
     1    10X,5HSTATE,10X,4HDOPT,10X,4HFOPT)
      DO 110 J=1,NSTATE
      WRITE(NW,111)STATE(J),DOPT(I,J),FOPT(I,J)
111   FORMAT(3X,3E14.5)
110   CONTINUE
100   CONTINUE
      RETURN
      END
```

```
      SUBROUTINE FACTRL(I,FCTL)
C *****************************************************
C THE ROUTINE CALCULATES THE FACTORIAL OF THE
C INTEGER I.
C    I    = SPECIFIED INTEGER
C    FCTL = COMPUTED FACTORIAL OF I
C           NOTE: FCTL IS A REAL NUMBER
C  I SHOULD BE POSITIVE.
C *****************************************************
      FCTL=1.0
C CHECK FOR I=0
      IF(I.EQ.0) RETURN
      DO 10 J=1,I
         FCTL=FCTL*FLOAT(J)
   10 CONTINUE
      RETURN
      END




      SUBROUTINE GAUSS(ISEED,STDEV,AVE,VAL)
C *********************************************************
C  THE ROUTINE COMPUTES A NORMALLY DISTRIBUTED
C  RANDOM NUMBER WITH A SPECIFIED MEAN AND STANDARD
C  DEVIATION.
C     ISEED  = MUST CONTAIN ON ENTRY INTEGER SEED
C              WHICH IS REPLACED ON EXIT BY A NEW
C              RANDOM SEED VALUE FOR USE IN SUB-
C              SEQUENT CALLS.
C     STDEV  = STANDARD DEVIATION OF THE REQUIRED
C              DISTRIBUTION.
C     AVE    = AVERAGE VALUE OF THE REQUIRED DIS-
C              TRIBUTION.
C     VAL    = COMPUTED VALUE OF THE RANDOM VARIABLE
C  USES THE SYSTEM ROUTINES RAND1 AND RAND TO
C  GENERATE UNIFORMLY DISTRIBUTED RANDOM NUMBERS
C  IN THE INTERVAL 0.0 TO 1.0 AND PROCESS THEM
C  USING THE CENTRAL LIMIT THEOREM.
C *********************************************************
      A=0.0
      DO 10 I=1,12
         CALL URAND(ISEED,P)
         A=A+P
   10 CONTINUE
      VAL=(A-6.0)*STDEV+AVE
      RETURN
      END
```

```
      SUBROUTINE GOLDEN(FUN,XMIN,DX,EPS,NW,XOPT,FMIN)
C **************************************************************
C THE SUBROUTINE OBTAINS THE MINIMUM OF A SPECIFIED FUNCTION
C OF A SINGLE VARIABLE AND THE CORRESPONDING ARGUMENT, BY
C THE METHOD OF GOLDEN SECTION OR FIBONACCI SEARCH.
C    FUN    = NAME OF SUBROUTINE SUPPLIED BY THE USER
C             TO DEFINE THE FUNCTION TO BE MINIMIZED
C             FOR EXAMPLE:
C                 SUBROUTINE COST(X,F)
C                 F=X*X-6.0*X+8.0
C                 RETURN
C                 END
C    XMIN   = SPECIFIED LOWER LIMIT OF SEARCH
C    DX     = SPECIFIED INITIAL INCREMENT OF COARSE SEARCH
C    EPS    = MINIMUM ACCEPTABLE INTERVAL OF UNCERTAINTY
C             FOR CONVERGENCE
C    NW     = OUTPUT CHANNEL FOR INTERMEDIATE PRINTOUT.
C             SET NW=0 TO SUPPRESS ALL PRINTOUT.
C    XOPT   = COMPUTED VALUE OF OPTIMUM ARGUMENT
C    FMIN   = COMPUTED VALUE OF FUNCTION MINIMUM
C THE ROUTINE WILL FIND ONLY THE FIRST (LOWEST) OF MORE
C THAN ONE STATIONARY POINTS.  MAY BE USED FOR MAXIMIZING
C BY SETTING OBJECTIVE FUNCTION ZERO.
C THE NAME OF THE ACTUAL SUBROUTINE CORRESPONDING TO FUN
C MUST APPEAR IN AN EXTERNAL STATEMENT IN THE CALLING
C PROGRAM.
C **************************************************************
C SET GOLDEN SECTION FACTORS
      FAC1=1.618033989
      FAC2=FAC1-1.0
C SET AND TEST POINTS 1 AND 2
      DX1=DX
      DX2=FAC1*DX1
      X1=XMIN
      X2=X1+DX1
      CALL FUN(X1,F1)
      CALL FUN(X2,F2)
      IF(NW.GT.0)WRITE(NW,100)X1,F1
      IF(NW.GT.0)WRITE(NW,100)X2,F2
   10 CONTINUE
C INITIAL COARSE SEARCH STARTS HERE
      X4=X2+DX2
      CALL FUN(X4,F4)
      IF(NW.GT.0)WRITE(NW,100)X4,F4
C IF FUNCTION INCREASING TERMINATE COARSE SEARCH
      IF(F2.LT.F4) GOTO 20
      X1=X2
      X2=X4
      F1=F2
      F2=F4
```

```
      DX1=DX2
      DX2=DX2*FAC1
      GOTO 10
   20 CONTINUE
C  START OF ITERATIONS TO REDUCE INTERVAL OF UNCERTAINTY
C  FROM X1 TO X4
      DX2=DX1
      DX1=DX2*FAC2
      X3=X4-DX2
      CALL FUN(X3,F3)
      IF(NW.GT.0) WRITE(NW,100)X3,F3
   40 CONTINUE
C  TEST FOR CONVERGENCE
      IF(DX1.LT.EPS) GOTO 90
C  COMPARE TWO INSIDE POINTS X2 AND X3 TO DECIDE
C  WHICH OUTSIDE SEGMENT TO DISCARD
      IF(F3.GT.F2) GOTO 30
C  DISCARD SEGMENT X1 TO X2
      X1=X2
      X2=X3
      F1=F2
      F2=F3
      GOTO 20
   30 CONTINUE
C  DISCARD SEGMENT X3 TO X4
      X4=X3
      X3=X2
      F4=F3
      F3=F2
      DX2=DX1
      X2=X1+DX1
      DX1=FAC2*DX1
      CALL FUN(X2,F2)
      IF(NW.GT.0) WRITE(NW,100)X2,F2
      GOTO 40
   90 CONTINUE
C  SEARCH CONVERGED, SELECT BETTER OF POINTS X2 AND X3
      FMIN=F2
      XOPT=X2
      IF(F2.LT.F3) RETURN
      FMIN=F3
      XOPT=X3
      RETURN
  100 FORMAT(10H IN GOLDEN,2F16.4)
      END
```

```
      SUBROUTINE HJMIN(VAR,DVAR,NVAR,FMIN,RHO,EPS,FUN,NW,NMAX)
C **************************************************************
C   FINDS THE MINIMUM OF A FUNCTION OF SEVERAL VARIABLES BY
C   THE HOOKE AND JEEVES PATTERN SEARCH METHOD.  EMPLOYS A
C   USER CREATED OBJECTIVE FUNCTION.
C     VAR    = ARRAY OF SIZE (NVAR) CONTAINING ON ENTRY
C              THE INITIAL APPROXIMATIONS FOR THE VARIABLES
C              AND ON EXIT THE OPTIMAL SET.
C     DVAR   = ARRAY OF SIZE (NVAR) CONTAINING INITIAL
C              VALUES OF THE VARIABLE INCREMENTS.
C     NVAR   = NO. OF VARIABLES (NVAR,LE.20)
C     FMIN   = COMPUTED VALUE OF FUNCTION MINIMUM.
C     RHO    = FACTOR BY WHICH VARIABLE INCREMENTS ARE
C              MULTIPLIED TO REFINE SEARCH. (0.0.LT.RHO.LT.1.0)
C     EPS    = FRACTIONAL PART OF DVAR(N) USED AS CRITERION
C              FOR CONVERGENCE.
C     FUN    = NAME OF USER CREATED SUBROUTINE TO COMPUTE
C              THE OBJECTIVE FUNCTION,  E.G.
C              SUBROUTINE FUN(X,F)
C              DIMENSION X(3)
C              F=.......
C              RETURN
C              END
C     NW       OUTPUT CHANNEL FOR PRINTOUT OF
C              INTERMEDIATE RESULTS, I.E. OBJECTIVE
C              FUNCTION AND VARIABLES AT EACH
C              BASEPOINT.  SET NW=0 TO SUPPRESS
C              ALL PRINTOUT
C     NMAX   = ON ENTRY, THE MAX. NO. OF OBJECTIVE FUNCTION
C              EVALUATIONS ALLOWED, AND ON EXIT THE ACTUAL
C              NUMBER USED.  A DIAGNOSTIC IS PRINTED IF
C              THE ALLOWABLE NUMBER IS EXCEEDED.
C   THE ACTUAL PARAMETER CORRESPONDING TO FUN IN THE CALLING
C   STATEMENT MUST APPEAR IN AN EXTERNAL STATEMENT IN THE
C   CALLING PROGRAM.
C **************************************************************
C   SET UP STORAGE AND WORKING ARRAYS.  THESE MUST BE
C   RE-DIMENSIONED IF MORE THAN 20 VARIABLES USED.
      DIMENSION VAR(NVAR),DVAR(NVAR)
      DIMENSION X(10),DX(10),XCRIT(10)
      EXTERNAL FUN
C   SET UP CONVERGENCE CRITERIA XCRIT()
      DO 10 IVAR=1,NVAR
         XCRIT(IVAR)=EPS*DVAR(IVAR)
   10 CONTINUE
      CALL FUN(VAR,FMIN)
      N=1
      IF(NW.GT.0)WRITE(NW,900)FMIN,VAR
  900 FORMAT(12H IN HJMIN F=,F16.4,/,(1X,5F10.3))
```

```
C     START EXPLORATION WITH NO PATTERN
 20   CONTINUE
      FN=FMIN
C     TRANSFER CURRENT VAR() TO WORKING ARRAY
      DO 30 IVAR=1,NVAR
         X(IVAR)=VAR(IVAR)
         DX(IVAR)=DVAR(IVAR)
 30   CONTINUE
C     CARRY OUT LOCAL EXPLORATION FOR NEW BASEPOINT
      NENTRY=1
      GOTO 140
 150  CONTINUE
      IF(NW.GT.0)WRITE(NW,900)FN,(X(J),J=1,NVAR)
C     TEST FOR NUMBER OF FUNCTION EVALUATIONS
      IF(N.GT.NMAX) GOTO 90
C     CHECK IF NO IMPROVEMENT IN OBJECTIVE FUNCTION
      IF(FN.GE.FMIN)GOTO 70
 40   CONTINUE
C     CARRY OUT PATTERN MOVE
      DO 50 IVAR=1,NVAR
C     FIRST, CORRECT SIGN OF DX() ELEMENTS TO
C     SPEED EXPLORATION
      IF((X(IVAR).GT.VAR(IVAR).AND.DX(IVAR).LT.0.0).OR.
     +    (X(IVAR).LE.VAR(IVAR).AND.DX(IVAR).GE.0.0))
     +    DX(IVAR)= -DX(IVAR)
         X1=VAR(IVAR)
         VAR(IVAR)=X(IVAR)
         X(IVAR)=2.0*X(IVAR)-X1
 50   CONTINUE
      FMIN=FN
      CALL FUN(X,FX)
      N=N+1
      IF(N.GT.NMAX) GOTO 90
      FN=FX
C     CARRY OUT LOCAL EXPLORATION FOLLOWING PATTERN MOVE
      NENTRY=2
      GOTO 140
 160  CONTINUE
C     OUTPUT NEW BASEPOINT IF REQUIRED
      IF(NW.GT.0) WRITE(NW,900)FN,(X(J),J=1,NVAR)
C     TEST NUMBER OF FUNCTION EVALUATIONS
      IF(N.GT.NMAX) GOTO 90
C     CHECK IF NEW BASEPOINT NO IMPROVEMENT
C     IF SO, ABANDON PATTERN
      IF(FN.GT.FMIN) GOTO 20
C     TEST IF POSITION VECTOR HAS CHANGED
      DO 60 IVAR=1,NVAR
         IF(ABS(X(IVAR)-VAR(IVAR)).GT.DVAR(IVAR)/2.0) GOTO 40
 60   CONTINUE
```

```
C  NO CHANGE IN POSITION VECTOR.  CHECK FOR CONVERGENCE
   70  CONTINUE
       DO 120 IVAR=1,NVAR
         IF(DVAR(IVAR).GT.XCRIT(IVAR)) GOTO 130
  120  CONTINUE
C  ALL VARIABLES SATISFY CONVERGENCE CRITERIA
C  RESET NMAX PRIOR TO EXIT
       NMAX=N
       RETURN
  130  CONTINUE
C  REDUCE INCREMENT SIZE
       DO 80 IVAR=1,NVAR
         DVAR(IVAR)=RHO*DVAR(IVAR)
         DX(IVAR)=RHO*DX(IVAR)
   80  CONTINUE
C  RESTART SEARCH WITHOUT PATTERN
       GOTO 20
   90  CONTINUE
       IF(NW.GT.0)WRITE(NW,901)N,NMAX
  901  FORMAT(37H IN HJMIN NO. OF FUNCTION EVALUATIONS,/,
      +        I4,13H GREATER THAN,I4,8H ALLOWED)
       RETURN
C
C  PROCEDURE FOR LOCAL EXPLORATION
C
  140  CONTINUE
       DO 220 IVAR=1,NVAR
         X(IVAR)=X(IVAR)+DX(IVAR)
         CALL FUN(X,FX)
         N=N+1
C  TEST IF +VE MOVE SUCCESSFUL
         IF(FX.LT.FN) GOTO 210
         DX(IVAR)=-DX(IVAR)
         X(IVAR)=X(IVAR)+2.0*DX(IVAR)
         CALL FUN(X,FX)
         N=N+1
C  TEST IF -VE MOVE SUCCESSFUL
         IF(FX.LT.FN) GOTO 210
         X(IVAR)=X(IVAR)-DX(IVAR)
         GOTO 220
C  IF SUCCESSFUL, UPDATE FN
  210    CONTINUE
         FN=FX
  220  CONTINUE
C  RETURN TO MAIN BODY OF ROUTINE
       GOTO(150,160) NENTRY
       END
```

```
      SUBROUTINE IBEAM(BM,SPAN,FY,B,D,WT,XXI)
C ***********************************************************
C     SELECTS A STEEL BEAM SECTION FROM THE FOLLOWING
C     TABLE, TO CARRY A SPECIFIED BENDING MOMENT
C     RESULTING FROM A SUPERIMPOSED LOAD.  THE DEAD LOAD
C     B.M. IS AUTOMATICALLY INCLUDED BY THE ROUTINE.
C
C     DESIGN USES A STRESS FACTOR OF 0.66.
C       BM    = IMPOSED BENDING MOMENT (FT.LB.)
C       SPAN  = SIMPLY SUPPORTED SPAN (FT.)
C       FY    = STEEL YIELD STRESS (P.S.I.)
C       B     = BREADTH (IN.) OF SELECTED SECTION
C       D     = DEPTH (IN.) OF SELECTED SECTION
C       WT    = WEIGHT PER FT (LB/FT) OF SECTION
C       XXI   = MOMENT OF INERTIA (IN**4) ABOUT
C               X-X AXIS OF SELECTED SECTION
C     IF LARGEST SECTION NOT STRONG ENOUGH ZERO VALUES
C     ARE RETURNED FOR B, D, WT AND XXI.
C ***********************************************************
      DIMENSION SECDAT(4,31)
      DATA ((SECDAT(I,J),I=1,4),J=1,14)/
     + 11.91,   3.97,   14.0,     88.0,
     + 12.00,   4.00,   16.5,    105.0,
     + 12.16,   4.01,   19.0,    130.0,
     + 13.72,   5.00,   22.0,    197.0,
     + 15.65,   5.50,   26.0,    298.0,
     + 15.84,   5.53,   31.0,    372.0,
     + 15.85,   6.99,   36.0,    446.0,
     + 16.00,   7.00,   40.0,    516.0,
     + 17.86,   7.48,   45.0,    704.0,
     + 18.00,   7.50,   50.0,    801.0,
     + 18.12,   7.53,   55.0,    890.0,
     + 20.20,   8.24,   62.0,   1327.0,
     + 23.71,   8.96,   68.0,   1814.0,
     + 23.91,   8.99,   76.0,   2096.0/
      DATA ((SECDAT(I,J),I=1,4),J=15,30)/
     + 24.09,   9.02,   84.0,   2364.0,
     + 26.69,   9.96,   84.1,   2825.0,
     + 26.91,   9.99,   94.0,   3267.0,
     + 29.64,  10.46,   99.0,   3989.0,
     + 29.82,  10.48,  108.0,   4461.0,
     + 30.00,  10.50,  116.0,   4919.0,
     + 32.86,  11.48,  118.0,   5886.0,
     + 33.10,  11.51,  130.0,   6699.0,
     + 35.55,  11.95,  135.0,   7796.0,
     + 36.00,  12.00,  160.0,   9739.0,
     + 36.32,  12.07,  182.0,  11282.0,
     + 36.48,  12.12,  194.0,  12103.0,
     + 35.88,  16.48,  230.0,  14988.0,
     + 36.24,  16.56,  260.0,  17234.0,
     + 36.50,  16.60,  280.0,  18819.0,
     + 36.72,  16.66,  300.0,  20290.0/
      NO=30
      X1=SPAN*SPAN/8.0
      DLBM=SECDAT(3,NO)*X1
```

```
         Z=(BM+DLBM)/(0.66*FY)
         Z=Z*12.0
         ZMAX=SECDAT(4,NO)*2.0/SECDAT(1,NO)
         IF(Z.LE.ZMAX) GOTO 10
         B=0.0
         D=0.0
         WT=0.0
         XXI=0.0
         RETURN
10       MIN=1
         MAX=NO
20       NEXT=(MIN+MAX)/2
         DLBM=SECDAT(3,NEXT)*X1
         Z=(BM+DLBM)/(0.66*FY)
         Z=Z*12.0
         ZNEXT=SECDAT(4,NEXT)*2.0/SECDAT(1,NEXT)
         IF(Z.LE.ZNEXT)GOTO 30
         MIN=NEXT
         GOTO 40
30       MAX=NEXT
40       IF((MAX-MIN).GT.1) GOTO 20
         B=SECDAT(2,MAX)
         D=SECDAT(1,MAX)
         WT=SECDAT(3,MAX)
         XXI=SECDAT(4,MAX)
         RETURN
         END
```

```
      SUBROUTINE INTER1(F,X,NPTS,XSPEC,FXSPEC)
C ********************************************************
C   FINDS BY LINEAR INTERPOLATION THE VALUE OF A FUNCTION
C   CORRESPONDING TO A SPECIFIED ARGUMENT.  FUNCTION IS
C   DEFINED BY A SET OF PAIRS OF VALUES.
C     F     = ARRAY OF SIZE NPTS HOLDING VALUES OF
C             FUNCTION
C     X     = ARRAY OF SIZE NPTS HOLDING CORRESPONDING
C             ARGUMENTS
C     NPTS  = NO. OF POINTS DESCRIBING FUNCTION
C     XSPEC = SPECIFIED ARGUMENT
C     FXSPEC = COMPUTED VALUE OF FUNCTION AT XSPEC
C   IF XSPEC NOT IN RANGE FUNCTION IS ASSUMED EXTENDED
C   INDEFINITELY WITH CONSTANT VALUES F(1) OR F(NPTS).
C   FUNCTION MUST BE UNIMODAL IN X.
C ********************************************************
      DIMENSION F(NPTS),X(NPTS)
      I=0
      IF(XSPEC.LE.X(1)) I=1
      IF(XSPEC.GE.X(NPTS)) I=NPTS
      IF(I.EQ.0)GO TO 10
      FXSPEC=F(I)
      RETURN
   10 MIN=1
      MAX=NPTS
   40 NEXT=(MIN+MAX)/2
      IF(XSPEC.GE.X(NEXT))MIN=NEXT
      IF(XSPEC.LT.X(NEXT))MAX=NEXT
      IF((MAX-MIN).GT.1)GO TO 40
      XFAC=(XSPEC-X(MIN))/(X(MIN+1)-X(MIN))
      FXSPEC=F(MIN)+XFAC*(F(MIN+1)-F(MIN))
      RETURN
      END
```

```
      SUBROUTINE JSTATE(STATE,NST,ST,JST)
C **************************************************************
C  THE ROUTINE DETERMINES WHICH ELEMENT OF AN ARRAY IS
C  REPRESENTED BY A PSECIFIC VALUE.  THE TOLERANCE IS
C  PASSED THROUGH A LABELLED COMMON BLOCK
C     COMMON /ACC/EPS
C
C     STATE = ARRAY (NST) CONTAINING THE ARRAY TO BE TESTED.
C     NST   = NO. OF ELEMENTS IN STATE()
C     ST    = SPECIFIED VALUE TO BE MATCHED.
C     JST   = COMPUTED INTEGER VALUE OF LOWEST ELEMENT OF
C             STATE() = ST +/- EPS.
C **************************************************************
      DIMENSION STATE(NST)
      COMMON /ACC/EPS
      JST=0
      DO 10 J=1,NST
      DIFF=ABS(STATE(J)-ST)
      IF(DIFF.GT.EPS) GOTO 10
      JST=J
10    CONTINUE
      RETURN
      END
```

```
      SUBROUTINE LPDATA(N,M,A,B,C,ITAB,NTAPE)
C ******************************************************************
C   ROUTINE TO INPUT DATA FOR USE IN ROUTINE SIMPLEX
C     N     = NO. OF VARIABLES (REAL, SLACK,AND ARTIFICIAL)
C     M     = NO. OF EQUALITY CONSTRAINTS
C     A     = ARRAY (M,N) OF STRUCTURAL COEFFS.
C     B     = ARRAY (M) OF STIPULATIONS
C     C     = ARRAY (N) OF COST COEFFS.
C     ITAB  = LOGICAL VARIABLE SET .TRUE. TO PRINT TABLES
C     NTAPE = NO. OF INPUT PERIPHERAL CHANNEL
C   ONLY N,M,AND NTAPE MUST BE DEFINED PRIOR TO CALL OF .LPDATA.
C ******************************************************************
      DIMENSION A(M,N),B(M),C(N)
      LOGICAL ITAB
      WRITE(6,101)
  101 FORMAT(32H ARE TABLES REQUIRED?...YES/NO)
      READ(NTAPE,502)ANS
  502 FORMAT(A1)
      ITAB=.FALSE.
      IF(ANS.EQ.1HY) ITAB=.TRUE.
      WRITE(6,602)N
  602 FORMAT(17H SUPPLY C(I),I =1,I2,9H...7F10.5,/)
      READ(NTAPE,503)(C(I),I=1,N)
  503 FORMAT(7F10.5)
      WRITE(6,603)N
  603 FORMAT(19H SUPPLY A(I,J),J= 1,I2,18H AND B(J)...7F10.5)
      DO 2 I=1,M
      WRITE(6,604)I
  604 FORMAT(10H FOR EQN.(,I2,1H),/)
      READ(NTAPE,503)(A(I,J),J=1,N),B(I)
    2 CONTINUE
      RETURN
      END
```

```
      SUBROUTINE MOFICR(B,D,AST,FC,FY,XMOFI)
C     ********************************************************
C     THE SUBROUTINE FINDS THE MOMENT OF INERTIA ABOUT THE
C     NEUTRAL AXIS OF A CRACKED TRANSFORMED SECTION OF
C     A SINGLY REINFORCED RECTANGULAR BEAM.
C     B     = SECTION BREADTH (IN)
C     D     = EFFECTIVE DEPTH (IN)
C     AST   = AREA OF TENSILE STEEL (SQ.IN)
C     FC    = CONCRETE COMPRESSIVE STRENGTH (PSI)
C     FY    = YIELD STRENGTH OF STEEL (PSI)
C     XMOFI = COMPUTED VALUE OF MOMENT OF
C             INERTIA (IN**4)
C     ********************************************************
      REAL N
      CALL NADPTH(B,D,AST,FC,FY,X)
      N=29000000.0/(57000.0*SQRT(FC))
      XMOFI=B*X*X/3.0 + N*AST*(D-X)**2.0
      RETURN
      END




      SUBROUTINE NADPTH(B,D,AST,FC,FY,X)
C     ********************************************************
C     THE ROUTINE FINDS THE NEUTRAL AXIS DEPTH IN A
C     RECTANGULAR SINGLY REINFORCED CONCRETE SECTION.
C     B     = SECTION BREADTH (IN)
C     D     = EFFECTIVE DEPTH (IN)
C     AST   = AREA OF TENSILE STEEL (SQ.IN)
C     FC    = CONCRETE COMPRESSIVE STRENGTH (PSI)
C     FY    = YIELD STRENGTH OF STEEL (PSI)
C     X     = COMPUTED NEUTRAL AXIS DEPTH (IN)
C     ********************************************************
      REAL N
      EC=57000.0*SQRT(FC)
      ES=29000000.0
      N=ES/EC
      A1=B/2.0
      B1=N*AST
      C1=-N*AST*D
      X=(-B1+SQRT(B1*B1-4.0*A1*C1))/(2.0*A1)
      RETURN
      END
```

```
      SUBROUTINE SIMPLEX(N,M,A,B,C,XOPT,ZMIN,ITAB,NBASIC,SIMCO)
C *****************************************************************
C     SOLVES (MINIMIZES) A LINEAR PROGRAMMING PROBLEM BY THE
C     STANDARD SIMPLEX METHOD. DATA MUST BE PRESENTED IN THE
C     FORM OF EQUALITY CONSTRAINTS COMPLETE WITH ANY SLACK, SURPLUS,
C     OR ARTIFICIAL VARIABLES. THE STIPULATIONS SHOULD BE POSITIVE,
C     DOES NOT CHECK FOR INFEASIBLE OR DEGENERATE CASES.
C     INFEASIBLE CASES ARE INDICATED BY NON-ZERO VALUES FOR
C     ONE OR MORE ARTIFICIAL VARIABLES IN THE OPTIMAL SOLUTION.
C       N      = NO. OF VARIABLES (REAL,SLACK AND ARTIFICIAL)
C       M      = NO. OF EQUALITY CONSTRAINTS
C       A      = ARRAY (M,N) OF STRUCTURAL COEFFS.
C       B      = ARRAY (M) OF STIPULATIONS (POSITIVE)
C       C      = ARRAY (N) OF COST COEFFS. MUST INCLUDE
C                ZEROS FOR SLACK VARIABLES AND LARGE VALUES FOR
C                ARTIFICIAL VARIABLES
C       XOPT   = ARRAY (N) OF COMPUTED OPTIMAL VALUES
C       ZMIN   = COMPUTED MINIMUM OBJECTIVE FUNCTION
C       ITAB   = LOGICAL VARIABLE SET .TRUE. IF TABLES TO BE
C                PRINTED
C       NBASIC = WORK STORE ARRAY (M). CONTAINS THE INTEGERS
C                DEFINING THE CURRENT BASIC SOLUTION
C       SIMCO  = WORK STORE ARRAY (N). CONTAINS THE SIMPLEX
C                COEFFS.
C     EXAMPLE:
C        DIMENSION A(225),B(15),C(15),X(15),NBASIC(15),SIMCO(15)
C        NTAPE=5
C        REWIND NTAPE
C        WRITE(6,10)
C     10 FORMAT(* SUPPLY N,M...2I5*)
C        READ(NTAPE,20)N,M
C     20 FORMAT(2I5)
C        CALL LPDATA(N,M,A,B,C,ITAB,NTAPE)
C        CALL SIMPLEX(N,M,A,B,C,X,ANS,ITAB,NBASIC,SIMCO)
C        WRITE(6,607)
C    607 FORMAT(1X,8HVARIABLE,4X,5HVALUE)
C        DO 6 I=1,N
C           WRITE(6,608)I,X(I)
C    608    FORMAT(1X,2HX(,I3,4H) = ,F12.3)
C    6   CONTINUE
C        WRITE(6,610)ANS
C    610 FORMAT(1X,10X,30HOBJECTIVE FUNCTION VALUE IS    ,F15.5)
C        END
C *****************************************************************
```

```
      DIMENSION A(M,N),B(M),C(N),NBASIC(M),SIMCO(N)
      DIMENSION XOPT(N)
      INTEGER PIVROW,PIVCOL
      LOGICAL ITAB
      IF(ITAB) CALL TAB1(N,M,A,B,C)
C     SET NBASIC TO ZERO
      DO 1 J=1,M
      NBASIC(J)=0
1     CONTINUE
C     CHECK EACH COLUMN OF THE A( ) MATRIX AND LOOK FOR
C     SINGLE ELEMENTS WITH VALUES OF +1.0
      DO 6 ICOL=1,N
C     SET COUNTER FOR NUMBER OF POSSIBLE UNIT ELEMENTS.
      NUNIT=0
      DO 2 JROW=1,M
      IF(A(JROW,ICOL).EQ.0.0) GO TO 2
C     IF ELEMENT NON ZERO BUT NOT +1.0 REJECT THIS COLUMN.
      IF(A(JROW,ICOL).NE.1.0) GO TO 6
C     POSSIBLE UNIT ELEMENT FOUND.
      NUNIT=NUNIT+1
      JUNIT=JROW
2     CONTINUE
C     TEST IF ONLY ONE POSSIBLE UNIT ELEMENT FOUND.
      IF(NUNIT.EQ.1) NBASIC(JUNIT)=ICOL
6     CONTINUE
C     TEST IF BASIC FEASIBLE SOLUTION FOUND.
      DO 7 I=1,M
      IF(NBASIC(I).GT.0) GO TO 7
      WRITE(6,102)
102   FORMAT(3X,*INITIAL BASIC FEASIBLE SOLN. NOT FOUND*)
      RETURN
7     CONTINUE
C     GET .Z. FOR INITIAL B.F.S.
      Z=0.0
      DO 8 I=1,M
      NVAR=NBASIC(I)
      Z=Z+B(I)*C(NVAR)
8     CONTINUE
      NTABLE=1
10    CONTINUE
      SCMIN=0.0
C     FIND .ZK. FOR NON-BASIC VARIABLES
C     AND CALCULATE SIMPLEX COEFFS. (CK-ZK)
      DO 16 K=1,N
      DO 12 I=1,M
      IF(K.EQ.NBASIC(I)) GOTO 16
12    CONTINUE
      SUM=0.0
      DO 14 I=1,M
      J=NBASIC(I)
      SUM=SUM + C(J)*A(I,K)
14    CONTINUE
      SIMCO(K)=C(K)-SUM
C     FIND LOWEST SIMPLEX COEFF. FOR MINIMIZATION
```

```
      C   AND NOTE PIVOT COLUMN
              IF(SIMCO(K).GE.SCMIN) GOTO 16
              SCMIN=SIMCO(K)
              PIVCOL=K
      16  CONTINUE
      C   SET SIMPLEX COEFFS. = 0.0 FOR BASIC VARIABLES
              DO 18 I=1,M
              NVAR=NBASIC(I)
              SIMCO(NVAR)=0.0
      18  CONTINUE
      C   CHECK FOR CONVERGENCE ... NO NEGATIVE SIMPLEX COEFFS.
              IF(SCMIN.GE.0.0) GOTO 38
              NTABLE=NTABLE+1
      20  CONTINUE
      C   LOOK FOR MINIMUM THETA = B(I)/A(I,K)
              THETMN=1.0E+10
              DO 26 I=1,M
              IF(A(I,PIVCOL).LE.0.0) GOTO 26
      22  CONTINUE
              THETA=B(I)/A(I,PIVCOL)
              IF(THETA.GT.THETMN) GOTO 26
      24  CONTINUE
      C   NOTE PIVOT ROW ... I TH VARIABLE LEAVES BASIS
              PIVROW=I
              THETMN=THETA
      26  CONTINUE
      C   CHECK FOR UNBOUNDED SOLUTION
              IF(THETMN.LT.1.0E+10) GOTO 27
              WRITE(6,100)
      100 FORMAT(* UNBOUNDED SOLUTION*)
              RETURN
      27  CONTINUE
      C   UPDATE BASIC FEASIBLE SOLUTION
              NBASIC(PIVROW)=PIVCOL
      C   REDUCE PIVOT ELEMENT A(PIVROW,PIVCOL) TO 1.0
      C   AND ELIMINATE ALL OTHER ELEMENTS IN PIVOT COLUMN
      C   BY ROW OPERATIONS
              DIV=A(PIVROW,PIVCOL)
              DO 28 I=1,N
              A(PIVROW,I)=A(PIVROW,I)/DIV
      28  CONTINUE
              B(PIVROW)=B(PIVROW)/DIV
              IF(ITAB) CALL TAB2(SIMCO,SCMIN,NTABLE,N,Z)
              DO 36 I=1,M
      C   OMIT PIVOT ROW
              IF(I.EQ.PIVROW) GOTO 34
              AIK=A(I,PIVCOL)
              DO 32 J=1,N
              A(I,J)=A(I,J) - A(PIVROW,J)*AIK
      32  CONTINUE
              B(I)=B(I) - B(PIVROW)*AIK
      34  CONTINUE
              IF(ITAB) CALL TAB3(N,M,NBASIC,A,B,I)
```

```
36      CONTINUE
C       UPDATE OBJECTIVE FUNCTION
        Z=Z+THETMN*SCMIN
        GOTO 10
38      CONTINUE
        IF(ITAB) CALL TAB2(SIMCO,SCMIN,NTABLE,N,Z)
C       ASSIGN OPTIMAL POLICY TO XOPT()
        DO 40 I=1,N
        XOPT(I)=0.0
40      CONTINUE
        DO 42 J=1,M
        I=NBASIC(J)
        XOPT(I)=B(J)
42      CONTINUE
        ZMIN=Z
        RETURN
        END
```

```
      SUBROUTINE SLABBM(SPAN,NO,WDL,WLL,BMMAX)
C ****************************************************************
C THE ROUTINE DETERMINES THE MAXIMUM +VE OR -VE BENDING
C BENDING MOMENT IN A ONE-WAY SPANNING SLAB FOR A SERIES
C (ONE OR MORE) OF EQUAL CONTINUOUS SPANS.
C    SPAN  = SPECIFIED SPAN
C    NO    = NO. OF EQUAL SPANS
C    WDL   = SPEC. UNIFORMLY DISTRIBUTED DEAD LOAD
C    WLL   = SPEC. UNIFORMLY DISTRIBUTED LIVE LOAD
C    BMMAX = COMPUTED MAXIMUM BENDING MOMENT
C UNITS MUST BE CONSISTENT THROUGHOUT, E.G. IF SPAN IS
C IN FEET AND WDL, WLL ARE IN LB/FT THEN BMMAX IS IN
C FT-LB.
C USER SHOULD PROVIDE APPROPRIATE LOAD FACTORS IN THE
C CALLING STATEMENT IF REQUIRED.
C ****************************************************************
      DIMENSION BMCDL(5),BMCLL(5)
      DATA BMCDL/0.125,0.125,0.100,0.107,0.105/
      DATA BMCLL/0.125,0.125,0.117,0.116,0.116/
      N=NO
C COEFFS. FOR MORE THAN 5 SPANS ARE SAME AS FOR NO=5
      IF(N.GT.5) N=5
      BMMAX=SPAN*SPAN*(BMCDL(N)*WDL + BMCLL(N)*WLL)
      RETURN
      END
```

```
      SUBROUTINE SLAB1W(SPAN,NO,WLL,FC,FY,RHO,DO,AST)
C ****************************************************************
C   THE ROUTINE DESIGNS BY ULTIMATE LOAD THEORY A ONE-WAY
C   SPANNING SINGLY REINFORCED CONCRETE SLAB FOR A SERIES
C   (ONE OR MORE) OF EQUAL SPANS.  END SUPPORTS ARE
C   SIMPLY SUPPORTED; INTERNAL SUPPORTS ARE FULLY CONTIN-
C   UOUS.  LOAD FACTORS OF 1.4 AND 1.7 ARE USED FOR DEAD
C   AND LIVE LOAD RESPECTIVELY.
C     SPAN   = SPECIFIED SPAN (FEET)
C     NO     = NUMBER OF EQUAL CONTINUOUS SPANS
C     WLL    = SPEC. UNIFORMLY DISTRIBUTED LIVE LOAD
C              (LB/SQ.FT)
C     FC     = CONCRETE COMPRESSIVE STRENGTH (PSI)
C     FY     = YIELD STRENGTH INSTEEL (PSI)
C     RHO    = SPECIFIED STEEL RATIO AST/(B*D)
C              NOT GREATER THAN 0.75*RHOB
C     DO     = COMPUTED OVERALL SLAB THICKNESS (INCHES)
C     AST    = COMPUTED STEEL AREA (SQ.IN.) PER FOOT.
C   SLAB DEPTH IS ROUNDED UP TO NEAREST 0.25 INCH WITH
C   1 INCH COVER.
C   USES ROUTINES SLABBM, SRSECT AND ASTBAL.
C ****************************************************************
C   ESTIMATE DEAD LOAD
      WDL=0.5*SPAN*SQRT(WLL)
   10 CONTINUE
      CALL SLABBM(SPAN,NO,1.4*WDL,1.7*WLL,BMMAX)
C   ALLOW FOR CAPACITY REDUCTION FACTOR
      RM=BMMAX/0.9
      CALL SRSECT(RM*12.0,FC,FY,RHO,12.0,D,AST)
C   ESTIMATE BAR DIAMETER ASSUMING SPACING = D AND ROUNDING
C   UP TO NEAREST 1/8 INCH
      DIA=SQRT(AST*D*4.0/(12.0*3.14))
      DIA=FLOAT(IFIX(DIA*8.0)+1)/8.0
C   ASSUME 1 INCH COVER
      DO=D + 0.5*DIA + 1.0
C   ROUND UP TO NEAREST 1/4 INCH
      DO=FLOAT(IFIX(DO*4.0)+1)/4.0
C   CHECK MINIMUM SLAB THICKNESS
      IF(DO.LT.3.5) DO=3.5
      WDL2=DO*12.5
      IF(ABS(WDL2-WDL)/WDL.LT.0.001) GOTO 20
      WDL=WDL2
      GOTO 10
   20 CONTINUE
      RETURN
      END
```

```
      SUBROUTINE SRMULT(B,D,AST,FC,FY,RM)
C *****************************************************
C THE ROUTINE DETERMINES THE RESISTANCE MOMENT OF A
C SINGLY REINFORCED RECTANGULAR CONCRETE SECTION BY
C ULTIMATE LOAD THEORY.
C    B    = SECTION BREADTH (IN)
C    D    = EFFECTIVE DEPTH (IN)
C    AST  = AREA OF TENSILE STEEL (SQ.IN)
C    FC   = CONCRETE COMPRESSIVE STRENGTH (PSI)
C    FY   = YIELD STRENGTH OF STEEL (PSI)
C    RM   = COMPUTED VALUE OF RESISTANCE MOMENT
C           IN INCH-LB.
C USES ROUTINE ASTBAL.
C *****************************************************
      CALL ASTBAL(FC,FY,BETA1,RHOB)
      RHO=AST/(B*D)
      IF(RHO.GT.RHOB) GOTO 10
      A=AST*FY/(0.85*B*FC)
      RM=AST*FY*(D-A/2.0)
      RETURN
   10 CONTINUE
      X=AST*87000.0/(0.85*BETA1*B*FC)
      C=-X/2.0 + SQRT(0.25*X*X + X*D)
      RM=0.85*FC*B*BETA1*C*(D-0.5*BETA1*C)
      RETURN
      END
```

```fortran
      SUBROUTINE SRSECT(BM,FC,FY,RHO,B,D,AST)
C ***************************************************************
C THE ROUTINE DETERMINES THE PROPORTIONS OF A RECTANGULAR,
C SINGLY REINFORCED CONCRETE SECTION FOR A SPECIFIED
C BENDING MOMENT.  THE SPECIFIED STEEL RATIO RHO MAY
C NOT EXCEED 0.75*RHOB.
C    BM    = SPECIFIED BENDING MOMENT (INCH-LB)
C    FC    = CONCRETE COMPRESSIVE STRENGTH (PSI)
C    FY    = YIELD STRENGTH IN STEEL (PSI)
C    RHO   = SPEC. STEEL RATIO (MAY BE OVERWRITTEN
C            BY THE ROUTINE)
C    B     = SPECIFIED SECTION BREADTH (INCHES);
C            IF ZERO ON ENTRY, THE SECTION IS PROPORTIONED
C            WITH B/D = 0.4, AND B IS OVERWRITTEN.
C    D     = COMPUTED EFFECTIVE DEPTH (INCH)
C    AST   = COMPUTED TENSION STEEL AREA (SQ.IN.)
C USES ROUTINE ASTBAL.
C ***************************************************************
      REAL MOD
      CALL ASTBAL(FC,FY,BETA1,RHOB)
C CHECK FOR MAXIMUM STEEL RATIO
      RHOMAX=0.75*RHOB
      IF(RHO.GT.RHOMAX) RHO=RHOMAX
      Q=RHO*FY/FC
      MOD=FC*Q*(1.0 - 0.59*Q)
C TEST IF B SPECIFIED
      IF(B.GT.0.0) GOTO 10
      D=(2.5*BM/MOD)**(1.0/3.0)
      B=0.4*D
      GOTO 20
   10 CONTINUE
      D=SQRT(BM/(B*MOD))
   20 CONTINUE
C COMPUTE STEEL AREA
      AST=RHO*B*D
      RETURN
      END
```

```
      SUBROUTINE TAB1(N,M,A,B,C)
C ***************************************************************
C ROUTINES TAB1, TAB2 AND TAB3 ARE INTENDED FOR USE WITH
C ROUTINE SIMPLEX TO OUTPUT THE TABLEAUX AT EACH ITERATION
C OF THE SIMPLEX PROCEDURE.
C REFERENCE SHOULD BE MADE TO THE LISTING OF SIMPLEX TO
C DETERMINE THE SIGNIFICANCE OF THE VARIABLES USED.
C ***************************************************************
      DIMENSION A(M,N),B(M),C(N)
      WRITE(6,601)
  601 FORMAT(///,20H THE INITIAL TABLEAU)
      WRITE(6,602)(I,I=1,N)
  602 FORMAT(16X,5I10)
      WRITE(6,603)(C(I),I=1,N)
  603 FORMAT(10H OBJ.FNCTN,10X,5E10.4)
      DO 10 I=1,M
        WRITE(6,604)B(I),(A(I,J),J=1,N)
  604   FORMAT(10X,6E10.4,/(20X,5E10.4))
   10 CONTINUE
      RETURN
      END

      SUBROUTINE TAB2(SIMCO,SCMIN,NTABLE,N,Z)
      DIMENSION SIMCO(N)
      WRITE(6,601)(SIMCO(J),J=1,N)
  601 FORMAT(10H SIMPLEX C,10X,5E10.4)
      WRITE(6,602)Z
  602 FORMAT(1X,10X,30HOBJECTIVE FUNCTION VALUE IS   ,F15.5)
      IF(SCMIN.GE.0.0) RETURN
      WRITE(6,603)NTABLE
  603 FORMAT(//,12H TABLEAU NO.,I6)
      WRITE(6,604)(I,I=1,N)
  604 FORMAT(10X,6HSTIP  ,5I10)
      RETURN
      END

      SUBROUTINE TAB3(N,M,NBASIC,A,B,I)
      DIMENSION NBASIC(M),A(M,N),B(M)
      WRITE(6,601)NBASIC(I),B(I),(A(I,J),J=1,N)
  601 FORMAT(4H X(,I2,4H)   ,6E10.4)
      RETURN
      END
```

```
      SUBROUTINE TGSORT(A,ITAG,N,M)
C *****************************************************
C THE ROUTINE PROCESSES AN ARRAY OF VALUES A(N) AND
C GENERATES AN INTEGER ARRAY ITAG(N) SUCH THAT THE
C SEQUENCE  A(ITAG(1)), A(ITAG(2)), A(ITAG(3))...
C IS IN ASCENDING OR DESCENDING ORDER
C   A     = ARRAY OF SIZE (N) CONTAINING ON
C           ENTRY THE VALUES TO BE SORTED.
C           A(N) IS UNCHANGED ON EXIT.
C   ITAG  = INTEGER ARRAY OF SIZE (N) CONTAINING
C           ON EXIT THE COMPUTED VALUES OF THE
C           TAGS (SUBSCRIPTS) IN ORDER.
C   N     = NUMBER OF ELEMENTS TO BE PROCESSED.
C   M     = +VE FOR DESCENDING ORDER
C           -VE FOR ASCENDING ORDER.
C FOR EXAMPLE IF A(8) CONTAINS THE VALUES
C   2.1  9.2  5.3  3.7  4.2  11.0  6.7  12.5
C THE ARRAY ITAG(8) WILL BE ASSIGNED THE VALUES
C    1    4    5    3    7    2    6    8
C *****************************************************
      DIMENSION A(N),ITAG(N)
      LOGICAL TIME1
      N1=N+1
      N2=N1/2
      DO 10 J=1,N
        ITAG(J)=-1
 10   CONTINUE
      DO 11 K=1,N2
        TIME1=.TRUE.
        DO 12 J=1,N
          IF(ITAG(J).GT.0) GOTO 12
          IF(.NOT.TIME1) GOTO 14
          TIME1=.FALSE.
          SMALL=A(J)
          BIG=A(J)
          JS=J
          JB=J
          GOTO 12
 14       IF(A(J).GT.SMALL) GOTO 13
          SMALL=A(J)
          JS=J
 13       IF(A(J).LE.BIG) GOTO 12
          BIG=A(J)
          JB=J
 12     CONTINUE
        L=N1-K
        ITAG(JB)=IABS(ITAG(JB))
        ITAG(JS)=IABS(ITAG(JS))
        IF(M.LT.0) GOTO 15
        ITAG(L)=ISIGN(JS,ITAG(L))
        ITAG(K)=ISIGN(JB,ITAG(K))
        GOTO 11
 15     ITAG(L)=ISIGN(JB,ITAG(L))
        ITAG(K)=ISIGN(JS,ITAG(K))
 11   CONTINUE
      RETURN
      END
```

```
      SUBROUTINE TREE(NA,NB,C,NUS,COPT,I,J,CJ,N,L)
C ****************************************************************
C   THE ROUTINE OPERATES ON A NETWORK COMPRISING A COMPLETE
C   PLANAR GRAPH OF DIRECTED OR UNDIRECTED LINKS THE COSTS
C   OF WHICH ARE SPECIFIED.
C   FOR ANY SPECIFIED ORIGIN (OR HOME) NODE THE ROUTINE
C   FINDS THE SHORTEST (I.E. MINIMUM COST) CHAIN (OR DIRECTED
C   PATH) TO ALL OR A SUBSET OF THE OTHER NODES.
C   FOR EACH NODE THE OPTIMAL UPSTREAM NODE IS ALSO FOUND
C   THUS ENABLING THE INTERMEDIATE NODES IN THE CHAIN TO BE
C   TRACED.
C      NA    = SPECIFIED ORIGIN OR HOME NODE.
C      NB    = SPECIFIED DESTINATION NODE.  SET ZERO IF ALL
C              NODES ARE TO BE PROCESSED.
C      C     = ARRAY (N,N) OF DIRECTED LINK COSTS.
C      NUS   = ARRAY (N) OF COMPUTED UPSTREAM OPTIMAL NODES.
C      COPT  = ARRAY (N) OF COMPUTED MINIMUM COST CHAINS.
C      I     = )
C      J     = ) WORK ARRAYS OF SIZE (L)
C      CJ    = )
C      N     = NO. OF NODES.
C      L     = NO. OF LINKS.
C   IT IS ASSUMED THAT ONLY ONE LINK BETWEEN ANY PAIR OF NODES
C   MAY EXIST BUT THAT THE COST IN EACH DIRECTION MAY DIFFER.
C   WHERE NO DIRECTED LINK EXISTS THE COST SHOULD BE SET
C   TO ZERO.
C ****************************************************************
      DIMENSION C(N,N),NUS(N),COPT(N),I(L),J(L),CJ(L)
      M=0
      BIGM=1.0E10
C   SET ARRAYS TO LARGE VALUE TO INDICATE NON-PROCESSED
C   STATE.
      DO 10 JJ=1,L
      CJ(JJ)=BIGM
10    CONTINUE
      DO 12 JJ=1,N
      COPT(JJ)=BIGM
12    CONTINUE
      COPT(NA)=0.0
      NUS(NA)=0
      K1=NA
C   MAIN LOOP STARTS HERE.
15    CONTINUE
      DO 20 K=1,N
C   SKIP IF NO DIRECTED LINK EXISTS.
      IF(C(K1,K).LE.0.0) GOTO 20
C   SKIP IF DOWNSTREAM NODE ALREADY MADE PERMANENT.
      IF(COPT(K).LT.BIGM) GOTO 20
      M=M+1
      I(M)=K1
      J(M)=K
      CJ(M)=COPT(K1) + C(K1,K)
20    CONTINUE
25    CONTINUE
```

```
C     SCAN WORK ARRAY FOR MINIMUM COST NODE.
      SMALL=BIGM
      DO 30 K=1,L
      IF(CJ(K).GE.SMALL) GOTO 30
      SMALL=CJ(K)
      IMIN=I(K)
      JMIN=J(K)
      KMIN=K
30    CONTINUE
      IF(SMALL.EQ.BIGM) RETURN
C     DELETE LINK (BOTH DIRECTIONS) FROM FURTHER CONSIDERATION.
      C(K1,JMIN)=-1.0
      C(JMIN,K1)=-1.0
      NUS(JMIN)=IMIN
      COPT(JMIN)=SMALL
C     DELETE CURRENT COST FROM FURTHER SCANNING SEARCHES.
      CJ(KMIN)=BIGM
C     DELETE ALL OTHER LINKS LEADING TO THIS NODE FROM
C     FURTHER CONSIDERATION.
      DO 40 K=1,L
      IF(J(K).NE.JMIN) GOTO 40
      CJ(K)=BIGM
40    CONTINUE
C     TERMINATE IF A SPECIFIED DESTINATION NODE HAS BEEN REACHED.
      IF(JMIN.EQ.NB) RETURN
      K1=JMIN
      IF(M.LT.L) GOTO 15
C     ALL CHAINS COMPUTED...CONTINUE SCANNING.
      GOTO 25
      END
```

```
      SUBROUTINE URAND(IY,P)
C ****************************************************************
C GENERATES A UNIFORMLY DISTRIBUTED RANDOM NUMBER IN THE
C INTERVAL 0.0 TO 1.0.  ALSO RE-GENERATES A NEW SEED
C VALUE FOR SUBSEQUENT CALL.
C    IY   = INTEGER CONTAINING ON ENTRY AN ARBITRARY
C           SEED VALUE.  A NEW RANDOM VALUE IS CONTAINED
C           ON EXIT.
C    P    = COMPUTED RANDOM VALUE UNIFORMLY DISTRIBUTED
C           IN THE INTERVAL  0.0 .LE. P .LE. 1.0
C BASED ON FORSYTHE, MALCOLM AND MOLER (1977) AFTER
C D.A.KNUTH (1969).
C ****************************************************************
      DOUBLE PRECISION HALFM,DATAN,DSQRT
      DATA M2/0/, ITWO/2/
      IF(M2.NE.0) GOTO 20
C IF FIRST ENTRY, COMPUTE MACHINE INTEGER WORD LENGTH
      M=1
   10 CONTINUE
      M2=M
      M=ITWO*M2
      IF(M.GT.M2) GOTO 10
      HALFM=M2
C COMPUTE MULTIPLIER AND INCREMENT FOR LINEAR
C CONGRUENTIAL METHOD
      IA=8*IDINT(HALFM*DATAN(1.D0)/8.D0) + 5
      IC=2*IDINT(HALFM*(0.5D0-DSQRT(3.D0)/6.D0)) + 1
      MIC=(M2-IC) + M2
C S IS THE SCALE FACTOR FOR CONVERTING TO FLOATING POINT
      S = 0.5/HALFM
C COMPUTE NEXT RANDOM NUMBER
   20 CONTINUE
      IY = IY*IA
C THE FOLLOWING STATEMENT IS FOR COMPUTERS WHICH DO NOT
C ALLOW INTEGER OVERFLOW ON ADDITION
      IF(IY.GT.MIC) IY=(IY-M2)-M2
      IY = IY + IC
C THE FOLLOWING STATEMENT IS FOR COMPUTERS WHERE THE
C WORD LENGTH FOR ADDITION IS GREATER THAN FOR
C MULTIPLICATION
      IF(IY/2.GT.M2) IY=(IY-M2)-M2
C THE FOLLOWING STATEMENT IS FOR COMPUTERS WHERE INTEGER
C OVERFLOW AFFECTS THE SIGN BIT
      IF(IY.LT.0) IY=(IY+M2)+M2
      P = FLOAT(IY)*S
      RETURN
      END
```

Index

activity, 277, 279, 305
activity times, 289
allocation process (DP), 215
alternative optimal solution (transportation method), 126
analogue models, 330
annual-cost comparison, 325
Antill, J. M., 308
Aoki, M., 199
Aris, R., 236
Armstrong-Wright, A. T., 308
artificial variables (LP), 83
Arora, T. S., 112
assignment method, 135–139
   crane problem, 136
   flowchart, 137
   Hungarian method, 136
   opportunity-cost table, 136
ASTBAL (subroutine), 335, 425, 427
Au, T., 112, 275, 355

balanced transportation problem, 115
Balinski, M. L., 152
BALNCE (subroutine), 407
Bandler, J. W., 199
bar chart, 276, 301
basic cells (transportation method), 119
basic feasible solution (LP), 70
basic solution, 69
basic variable, 70
basis (LP), 70
Battersby, A., 308
Bayes' formula, 370, 371
Bayes' strategy, 361, 362, 375
Bayes' theorem, 369, 371
Bellman, R., 236
Benjamin, I. R., 379
Berge, C., 275

big 'M' method (LP), 85
big 'M' method flowchart (LP), 87
BISECT (subroutine), 161, 164, 425, 428, 429
Biswas, A. K., 275
Blanchard, B. S., 27

calculus, 26, 28
canonical form (LP), 60, 73
capacity problems (networks), 256, 258–260
capital recovery factor, 315
cash flow diagram, 309
CDF cumulative distribution function, 340
chain flow (networks), 268, 270
circular flow (networks), 268
COL conditional opportunity loss, 357, 358, 363
COMBIN (subroutine), 265, 425, 430, 431
compound interest, 312
compound interest factors, 311
conditional opportunity loss (COL), 357, 358, 363
conditional probability, 360
constraints, 23, 398
   equality, 38, 187, 190
   inequality, 43, 187, 192
continuous variable, 347
convex sets, 67
Cornell, C. A., 379
COST (subroutine – chapter 1), 21
COST (subroutine – chapter 12), 408, 410, 411
cost coefficients, 59
CPM (critical path method), 276, 282, 296, 299
CPM (subroutine), 296, 297, 425, 432, 433
crane problem (assignment method), 136

crash situation, 302, 303–305
creativity, 1
critical activity, 288, 295, 301
critical path analysis, 276–308
critical path method (CPM), 276, 282, 296, 299
critical path, 286, 290, 303
cumulative distribution function (CDF), 340
CURFIT (subroutine), 425, 434, 435
cut-set algorithm (networks), 261, 265

Daellenbach, H. G., 112, 153, 198, 236
Dantzig, G. B., 112, 152
decision analysis, 356
decision trees, 363, 364, 374, 375
decision variables, 22
  dynamic programming, 206
declining-balance depreciation, 320
degeneracy flowchart (LP), 87
degeneracy (simplex method), 86
degeneracy (transportation method), 134
deNeufville, R., 27, 255
dependent float, 300
depreciation, 320
design criteria, 3
design methodology, 2
destinations (transportation method), 113
deterministic models, 25, 330
deterministic problem, 356
direct cost, 302
direction vector methods (non-linear programming), 180
direct search methods (non-linear programming), 164, 178
discrete (integer) variable, 348
DISTN1 (subroutine), 346, 349, 393, 425, 436
DISTN2 (subroutine), 346, 349, 394, 425, 437
distributed parameter models, 25
DP (dynamic programming), 201
Dreyfus, S. E., 236
dual (LP), 89, 104
duality (LP), 104
dual problem (networks), 248, 249
dual variables (MODI-transportation method), 130
Dublin, S. W., 329
dummy activity (networks), 280, 281
dummy destinations (transportation method), 128
dummy origins (transportation method), 128

dynamic models, 25, 338, 347
dynamic programming, 26, 200, 237, 246, 251
  allocation processes, 215
  decision variables, 206
  input state variables, 205
  output state variables, 205
  pipeline network problem, 201, 208, 229
  policy, 207
  problem definition flowchart, 214
  resource allocation problem, 218, 225
  second-best policies, 232
  specified state variables, 207
  stage, 204
  stage return function, 206
  state, 204
  state transformation function, 205
  state variables, 205
  waste-water treatment example, 220, 225
DYNAM (subroutine), 224, 425, 438, 439
DYNSOL (subroutine), 224, 425, 440, 441

earliest event time (CPM), 283, 284, 286–288
earliest finish time (CPM), 289
earliest start time (CPM), 289
economics, 16, 309
EMV expected monetary value, 366, 367
equality constraints, 38
events (networks), 278, 282, 292, 359
example 2.1, 28
example 2.2, 29
example 2.3, 30
example 2.4, 30
example 2.5, 30
example 2.6 (traffic in road tunnel), 32
example 2.7 (steel plate storage tank), 33
example 2.8 (open-topped reservoir), 36
example 2.9 (sedimentation tank), 40
example 2.10 (precast concrete operations), 45
example 2.11 (traffic flow), 50
example 9.1, 318
example 9.2, 319
example 9.3, 321
example 9.4, 322
example 9.5, 323
example 9.6, 324
example 9.7, 326
example 9.8, 327
example 10.1 (simulating the carrying capacity of a beam), 331
example 10.2 (minimum-cost design of R.C. beam), 333

example 10.3 (minimum cost design of floor), 337
example 10.4 (probability of failure of a R.C. beam), 339
example 10.5 (maintenance problem), 349
excavation equipment problem (integer programming), 140
excess variables, 83
expected monetary value (EMV), 366, 367
expected opportunity lost, 362
expected pay off, 361
expected values, 338, 365–367, 373

Fabrycky, W. J., 27
FACTRL (subroutine), 425, 442
false-position method (non-linear programming), 159
FEASBL (subroutine), 224–231
Fiacco, A. V., 199
Fibonacci constant, 165
fixed shipment route (transportation method), 129
flow chains, 260
flow variables, 239
Forsythe, G. E., 425
Fox, R. L., 112
free float (CPM), 288
functions, SRMULT, 333, 336, 342, 343

Gallagher, R. H., 112
Gantt chart (networks), 276
Garfinkel, R. S., 153
Gass, S. I., 112
GAUSS (subroutine), 341, 425, 442
Gauss–Jordan method (LP), 78
geometric series (economics), 318
George, J. A., 112, 153, 198, 236
Ghouila-Houri, A., 275
GOLDEN (subroutine), 167, 169, 174, 422, 423, 425, 443, 444
golden section method (non-linear programming), 165
Gomory, R. E., 152
Gomory's cutting plane algorithm (integer programming), 139, 141
Gomory's cutting plane algorithm flowchart (integer programming), 142
gradient methods (non-linear programming), 158, 171
gradient series (economics), 316
graphical method (integer programming), 140
graphical method (LP), 63

graph theory (networks), 238
Greenberg, N., 153

Hadley, G., 112, 153
Hardy ready-mix concrete company example, (transportation method), 114
Hardy-Cross, 240, 242
Haug, E. J., 112
hessian matrix, 35
Heyman, J., 424
Hitchcock, F. L., 152
HJMIN (subroutine), 181, 406–408, 425, 445–447
Hoel, L. A., 355
Hooke and Jeeves algorithm flowchart, 182
Hooke and Jeeves pattern search (non-linear programming), 171, 180
Hungarian method of assignment, 136

IBEAM (subroutine), 338, 425, 448, 449
implicit precedence matrix, 280
improvement index (transportation method), 120
indirect cost, 302
inequality constraints, 43
inflation, 318
initial basic feasible solution (LP), 74
input state variable (DP), 205
integer programming, 139–145
    excavation equipment example, 140
    Gomory's cutting plane algorithm, 139, 141
    graphical solution, 140
interval halving (non-linear programming), 160, 163
INTER1 (subroutine), 346, 425, 450
isocost lines (LP), 66
iterative method, 255

Jacobs, C. L. R., 236
James, L. D., 329
JSTATE (subroutine used by DYNAM and DYNSOL), 426, 451

Kimmett, D., 199
Kirchhoff's laws, 240, 241, 244, 247, 257, 258
Koopmans, T. C., 112, 152

labelling technique (networks), 265, 266–268, 270
lagrangian multipliers, 39, 43, 90, 398
latest event time (CPM), 285–287, 291

latest finish time (CPM), 289
latest start time (CPM), 289
least cost, 302
least squares, 300
Lee, R. R., 329
linear continuous model, 417
linear models, 24
linear programming, 26, 57–153, 237, 245–248, 255
  artificial variables, 83
  assignment method, 135–139
  basic feasible solution, 70
  basic solution, 69
  basic variable, 70
  basis, 70
  big 'M' method, 85
  big 'M' method flowchart, 87
  canonical form, 60, 73
  convex sets, 67
  cost coefficients, 59
  degeneracy, 86
  degeneracy flowchart, 87
  dual, 89, 104
  duality, 87, 104
  excess variables, 83
  general form, 59
  graphical method, 63
  initial basic feasible solution, 74
  integer programming, 139–145
  isocost lines, 66
  non-basic variables, 70
  non-convex sets, 67
  non-negativity condition, 59
  objective function, 59
  pivot, 78
  plant location example, 61
  portal frame problem, 58, 100, 413–424
  primal, 89, 104
  sensitivity analysis, 94
  simplex criterion, 75
  simplex method, 68, 72
  simplex method flowchart, 82
  slack variable, 73
  stipulations, 59
  structural coefficients, 59
  transportation method, 113–135
Lockyer, K. G., 308
Loomba, N. P., 152
loop incidence matrix, 258, 270, 271
LP (linear programming), 57
LPDATA (subroutine used in SIMPLEX), 100, 426, 452
lumped parameter models, 25

Malcolm, M. A., 425
marginal values (LP), 91
Marks, D. H., 27, 355
mathematical model, 21
maximin criterion, 356–358, 363
maximization (transportation method), 128
maximum flow (networks), 261, 268, 270
maximum flow-minimum cut theorem (networks), 258, 265
McCormick, G. P., 199
measures of effectiveness, 5
methods of calculus, 26, 28
minimax criterion, 358, 363
minimum cost (networks), 244, 246, 252
minimum cost rule (transportation method), 118
minimum cost rule flowchart (transportation method), 118
minimum cut (networks), 260, 261
minimum route, 249, 251, 272
minimum weight design, 58, 100, 413–424
modelling, 330
models
  analogue, 330
  deterministic, 330
  probablistic, 330, 338
  symbolic, 330
MODI, modified distribution method (transportation method), 120, 130
modified distribution method flowchart (transportation method), 131
MOFICR (subroutine), 336, 425, 453
Moler, C. A., 425
monetary units, 367
Monte Carlo procedures, 338, 349
Moy, S. J., 112, 424
MTHVAL (subroutine), 350, 352

NADTH (subroutine), 336, 426, 453
Neal, B. G., 112
Nemhauser, N., 153, 236
network cuts, 256
Newton-Raphson method, 48, 158, 161, 240, 242
Nicholls, R. L., 112, 148, 255, 275
node incidence matrix (networks), 257, 258, 261, 271
non-basic cells (transportation method), 119
non-basic variable (LP), 70
non-convex sets, 67
non-linear continuous model, 421, 422

non-linear discrete model, 422–424
non-linear models, 24
non-linear programming, 26, 154–199
   design of a short column, 167
   direction vector methods, 180
   direct search methods, 164, 178
   equality constraints, 190
   false-position method, 159
   fixed step size (gradient methods), 173
   golden section method, 165
   gradient methods, 158, 171
   Hooke and Jeeves flowchart, 182
   Hooke and Jeeves pattern search, 171, 180
   inequality constraints, 192
   interval halving, 160, 163
   Newton-Raphson method, 158, 161
   optimization of a sedimentation process, 161
   optimum design of a R.C. tank, 183
   penalty term, 189
   random search, 179
   secant method, 159
   tabulation (direct search), 179
   transitional penalty term, 194
   variable step size (gradient methods), 174
non-negativity constraints (LP), 59
non-uniform series cash flows, 316
normal duration, 302, 303
normal probability density, 339
northwest corner rule (transportation method), 116
northwest corner rule flowchart (transportation method), 118

objective function, 22, 249
objectives, 2
O'Brien, J. J., 308
opportunity-cost table (assignment method), 136
optimal policy (DP), 207
optimization, selection of a, 381
origins (transportation method), 113
output state variable (DP), 205

pay off table, 356, 357
penalty term (non-linear programming), 189
pipeline network problem (DP), 201, 208, 229
pivot (LP), 78
plant location example, 61
policy (DP), 207

portal frame problem (LP), 58, 100
posterior probability, 368–370, 373, 375
potential variables, 239, 240
present value, 312
present value analysis, 322
primal (LP), 89
primal-dual graphs (networks), 270, 271
primal problem (networks), 248, 272
principal of optimality (DP), 207
prior probability, 368–370, 373, 375
probabilistic models, 330, 338
program TANKEX, 409–410
program TST, 423
prohibited shipment route (transportation method), 129
project appraisal techniques, 322
pseudorandom numbers, 341

random search (non-linear programming), 171
Read, M. B., 275
relocatable libraries, 336
resource allocation (CPM), 297, 298
resource allocation problem (DP), 218, 225
resource levelling (CPM), 300
RETFUN (subroutine), 224–231
Riggs, J. L., 329
rim requirements (transportation method), 116

salvage value, 320
Sanders, M., 329
scheduling problems, 282
secant method (non-linear programming), 159
selection of optimization method, 381
sensitivity analysis, 94
   cost coefficient changes (LP), 94
   stipulation changes (LP), 97
Seshu, S., 275
shadow prices, 45
Shane, R. M., 355
Siddall, J. N., 198, 425
SIMPLEX (subroutine), 100, 426, 454–457
simplex criterion, 75
simplex method (LP), 68, 72
simplex method flowchart, 82
simplex program (LP), 98
simulation, 26, 330, 391
sinking fund, 313
SLABBM (subroutine), 338, 426, 458
SLAB1W (subroutine), 338, 426, 459

slack variable (LP)
  linear programming, 73
  networks, 249, 282
Smith, A. A., 199, 275
specified state variables (DP), 207
SRMULT (subroutine), 333, 336, 342, 343, 426, 460
SRSECT (subroutine), 338, 426, 461
Stafford, J. H., 27, 355
stage return function (DP), 206
stage (DP), 204
Stark, R. M., 112, 198, 255, 275
state (DP), 204
state transformation function (DP), 205
state variables (DP), 205
static models, 25, 338, 339
Stelson, T. E., 112, 275
stepping stone method (transportation method), 120
stepping stone method flowchart (transportation method), 121
stipulations, 59
stone cells (transportation method), 120
straight-line depreciation, 320
structural coefficients, 59
subroutine ASTBAL, 335, 425, 427
subroutine BALNCE, 407
subroutine BISECT, 161, 164, 425, 428, 429
subroutine COMBIN, 265, 425, 430, 431
subroutine COST, 21
subroutine COST (chapter 12), 408, 410, 411
subroutine CPM, 296, 297, 425, 432, 433
subroutine CURFIT, 425, 434, 435
subroutine DISTN1, 346, 349, 393, 425, 436
subroutine DISTN2, 346, 349, 394, 425, 437
subroutine DYNAM, 224, 425, 438, 439
subroutine DYNSOL, 224, 425, 440, 441
subroutine FACTRL, 425, 442
subroutine FEASBL, 224-231
subroutine GAUSS, 341, 425, 442
subroutine GOLDEN, 167, 169, 174, 422, 423, 425, 443, 444
subroutine HJMIN, 181, 406-408, 425, 445-447
subroutine IBEAM, 338, 425, 448, 449
subroutine INTER1, 346, 425, 450
subroutine JSTATE (used by DYNAM and DYNSOL), 426, 451
subroutine LPDATA (used prior to SIMPLEX), 426, 452
subroutine MOFICR, 336, 425, 453
subroutine MTHVAL, 350

subroutine NADTH, 336, 426, 453
subroutine RETFUN, 224-231
subroutine SIMPLEX, 100, 426, 454-457
subroutine SLABBM, 338, 426, 458
subroutine SLABIW, 338, 426, 459
subroutine SRMULT, 333, 336, 342, 343, 426, 460
subroutine SRSECT, 338, 426, 461
subroutine TAB1 (used in SIMPLEX), 100, 426, 462
subroutine TAB2 (used in SIMPLEX), 100, 426, 462
subroutine TAB3 (used in SIMPLEX), 100, 426, 462
subroutine TGSORT, 393, 394, 406, 463
subroutine TRANSF, 224-231
subroutine TREE, 246, 253, 254, 272, 426, 464, 465
subroutine URAND, 426, 466
subroutine WEIGHT, 423
subroutine library, 334
symbolic models, 330
system, notion of a, 7
systems engineer, 8
systems engineering, 7

TAB1 (subroutine used by SIMPLEX), 100, 426, 462
TAB2 (subroutine used by SIMPLEX), 100, 426, 462
TAB3 (subroutine used by SIMPLEX), 100, 426, 462
tabulation (non-linear programming), 179
TANKEX (program), 409
TGSORT (subroutine), 393, 394, 406, 463
Thirstville project, 12
Thomann, R. V., 275
total float (CPM), 286, 287, 304
transitional penalty term (non-linear programming), 194
traffic networks, 244, 256
TRANSF (subroutine) 224-231
transportation matrix, 116
transportation method, 113-135
  alternative optimal solution, 126
  balanced problem, 115
  basic cells, 119
  degeneracy, 134
  destinations, 113
  dual variables (MODI), 130
  dummy destinations, 128
  dummy origins, 128
  fixed shipment route, 129

transportation method *contd.*
  flowchart, 117
  Hardy ready-mix concrete company example, 114
  improvement index, 120
  maximization, 128
  minimum cost rule, 118
  minimum cost rule flowchart, 119
  MODI, modified distribution method, 120, 130
  MODI flowchart, 131
  non-basic cells, 119
  northwest corner rule, 116
  northwest corner rule flowchart, 118
    origins, 113
    prohibited shipment route, 129
  rim requirements, 116
  stepping stone method, 120
  stepping stone method flowchart, 121
  stone cells, 120
  transportation matrix, 116
  unblanaced problem, 128
  water cells, 120
transportation problem
  linear programming, 113
  networks, 237, 244, 246
travelling salesman problem, 254, 255
tree network, 239, 251–253
TREE (subroutine), 246, 253, 254, 272, 426, 464, 465
TST (program), 423

Tufgar, R. H., 175
Turban, E., 152

unbalanced transportation, 128
unconditional probability, 359
uniform series compound interest, 313
uniform series present value, 314
URAND (subroutine), 426, 466
utility, 367, 368

variance, 338

waste-water treatment example (DP), 220, 225
water cells (transportation method), 120
WEIGHT (subroutine), 423
weight function (for portal frame), 416–418
White, D. J., 236
Woodhead, R. W., 308
Woods, D. R., 329
worked examples (chapter 12)
  Fassbuck Trading Co. (*see also* Thirstville project), 380–387
  parking-lot problem, 387–395
  design of an insulated pressure vessel, 395–400
  new water supply (*see also* Thirstville project), 400–412
  minimum weight of portal frame, 413–424

Zienkiewicz, O. C., 112